Crowns of Glory,
Tears of Blood

Crowns of Glory, Tears of Blood

THE DEMERARA SLAVE REBELLION OF 1823

Emilia Viotti da Costa

New York Oxford
OXFORD UNIVERSITY PRESS
1994

Oxford University Press

Oxford New York Toronto
Delhi Bombay Calcutta Madras Karachi
Kuala Lumpur Singapore Hong Kong Tokyo
Nairobi Dar es Salaam Cape Town
Melbourne Auckland Madrid

and associated companies in
Berlin Ibadan

Published by Oxford University Press, Inc.
200 Madison Avenue, New York, New York 10016

Library of Congress Cataloging-in-Publication Data
Costa, Emilia Viotti da.
Crowns of glory, tears of blood : the Demerara Slave Rebellion
of 1823 / Emilia Viotti da Costa.
p. cm. Includes bibliographical references and index.
ISBN 0-19-508298-2
1. Guyana—History—1803-1966.
2. Slavery—Guyana—Insurrections, etc.
3. Smith, John, 1790-1824. I. Title.
F2384.C67 1993
988.1—dc20 92-38696

9 8 7 6 5 4 3 2 1

Printed in the United States of America
on acid-free paper

For Jack

ACKNOWLEDGMENTS

During the past ten years, the leave policies of Yale University, a year's fellowship at the National Humanities Center in 1984–1985 (thanks to a grant from the Centers for Advanced Study Program of National Endowment for the Humanities), and a year at the Institute for the Advanced Study at Princeton in 1989–90 (thanks to grants from the Ford Foundation and the John D. and Catherine T. MacArthur Foundation), provided me not only with the opportunity to exchange my ideas with a select group of bright and committed scholars, but also with the peace of mind I needed to carry on my research. I came away from both places with a feeling that I had discovered what a scholar's paradise might look like, and I am grateful for that.

I also want to acknowledge the help I have received from other institutions and their devoted personnel: at Yale, the Sterling Library, the Mudd Library, and the Divinity School Library and Archive, where I found an extraordinary rich collection of microfiches containing documents from the London Missionary Society Archive, and a complete collection of the *Evangelical Magazine* and other missionary publications; in Guyana, the University of Guyana Library, the Public Library in Georgetown, and the Guyana National Archives, where exceptionally dedicated people struggle every day against all odds to maintain the records of their past; and in England, the Public Record Office at Kew, an astonishing repository of documents concerning British imperial activities around the world.

I want to thank my editor at Oxford University Press, Leona Capeless, for her careful and sympathetic reading of the book. To Richard Dunn, Edmund Morgan, David Montgomery, Janaina Amado, Bela Feldman, and most of all to Eric Arnesen, who were gracious enough and patient enough to read and criticize earlier versions of the manuscript, I will be forever grateful. Ira Berlin offered valuable criticism of a paper, "Slaves and Managers in the Age of Revolution: Sugar in the East Coast of Demerara 1807–1832," that I presented at a 1989 conference on Cultivation and Culture, at the University of Maryland. Sections of that paper were incorporated into this book. Equally useful were Barry Gaspar's comments on another paper I presented at a 1991 conference on Slavery and Freedom in Comparative Perspective, at the University of California at San Diego, under the title "From All According to Their Abilities, to All According to Their Needs: Day-to-Day Resistance and Slaves' Notions of Rights in Guyana, 1822–1832," parts of which were integrated into Chapter Two of this book.

I am also grateful to several Guyanese scholars, particularly Tommy Payne, Ana Benjamin, Simon Carbury, and Hazel Woolford, whose hospitality made enjoyable my stay in Guyana in 1991. The generous treatment I was given by Guyanese people of every color, including the few whites whom I met there, only confirmed something I always had known: the idea that racial conflict is inescapable is nothing more than a myth kept alive by powerful interests.

No acknowledgment can fully register an author's debt to her predecessors. But a book is always the product of a conversation among scholars. This book is no exception. If it were not for the extraordinarily rich literature on slavery and abolition, produced particularly in the United States, the Caribbean, Canada, and Great Britain, this book could not have been written the way it was.

In my long journey in search of the past, I daily shared with Raymond Jackson Wilson my findings, my ideas, my excitements, and my disappointments. I always benefited from his comments, criticisms, and editorial assistance. His sharp mind, his wit, his intellectual sensitivity and curiosity, and his almost infinite patience, made the journey pleasant and fruitful. No one has contributed as much as he to making this book possible.

CONTENTS

ILLUSTRATIONS

INTRODUCTION

From the moment the first Europeans arrived from across the ocean, Guyana became a land of wild dreams and bitter realities. The El Dorado of the early days of the "discovery," the prey of pirates and buccaneers, of Europeans in quest of wealth and power, the booty of European nations competing for supremacy over lands and peoples around the world, was turned into a colonial producer of tropical commodities, a land of masters and slaves. Officially incorporated into the British Empire at the beginning of the nineteenth century, the colonies of Demerara-Essequibo and Berbice became famous for their sugar.

In 1823, Demerara was the setting for one of the greatest slave uprisings in the history of the New World. Ten to twelve thousand slaves rose up in the name of their "rights." The rebellion started on plantation *Success,* which belonged to John Gladstone (father of the future British prime minister). It spread to about sixty plantations that lay in an intensively cultivated strip of land known as the East Coast, reaching for some twenty-five miles along the sea, eastward from the mouth of the Demerara River. The rebels were quickly and brutally repressed. More than two hundred were killed outright. Many were brought to trial, and a number were hanged—to the accompaniment of all the pomp and circumstance the colony could muster. John Smith, an evangelical missionary who had come from Britain to Demerara in 1817 to preach to slaves, was accused of being the instigator of the rebellion. He was tried by a court-martial, and condemned to death. This book tells their story.

Crises are moments of truth. They bring to light the conflicts that in daily life are buried beneath the rules and routines of social protocol, behind the gestures that people make automatically, without thinking of their meanings and purposes. In such moments the contradictions that lay behind the rhetoric of social harmony, consensus, hegemony, or control are exposed. That is precisely what happened in 1823 in Demerara. The slave uprising showed clearly where the lines of loyalty were drawn. It forced people to take sides and to make their commitments clear. It revealed the notions and feelings that created bondings and identities, or that set people against each other. It laid bare the motivations and rationalizations used by different groups in their social interaction. It made public, for a moment at least, the slaves' secret life. It removed the mask of benevolence and exposed in its nakedness all the brutality of the masters' power and brought into the open their growing opposition to the British government.

The men and women who lived through the rebellion and its aftermath could define the events only in very immediate and emotional terms. For the Smiths, John and his wife Jane, for their fellow missionary John Wray, and a few others who identified with their mission, the cause of rebellion was unmitigated oppression, and the responsibility for the tragic events belonged to Governor John Murray and to planters like Michael McTurk, who were doing the work of the Devil. By contrast, the governor, McTurk, and most planters saw the missionaries themselves as the main culprits. Missionaries couched their understanding of the rebellion in terms such as "sin," "greed," "tyranny," and "oppression." Planters, colonial authorities, and the local press spoke of what happened in terms of "treachery," "deceit," "defiance," and "fanaticism." Both sides talked as though such categories had lives of their own and, like malignant spirits, could make history by possessing men and women. Both sides searched their past experience for whatever might validate their actions, and demonstrate their truth. When they tried to go beyond the immediacy of their experience, the missionaries talked about the evils of the slave "system," while planters and authorities blamed dissenters, abolitionists, the British press, and members of Parliament who had lent ears to those who favored emancipation.[1]

The early impressions of the rebellion were inscribed in the many pages some of the participants wrote and in the books, pamphlets, magazine articles, and government papers they published. Their writings would later be carefully gathered and preserved by the institutions to which each side appealed, creating an impressive record that, with time, would grow with the addition of documents generated in Britain, where the events in Demerara had powerful repercussions.

Predictably, whites monopolized the historical record. Both sides in the debate saw the slaves as ciphers: as men and women who had risen either because they had been manipulated by devious missionaries, or because they had been victimized by godless planters and managers. Neither side acknowledged that slaves too had a history of their own, a history they were not allowed to tell until they were arrested and brought to trial—and even then only within severe constraints. Neither side was capable of producing a narrative that could encompass the experience of the other. This does not mean that their versions of the rebellion should be dismissed. The stories people told bespoke their individual experience and their dreams and nightmares. Their narratives revealed their perceptions, the way they organized their experience, and responded to others. Their stories were articulated within a frame of reference and a language which was both constituted by and constitutive of their experience.

People's self-definitions, their narratives about themselves and about others, however significant, are not enough to characterize them, or account for their experience, much less to explain a historical event. Stories people tell have a history their words and actions betray but that their narratives do not immediately disclose; a history that explains why they choose the words they use, say what they say, and act as they do; a history that explains the specific meanings behind the illusory universality their words suggest—a history they often are unaware of. Their utterances are not simply statements about "reality," they are commentaries on their present experiences, memories of a past willed to them by their ancestors, and anticipations of a future they wish to create.

The narratives planters, missionaries, and royal authorities produced expressed the positions from which they spoke, their class, religion, ethnicity, status, gender, and the role each played in society. Such categories, however, are historically constructed, not immutable and primordial essences from which people's ideas and behavior can be deduced. They mean different things at different times and places. To be a planter or a slave in Demerara in 1823 when the British government, under pressure from abolitionists, was taking measures to "ameliorate" the slaves' conditions of living in preparation for their future emancipation, was not the same as living there fifty years earlier, when slavery appeared to be a stable institution. To be a minister in England was one thing, to preach to slaves in Demerara was another. To preach to whites and free blacks in town was a quite different experience from preaching on plantations and living among slaves. To be born in Africa and then be transported to the New World and sold as a slave, was not the same as being born a slave in Demerara. And to be a woman was to have problems and opportunities that her male counterparts did not have.

Identities, language, and meanings are products of social interaction, which takes place within a specific system of social relations and power, with its own rituals, protocols, and sanctions. The material conditions of peoples' lives, the way human and ecological resources are utilized and distributed, the concrete ways power is exerted, are as important in shaping identities, defining language, and creating meanings, as the social codes that mediate experience or the conventions used to define what is real. In fact, material conditions and symbolic systems are intimately connected.

Slave plantation societies had much in common. The life of a slave in Demerara was in many ways similar to the life of a slave in Cuba, the southern United States, or Brazil. But there were also significant differences, depending on the nature of the crop, the degree of technological development, the layout and sizes of the plantations, the percentage of slaves and free blacks in the total population, the demographic profile of the slave population, the slaves' places of origins and culture, the characteristics of the planter class (whether absentee or resident for example), and the religious, political, and administrative institutions they created. All these conditions changed over time. More important, plantations everywhere produced primarily for an international market, and this exposed them to contacts of all sorts with the outside world. As in the United States, Brazil, or any other European colonies, the lives of the men and women who lived in Demerara—slaves, managers, masters, missionaries, and royal authorities—were not shaped only by local constraints. The outside world impinged upon them everyday. Political struggles in Britain had as much an impact on their lives as the fluctuations of the market, the decisions taken by the British government, and changing notions about religion, wealth and labor, crime and punishment, literacy and education, trade and Empire, citizenship and government. Since they were not passive bearers of ideology, missionaries, colonists, slaves, and royal authorities would create their own scripts out of the available discourses and the material provided by their own past and present experiences. Yet they all were trapped in a process that in great part escaped their control.

The uneven development characteristic of the modern world was creating a profound contradiction between the colony and the mother country. While one was ever more dependent on slave labor, the other had increasingly become a land of "free" laborers. During the eighteenth century the redefinition and expansion of imperial domination, the massive enclosures, the development of commerce, trade, and manufactures in Britain had all undermined the traditional social basis of the gentry's authority, and put to the test their ideology of deference and patronage. The challenge from below, the factionalism of rulers divided for reasons of interest, purpose, and conviction between those who supported tradition and those who preached reforms, and the ensuing political struggles—aggravated by debates over the French Revolution and the war against France—opened the doors for new notions and policies. The new trends would not only challenge the authority of masters in Demerara (as elsewhere in the British Caribbean), but also bring into question the whole system of slavery, fanning the slaves' hope for emancipation precisely at a time masters were intensifying the rhythm of labor on plantations. Outside this larger context it would be impossible to understand fully the slave uprising, the missionaries' actions, and the colonists' responses.

But there is another side to the story. Demerara society was also changing from within. There was growing confrontation between masters and slaves. Torn from kin-centered or tributary societies, with their rules, norms, and decorums, slaves had been forced to redefine their identities in slavery—though not merely as slaves. From scripts brought from their pasts, modified by their new condition and environment, slaves wove new narratives about the world, created new forms of kinship, and invented new utopias. They did not try simply to re-create their past, but to control their present and shape their future. In their day-to-day interactions with masters and missionaries, they appropriated symbols that originally were meant to subject them and wrought those symbols into weapons of their own emancipation. In this process they not only transformed themselves and everyone around, but they also helped to shape the course of history.

Missionaries stepped into an already tense situation which their presence would only complicate. Seeing themselves as instruments of divine providence, they had come from England imbued with their sense of mission, driven by their own convictions, full of certainties about the way society should be, and determined to change it to meet their ideal of the people in Christ. They arrived ignorant of the notions of propriety, the rules, rituals, and sanctions that regulated the relations between masters and slaves. They met a "reality" clothed with signs and symbols to which they were blind, a "reality" they could only assess through their own codes. Not surprisingly, they violated many rules, and provoked the resentment and irritation of masters and managers.

John Wray and John Smith were sent to Demerara to convert the "heathen," but this abstract notion did not prepare them to deal with their flocks. They expected to meet ignorant "babes" waiting to be saved, but instead they found a people whose system of meanings they ignored and often took at face value, men and women of flesh and blood, seasoned in their struggles against managers and masters. Convinced of the superiority of their European culture and their religion, missionaries were torn between two contradictory impulses: one that led them to emphasize the slaves' otherness, and another that compelled them to assert the universality of the faithful and to recognize the slaves as brethren in Christ. They went to Demerara with the notion that slaves were savages to be civilized. But they soon discovered "humanity" in the slaves and savagery in people of their "own kind." Their chapel created a space where slaves from different plantations could legitimately assemble to celebrate their humanity and their equality as God's children. Slaves appropriated the missionaries' language and symbols, and turned their lessons of love and redemption into promises of freedom. Incensed by rumors of emancipation and convinced they had allies in England, the slaves seized the opportunity to take history into their own hands. How and why they did it are two of the questions this book tries to answer.

The slaves' rebellion and John Smith's trial had important echoes and consequences in Britain, where evangelicals, abolitionists and anti-abolitionists took sides for or against the missionary, in meeting halls, in the press, and in Parliament. The rebellion, the trial, and the ensuing debates generated many documents. Missionaries' diaries and their voluminous correspondence—registering day-to-day life on plantations and the missionaries' interaction with slaves, masters, managers, and local authorities—were carefully preserved by the missionary societies. The London Missionary Society's board of director's meetings, minutes of the meetings of committees in charge of selecting missionaries, and the candidates' papers, the *Evangelical Magazine* and other missionary journals describing the progress of the missions—all offer valuable insight into the training of missionaries and their work. The Colonial Office dispatches and records, the governors' correspondence, the letters from merchants, planters, soldiers, and militiamen who participated in the repression were all preserved in the Public Record Office. There, too, were kept the trials' proceedings, and copies of the minutes of the Demerara Court of Policy, Court of Criminal Justice, and the books of the Fiscals and Protectors of Slaves, which registered "offences" committed by slaves, their punishments and complaints, before and after the rebellion. Several parliamentary inquiries and Blue Books published by order of the House of Commons, containing precious statistics and other information about the colony, found their way into libraries all over the world. Although Demerara had a small population of about 2,500 whites, who lived surrounded by an equal number of free blacks and 77,000 slaves, the colony had three newspapers, which together with the annual almanacs and guides, travelers' accounts, and agricultural manuals provide a vivid and detailed picture of life in the colony.

Such a variety and abundance of sources has allowed me to adopt a narrative strategy somewhat reminiscent of the "polyphonic novel," and to tell the story of the rebellion from multiple points of view—without, however, giving up the privileges and responsibilities of a narrator. I have tried to bring together a macro- and a micro-historical approach. This decision was born out of a conviction that it is impossible to understand one without the other. History is not the result of some mysterious and transcendental "human agency," but neither are men and women the puppets of historical "forces." Their actions constitute the point at which the constant tension between freedom and necessity is momentarily resolved.

We have become so habituated to seeing history as product of reified historical categories, to talking about "variables" and "factors," to dealing with abstractions such as capitalism, abolitionism, evangelicalism, and the like, that we often forget that history is made by men and women, even though they make it under conditions they themselves have not chosen. In the last instance, what matters is the way people interact, the way they think about the world and act upon it, and how in this process they transform the world and themselves.

As historians we understand that history never repeats itself. But we transform historical events into metaphors and see universality in uniqueness. Otherwise, history would be a museum of curiosities, and historians nothing but antiquarians. The slave rebellion of 1823 and the Reverend Smith's predicament have universal value. They remind us of the many missionaries and lay people who, imbued with a sense of mission, a deep commitment to human brotherhood, and a strong passion for justice, became scapegoats in other times and other places. They remind us also that the slaves' struggle for freedom and dignity continued to be re-enacted under new guises and new scripts long after "emancipation." That is what makes the story of John Smith and the Demerara slave rebellion worth telling.

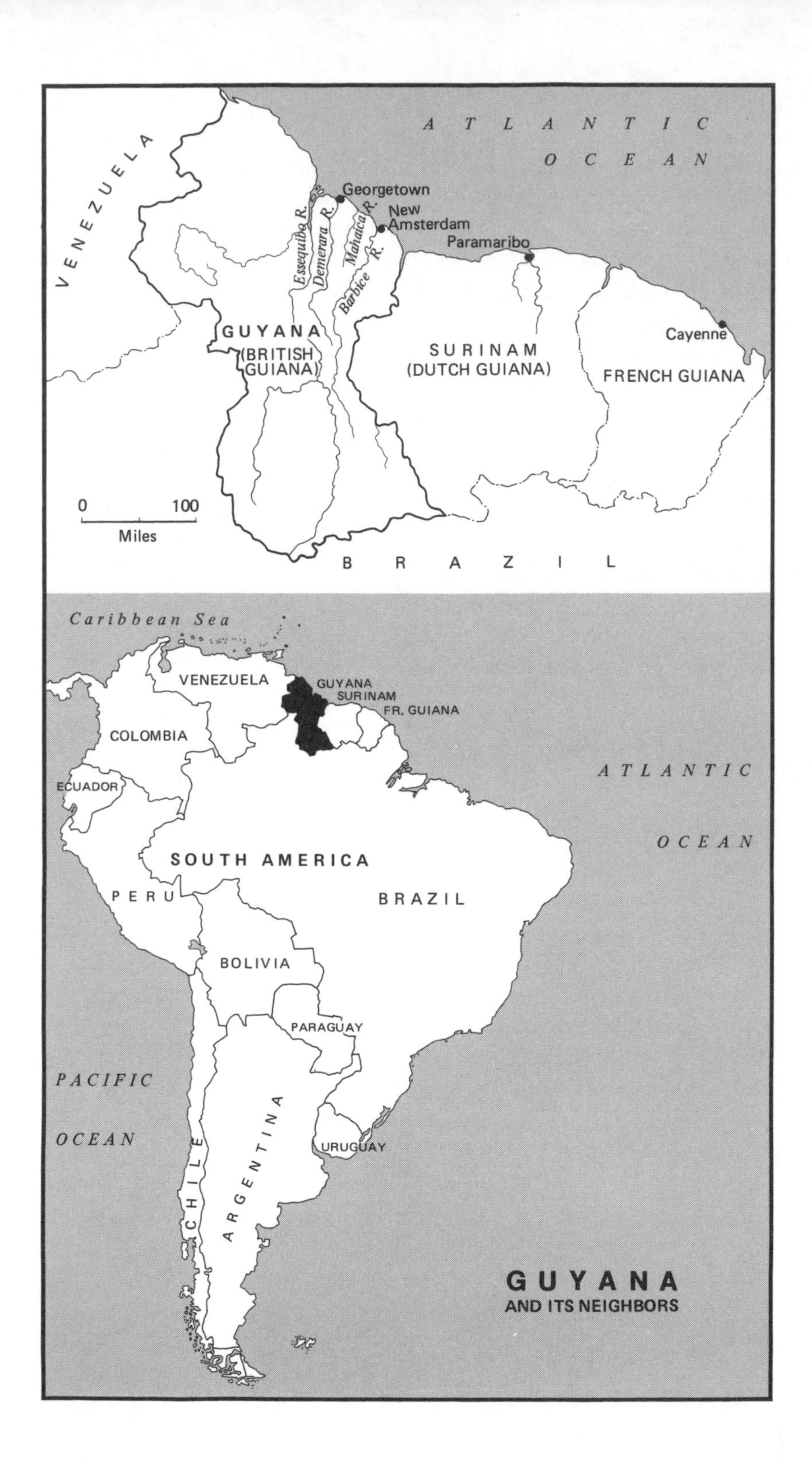
ATLANTIC
OCEAN
VENEZUELA
Georgetown
New
Amsterdam
Paramaribo
Essequibo R.
Demerara R.
Mahaica R.
Berbice R.
GUYANA
(BRITISH
GUIANA)
SURINAM
(DUTCH GUIANA)
Cayenne
FRENCH GUIANA
0
100
Miles
BRAZIL
Caribbean Sea
VENEZUELA
GUYANA
SURINAM
FR. GUIANA
COLOMBIA
ECUADOR
ATLANTIC
OCEAN
SOUTH AMERICA
PERU
BRAZIL
BOLIVIA
PARAGUAY
PACIFIC
OCEAN
URUGUAY
CHILE
ARGENTINA
GUYANA
AND ITS NEIGHBORS

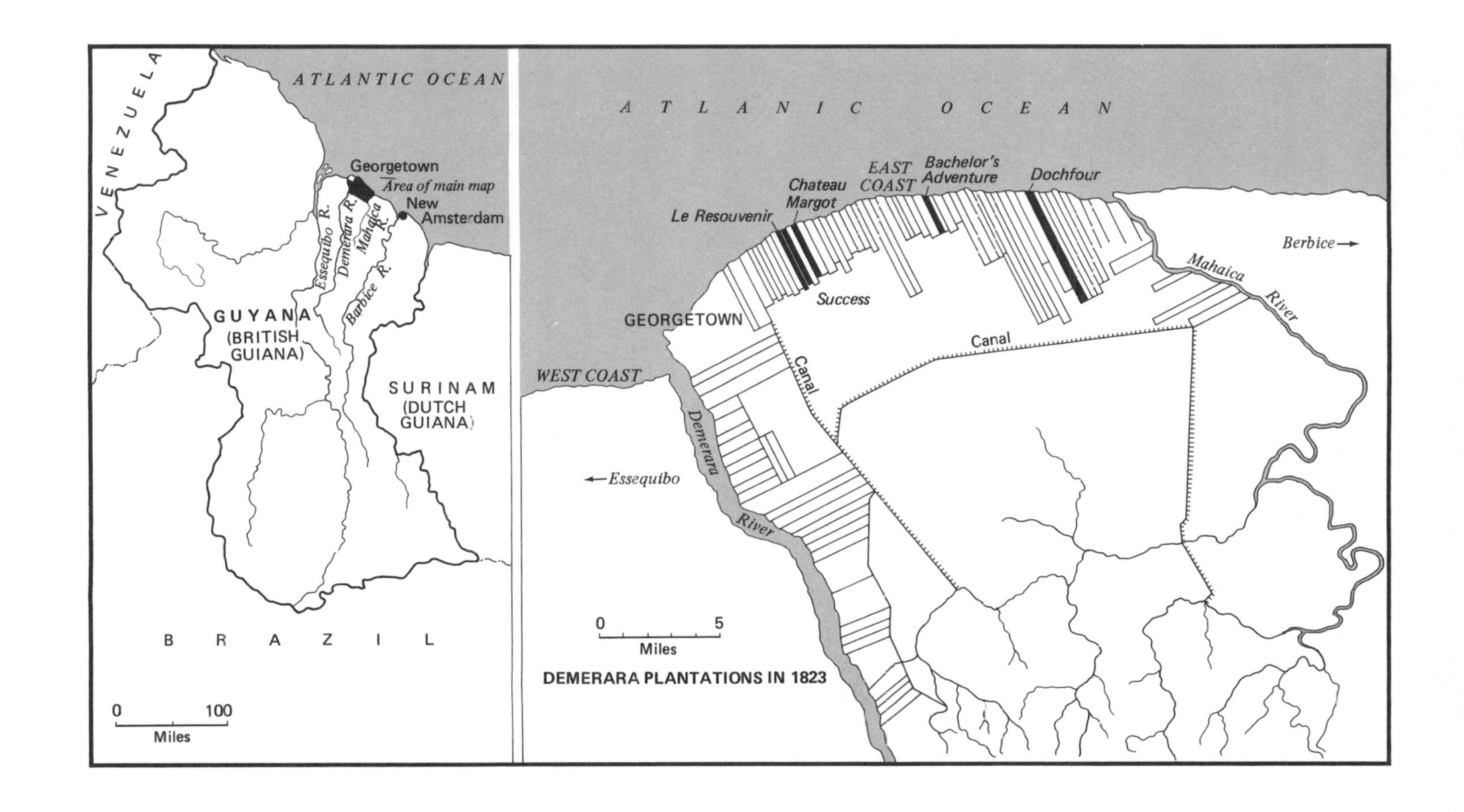
VENEZUELA
ATLANTIC OCEAN
Georgetown
Area of main map
New Amsterdam
Essequibo R.
Demerara R.
Mahaica R.
Barbice R.
GUYANA
(BRITISH GUIANA)
SURINAM
(DUTCH GUIANA)
BRAZIL
0 100
Miles
ATLANIC OCEAN
EAST COAST
Bachelor's Adventure
Dochfour
Chateau Margot
Le Resouvenir
Success
GEORGETOWN
Berbice
Mahaica River
Canal
Canal
WEST COAST
Demerara River
Essequibo
0 5
Miles
DEMERARA PLANTATIONS IN 1823

Crowns of Glory,
Tears of Blood

CHAPTER ONE

Contradictory Worlds: Planters and Missionaries

Every time is a time of change—but some are more so than others. Every time is a time of conflict—but there are historical moments when the scattered and individual conflicts and tensions that characterize day-to-day experience suddenly coalesce into a wider and encompassing phenomenon that threatens the "social order." At such moments, long-standing individual grievances are transformed into an overall critique of the system of power. Elites' assumptions about the world are challenged. What was once moral becomes immoral; what was right becomes wrong; what was fair becomes unfair. New discourses about society give consistency and organization to once-fragmented "revolutionary" notions, and claim the status of truth. These are dangerous times; these are exciting times: times of heroes and martyrs, of heresies and orthodoxies, of revolution and repression. Some will risk their lives in the name of the world being born, others to defend the world that is dying. Such are times of revolution. But when ruling groups appropriate radical discourses, purge them of their radical content, and manage through reform, cooptation, and repression to release some of the pressures from below—while creating new power blocs or coalitions—social revolutions are sometimes averted. If there is a price to pay for revolutionary upheavals, there is also a price to pay for reform and accommodation. Among the rulers, those who cannot make the leap will be left behind; among the ruled a few will benefit, but many more may see no fundamental change. For them, what changes during such moments of history is the system of exploitation and repression. It was in such a time that John Smith and his fellow missionaries lived.

Throughout the eighteenth century there had been talk of reform in Britain. But the first significant blow to the traditional social order was the independence of the American colonies, which triggered an intense debate over the notion of citizenship and brought into question the system of monopolies and privileges that had characterized the relations between European nations and their colonies. The second was the French Revolution, which brought into question relations between state and society, rulers and ruled. Events in France acted as a catalyst. They reinvigorated a libertarian and equalitarian rhetorical tradition that reached far back into the English history of Levellers, Dissenters, and Commonwealthmen. Old aspirations, debates, fears, and tensions found a new language and arguments, and a changed balance of forces.[1] English society was suddenly polarized, divided between those who hailed the French Revolution as the end of all tyranny and corruption and those who saw it as the beginning of anarchy and chaos. Finally, the Haitian Revolution, by challenging the power masters had over slaves, brought to the forefront the question of slavery. The powerful symbolic meaning of the three revolutions can only be understood with reference to the profound social and economic transformations taking place in Britain and the simultaneous redefinition and expansion of its empire.[2]

Few people living in Britain between the 1780s and the 1830s could have remained indifferent to political debate. The themes of equality, representation, freedom, tyranny, monopolies, corporate privileges, corruption, and representation were debated in every corner of England. These powerful notions challenged an order based on deference, rank, and patronage. Longtime grievances and resentments found expression in scores of books, pamphlets, and broadsides, perhaps most notably in Tom Paine's *Rights of Man*. Published in 1792, it rapidly gained tremendous popularity, selling 200,000 copies in six months.[3] Middle-class radical societies agitating for reform and the rights of man sprang up everywhere. Radical ideas found fertile ground among the growing urban populations, particularly in industrial centers such as Manchester, Sheffield, and Birmingham. Playing to the growing literacy among the poor, the radical press turned out a flood of pamphlets criticizing English political institutions and stimulating political debate.[4] The campaign for the repeal of the Test and Corporation Acts propelled dissenters to the forefront of popular mobilization.[5] Soon, however, conservatives were closing ranks to defend the establishment. Edmund Burke's *Reflections on the Revolution in France* (1790) was their source of inspiration. Well-orchestrated "Church and King" mobs were confronting radicals and dissenters. Spies were infiltrating radical associations. And to counteract the arguments of the reformers, conservatives were turning out their own popular tracts. Hannah More's *Village Politics: A Dialogue Between Jack Anvill the Blacksmith and Tom Hod the Mason* was a great success, and her pamphlets were outselling even Paine's.[6]

All over England working men and women came together to discuss political and social issues and to make their voices heard. Prominent among the issues was the abolition of the slave trade. William Wilberforce's motion to end the trade, brought before the House of Commons in 1789, aroused great public attention. The bill was delayed by parliamentary maneuvers of the opponents and eventually failed. When he tried in 1791 to introduce another motion to end the slave trade, Wilberforce was again defeated. A year later, backed by more than five hundred petitions from all over the country, he managed to carry his motion in the Commons, but the bill failed in the House of Lords.[7] The abolitionists turned then to a new strategy, stressing the horrors of slave labor and advocating a consumer boycott of West Indian sugar and rum. Their appeals to the public had a great impact.[8] By the time Smith and his fellow missionaries had reached the age when children start looking beyond the boundaries of their households, the abolition campaign had gained the heart of the "common man." Local abolitionist committees were created in towns and cities like Birmingham, York, Worcester, Sheffield, Leeds, Norwich, Northampton, Exeter, and Falmouth. Their members were manufacturers, tradesmen, doctors, clergymen, lawyers, clerks, and artisans.[9] Laboring men and women were being recruited in increasing numbers to support abolition.[10] From 1788 to 1791 the number of signatures on petitions for the abolition of the slave trade rose from 60,000 to 400,000. In Manchester, 20,000 people out of a total population of 60,000—virtually all the adults in the city—signed one petition, evincing both the abolitionists' capacity for mobilization and the petitioners' understanding that the revolutionary message of freedom and equality was a universal one.[11] As Thomas Hardy, one of the radical leaders, put it, the rights of man were not confined to England but "extended to the whole human race, black and white, high and low, rich or poor."[12] Abolition was firmly linked to the issue of reform at home. In the minds of many "ordinary" people, the abolition of the slave trade seemed to be tied to democratic principles, and the freedom of slaves to the rights of free men.

A wave of radicalism that appeared to find such widespread support among artisans and laborers was bound to alarm conservatives. To them it all seemed—as Charles Dickens would later put it—like an "obliteration of landmarks, and opening of floodgates, and cracking of the framework of society." In their opinion, "the country was going into pieces."[13] To them popular mobilization was a threat, and they were quick to respond with a series of repressive measures aimed at controlling public opinion and restraining grass-roots organizations. William Pitt, who five years earlier had supported a motion for the abolition of the slave trade, was by 1792 already negotiating with Robert Dundas to curb the radicals. Even Wilberforce, who had initially appealed to popular support, seemed to worry when it came with such vigorous enthusiasm.[14]

As early as May 1792 the government was encouraging magistrates to control "riotous" meetings more rigorously. Soon, the tragic direction taken by the French Revolution gave new arguments to the conservatives and put many radicals on the defensive. The execution of Louis XVI and the Massacres of September seemed to confirm the worst predictions of the enemies of the Revolution. And when Britain declared war on France in 1793, criticism of the social and political order could be labeled treason, and adepts of the French Revolution could be seen as traitors.

Repression was the principal weapon used by conservatives to contain the wave of radicalism. In 1794 Parliament suspended habeas corpus, so that "riotous people" could be tried without difficulties. And in 1795, after crowds took to the streets asking for bread and peace and hissed the king at the opening of Parliament, two new acts were passed. The Seditious Meetings Act prohibited meetings of more than fifty people and forbade lectures outside academic walls without permits. The Treasonable Practices Act defined the law of treason in stricter terms. These repressive measures were followed by a tightening of censorship. Radical leaders, editors, and authors were put on trial one after another. Not surprisingly, in the face of such repression, the abolitionist movement receded. And motions favoring reform presented to Parliament were defeated again and again. In the eyes of many, there was no difference between reform and revolution.

For the opponents of slavery, at least, the situation brightened somewhat when Napoleon tried to restore slavery in Haiti. Abolitionism became again an "acceptable" position in Britain, and the issue of the abolition of the slave trade was reopened. The Abolitionist Committee was reactivated in 1804, and in 1806 it fostered a campaign aimed at both voters and lawmakers. The issue was again debated in Parliament and in the press. In 1805 the British government issued an order in council prohibiting slave trading to the newly captured colonies. And finally in 1807, the House of Commons passed a bill making it illegal for any British ship to be involved in the slave trade after January 1, 1808. Three weeks after the King's assent had been given to the abolition of the slave trade, a group of high personalities under the patronage of the Duke of Gloucester and including the Bishop of London, George Canning, Henry Brougham, Thomas Clarkson, Thomas Babington Macaulay, William Pitt, William Wilberforce, Granville Sharp, James Stephen, and others founded the African Institution with the explicit purpose of monitoring the measure and promoting "civilization" in Africa.[15] Still, although voices in favor of reform were again being heard, radicals continued to be kept under tight control.

Repression weakened the radical movement, but did not erase the issues that had given rise to it. Curtailed in its public manifestations, British radicalism continued for the next thirty years to run underground, emerging here and there in different forms, from food riots to petition campaigns. From the Luddites of 1811 to Peterloo in 1819 to the Chartists of 1830, there was a continuous flow.[16] War expenditures, the Napoleonic blockade and the ensuing economic crisis, together with the decline of British trade, inflation, depreciation of the pound, bankruptcies and stoppages, successive bad harvests—all kept radicalism alive. In London and in the new industrial centers, people took to the streets again and again to protest the hardships imposed on them first by the war then by rapid economic and social change. And the hardship, the repression, and the protests all gave new strength to evangelicalism.

It was in this atmosphere of revolution and repression, intense class polarization, and social and economic change—which characterized the earlier stages of the industrial revolution—that John Smith had come of age. Like many others of his generation, he found in evangelicalism an antidote to the anxieties and confusions these processes had unleashed. The evangelical language of universal brotherhood carried the promise of bridging the gap between rich and poor, the powerful and the powerless, without the pains and costs of a violent revolution.[17]

During the French wars, there was a dramatic increase in the Methodist following. This coincided with a decline of revolutionary fervor among all the nonconformist sects.[18] It is possible to argue, as E. P. Thompson has, that this new religious movement was an attempt by its leaders to tame the radical impulse, that they aimed at fighting the enemies of the established order and at raising the standards of public "morality" by promoting loyalty in the middle ranks, and subordination and "industry" in the lower orders of society. That may have been the purpose of most of the Wesleyan bureaucrats of the day. The Wesleyan hierarchy may have disapproved of popular tumults and may have hoped to foster in their churches notions that might provide what Thompson calls "the psychic component of the work discipline of which manufacture stood most in need." Indeed there is much evidence that this was so.

But there is another side to the story. A bureaucracy's wish to impose orthodoxy is sometimes defeated by the variety of human experience of its constituency. This may explain the splits within Methodists, the appearance first of the New Connexion in 1797 and a few years later of the Primitive Methodists, who organized their churches in a more democratic way.[19] "As each group broke away from the Wesleyan Connexion (fast becoming a rigid institution), more pronounced forms of popular religion appeared. . . . Cottage religion became a recognized working-class alternative to the religion of the church."[20] With this came new and more radical interpretations of the Bible—interpretations that were carried to all corners of Britain by a growing number of itinerant preachers recruited among the popular classes.[21]

People always translate cultural messages into the terms of their own experience. A sermon may mean one thing to the preacher, but quite another to a congregation. And all the members of a congregation may not hear the same sermon the same way. It may mean one thing to the rich, another to the poor. Two preachers of the same denomination may extract from the same passage in the Bible two different lessons. Besides, the Biblical message is itself ambiguous. It can teach subservience, but it can also justify rebellion. Contradictions and ambiguities in the Biblical texts, and their profoundly metaphorical character, leave space for multiple uses and interpretations. Like any other message—perhaps even more so—the religious message is eminently symbolic, and its symbols will be decodified with reference to people's experience as a whole. When oppressed men and women, imbued with notions that make them see the world as a battlefield where God's soldier struggles against Satan and his followers, get together, no one can tell where their struggle will end. They may want to return to the past (as they imagine it); they may hold onto what they have in the present, or they may try to leap into a utopian future. They may follow anyone who claims to be a savior—as often happens in millenarian movements—but they can also come to question the legitimacy of their rulers.[22]

The whim or will of ordinary people could easily defeat the hegemonic purposes of the Methodist bureaucracies—even more so because there was a fundamental contradiction between their message of human brotherhood and the increasing social confrontation brought about by economic and social change. The British ruling classes may in the end have benefited from the ethic preached by the dissenters and evangelicals. But the artisans and laborers who flocked to the Methodist churches had motives of their own, particularly those who joined the more radical New Connexion or the Primitive Methodists, and those—like Smith—who in increasing numbers joined other nonconformist evangelical sects. The concept of calling, the values of freedom, self-discipline, self-reliance, frugality, and sobriety that the Methodists and other sectarians and evangelical groups preached had a strong appeal to broad sectors of the laboring classes. And the apparent deradicalization and submission of the poor—which Thompson attributes to the spread of Methodism—may have been more a product of their sense of helplessness and their fear of repression than of their religion.[23]

Artisans, laborers, or shopkeepers who were unhappy with the social order had two choices: they could follow the secular radicals or join the nonconformist evangelicals. Radicals raised the issue of citizenship, placed all evil in the social order, and called on men and women to struggle to reform society. They spoke to the workers' minds. Evangelical dissenters turned the relations between oppression and evil upside down, and made evil the source of oppression. They told people to reform their souls and promised that a reform of the soul would bring a new social order. They spoke to people's hearts. It is easy to imagine that to many people the task of reforming society may have seemed at this point more overwhelming and less rewarding than the task of reforming their own souls, particularly since they believed that in this effort they would have God on their side. The radicals' and evangelical dissenters' discourses about the world presented themselves as alternative views and practices. But sometimes there were surprising convergences.[24] When evangelicals identified piety with one social group and sin with another, and incorporated into their discourses notions about the rights of man, they could easily sound like radicals. And it was not by chance that many abolitionists and trade-union leaders were evangelicals and dissenters.

It is easy to understand the appeal that Methodists and other nonconformist evangelical sects had among laboring people in England during the years of war, repression, and economic change.[25] They may have helped workers escape their anxieties, preserve their radicalism under "acceptable" forms, or defend themselves against the destructive forces of the market. They may have helped workers protect themselves and their families—through education, self-discipline, thrift, and sobriety—from unemployment, starvation, drunkenness, and prostitution. To women, the evangelicals may have offered a means of keeping their families together, their husbands and children away from the tavern and the many "seductions" of the city. Evangelicals may have helped people to preserve a sense of dignity and worth in a world where humiliation was a daily experience. They may have given to those who felt powerless a sense of personal strength, and to those who despaired they may have given hope. For migrant workers—and they were numerous in England at this point—evangelical nonconformist churches may have offered community and support in an otherwise alien and hostile world.[26] Nonconformists challenged the Church of England and (implicitly or explicitly) the power and authority it represented. So when men like Smith went to other parts of the world they carried with them a message of freedom, equality, and brotherhood, and a sense of fairness, all of which could easily be turned against the established order. This would be particularly true in slave societies, where the ethic implicit in this new evangelicalism seemed not only out of place, but profoundly disruptive. No wonder most plantation owners opposed nonconformist evangelical missionaries and perceived them more as a threat than as a means of social control. But even in England, many people habituated to traditional means of social control and to an ethic of rank and patronage felt that evangelical dissenters were subverting the social "order."

The diatribes of an irate country parson from Ipswich—cited by Thompson—bear a remarkable resemblance to those of Demerara colonists against the missionaries and disclose the class nature of their fear and outrage. He complained that the field laborers converted to Methodism spread the idea that "Corn and all other fruits of the earth, are grown and intended by Providence, as much for the poor as the rich." They were less content with their wages, less ready "to work extraordinary hours as the exigencies of their masters might require." Worse, instead of "recuperating" themselves for the next week's labor, they exhausted themselves on Sundays by walking miles to hear a preacher. On week nights, instead of going straight to bed, they wasted fire and candles singing hymns—something the parson had been horrified to see "in some of the poorest cottages as late an hour as nine."[27]

Such fears and irritation were not unrealistic. In a time of profound class polarization, the language of universal brotherhood, betrayed by day-to-day experience, was potentially subversive (even more so in a slave society). The brotherhood it postulated, even if intended to reduce social tensions, could in fact aggravate them. Depending on political "praxis" as a whole, what might have started as a process of alienation could in the long run lead to emancipation. The evangelical message gave the oppressed a code with which to judge their oppressors. Its claims to universality undermined assumptions behind social differences and provided a utopia against which to judge the world. It was not by chance that many evangelical missionaries—particularly dissenters who gave a great autonomy to their congregations—like the Congregationalists, for example—were perceived by sectors of the ruling classes as a threat. Demerara colonists were not alone when they suspected that the Gospel the missionaries preached could be turned by the slaves against their masters.

The point of view of most of the colonists was brilliantly defined in an article in the *The Essequebo and Demerary Royal Gazette,*[28] just a few months after the arrival in Demerara of two missionaries from the London Missionary Society. The author condemned "the precarious preachers of a pretended enlightening doctrine, who announced equal rights, universal liberty," and who intended to make all men into one happy family. These preachers might have the best intentions, and "they spoke as philosophers," but they did not anticipate the possible and "unfortunately fatal consequences of their beautiful but premature principles." Confusion, patricide, persecution, and even destruction of the balance of nations, and "the butchering of millions of men" would follow from these principles: that all men being equal had equal rights; that the distinction of rank had ceased to exist; that understanding, virtue, and merit alone could justify an eminent place in society. So "dark and so terrible" seemed the consequences of these doctrines that the author of the article—who styled himself "an honest colonist"—felt the need to state that there was a fundamental incompatibility between Christianity and slavery. "He that chooses to make negroes Christians let him give them their liberty, let him not make them unhappier, let him not expose the Society of which he makes a part, to the illusions of his exalted brain." If the "negroes" arriving in the colony were unspotted children of nature, then it would be possible to instruct them in the noble truths of the Gospel. But who were these imported "negroes"? "A parcel of criminals who having been guilty of theft, murder, and other crimes in their own country . . . or prisoners of wars, who having lived a life of war, cruelly and dissolute, [were] not fit for a regular government based on moral persuasion" and had to be guided by corporal punishment.[29] The article vividly displayed most colonists' biases and concerns and their apprehension of the arrival in the colony of missionaries from the London Missionary Society.

Mistrust of Protestant dissenters had a long tradition in British history that not even the Toleration Act of 1689, revised in 1711 and 1812–13, had succeeded in erasing completely.[30] Hostility in the colonies against Protestant dissenters echoed the complaints of some conservative groups in the Church of England and in Parliament,[31] who continued to see them as "subversive" elements, and as late as 1811 were still trying to restrict their activities.[32] But such people were losing ground in Parliament. Lord Sidmouth's attempt to curtail the "abuses" of the Toleration Act, particularly the recruitment for the ministry of "improper" people such as "blacksmiths, chymney-sweepers, pig-drovers, pedlars, coblers and others of the same sort," was defeated in 1811 after a severe political struggle. And although Protestant dissenters continued to suffer several legal disabilities until 1828, the tendency, both in the government and in the Church of England, was to favor evangelical missionaries and their work among the slaves. But in the colonies nonconformist evangelical missionaries would meet growing resistance from planters and local authorities.[33]

Demerara planters did not draw only on British traditions to justify their opposition to the presence of evangelical dissenters among them. They also found reassurance in other Caribbean colonies where the same hostility existed. Everywhere in the Caribbean, nonconformist evangelical missionaries were under attack in the early decades of the nineteenth century. Demerara newspapers constantly reproduced articles from Jamaica, Barbados, Trinidad, or other islands, denigrating missionaries. The newspapers lumped them all together under the label "Methodists" and stressed their connections with abolitionists and the African Institution.[34] Colonists everywhere raised obstacles to the missionaries' work.[35] But they had to contend with the British government, which seemed intent on supporting the missionaries. In Jamaica, a Consolidated Slave Act went so far as to provide that religious instruction for slaves should be confined to the doctrines of the Established Church, and forbade Methodists and other "sectarians" to instruct their slaves or allow them in their chapels. The Privy Council, however, disallowed the provision as contrary to the principles of toleration prevailing in Britain.[36] In an 1811 circular to the crown colonies, Lord Liverpool, Secretary for War and Colonies, made it clear that the government favored the religious "instruction" of slaves.

Colonists resented such pressures, and at one time or another Bermuda, Anguilla, Jamaica, Tobago, and Saint Vincent all passed laws restricting missionary activities. But their royal governors could only obey instructions from London. The conflicting views of colonists and the British government placed governors in a difficult position, specially when they became slaveholders themselves. In 1816, Governor Woodford of Trinidad in a letter to Lord Bathurst admitted that if it were not for the pressure from England he would have expelled the missionaries from the colony: "I should have been much inclined long since, to have sent from the island the missionary and the Methodist preacher, if the general expression in England in their favor had not induced me to permit a continuance of their residence here. From the intimacies which they contract with the negroes, their presence is always a source of uneasiness."[37]

Throughout the Caribbean, what colonists seemed to be most offended by was the evangelical missionaries' "democratic" manners, their rhetoric of equality, and the degree of liturgical autonomy they gave to the slaves.[38] To the colonists, the ideas the missionaries brought were a threat to the social "order." It was because of such ideas—they thought—that slaves had risen in Haiti.[39] The colonists were convinced (not without reason) that the missionaries had connections with abolitionists, and that they were part of the increasingly powerful lobby in favor of emancipation. Joseph Marryat, perhaps the most eloquent spokesman for the West Indies in Parliament, gave frequent and free expression to the suspicion that missionaries constituted a powerful political network, ready to manipulate politics to achieve their goals.[40]

The abolition of the slave trade was a serious blow to the colonists—both to their pride and to their pockets.[41] During the campaign for abolition in England, slavery had been portrayed as the source of all evils, and the planters as godless tyrants ready to commit every sort of atrocity. Colonists resented both the abolitionists and the evangelical groups that supported them. Demerara planters were particularly hostile to missionaries from the London Missionary Society. They suspected the society had connections with Wilberforce, the African Institution, and other abolitionist groups in England.

(In truth, Joseph Hardcastle, the London Missionary Society's treasurer, was a close friend of well-known abolitionists like Thomas Clarkson, Granville Sharp, and William Wilberforce, and "a stout co-operator with them in every movement for the amelioration of the condition of the slave." And the LMS did indeed resort to Wilberforce—whenever missionaries confronted opposition in the colonies—and gained his support.)[42] Not surprisingly, most colonists saw missionaries as spies, always ready to send home tales that reinforced the worst prejudices in England against slaveowners.

Most colonists were convinced that giving religious instruction to slaves, teaching them to read, treating them as equals, calling them "brethren"—and so abolishing the social distinctions and protocols which in day-to-day experience reasserted the power masters had over slaves—would sooner or later lead slaves to rebel. Not even the constant reassurances from missionaries that they were careful to teach slaves submission to their masters could persuade colonists to see missionaries as trustworthy allies. Managers and masters wanted to have total power over their slaves and resented anything they feared might weaken their control. They disliked the missionaries' sabbatarianism, which interfered with the work discipline on the plantations, and were even more hostile to the idea of slaves going freely at nights to attend religious meetings that might be miles away. And they were exasperated whenever missionaries spoke for the slaves or against managers and masters. Equally annoying to the colonists were the missionaries' modest social origins and the atmosphere of intimacy they tried to create in their congregations, disregarding racial and social boundaries. This was made clear in a 1813 article published in the *Royal Gazette,* under the title "West Indies Methodists":

> Our colonies are now inundated with canting hypocritical tailors, carpenters, tinkers, cobblers, etc. who, too lazy to work for an honest livelihood in the Mother Country, and charmed with the idea of living in ease and luxury abroad, found it very convenient to become converts to the new light and volunteer to teach the Gospel without the ability to spell one of its verses. It was not until a conspiracy of a considerable extent was discovered, and which in a short time would have given a negro King to Jamaica, that the Assembly of that island thought proper to inquire into the cause of the apparent insubordination among the negroes. In the crowded Methodist chapel in Kingston, it is quite common to hear a fellow hold forth two hours, to a congregation of negroes and coloured people, in a jargon not otherwise intelligible to his audience than that we are all brethren, and on terms of perfect equality. Their love-feasts, in which Blacky was admitted to eat bread with Massa Parson, at once upset all colonial discipline and destroyed that respect for the Whites, which alone could ensure order and tranquility to the island. Under pretence of preaching and praying, members assembled and held meetings from three

> or four o'clock in the morning until day-light, and again after sun-set. At these a negro or mulatto, who did not know a letter of the alphabet, would mount a chair, and imitate, with tolerable accuracy, the gesture and grimace of the White Preacher, to the great edification and amusement of his audience.

All this seemed profoundly wrong and threatening to the author, who exhibited the interlocking race and class prejudice common to most colonists.[43]

Methodism—which in the English context E. P. Thompson assessed as a conservative ideology aiming at deradicalizing workers and domesticating their labor—appeared to the colonists as profoundly subversive. What for the missionaries was a means of social control for most colonists was revolutionary ferment. In England it may have been desirable, from the point of view of some sectors of the dominant classes, to train free laborers in the ethic of self-reliance and self-discipline so that they could become "their own slave drivers," as Thompson aptly put it.[44] In a colony, however, where there were *real* slaves, the only possible form of discipline colonists could envisage was that of the driver himself. When the Demerara colonists opposed Smith and his fellow missionaries, the colonists were reacting not only to what the missionaries preached but to what they represented. And the colonists were not wrong when they recognized that the missionaries were preaching a new way of seeing the world, a way more suitable to a society of free laborers, a way that could undermine the moral foundations of colonial society.[45]

Subversion was certainly not the conscious intention of the London Missionary Society or its missionaries. In fact, the society explicitly urged the men and women it sent into the world not to meddle with political affairs, and to teach slaves to obey their masters. Missionaries were told to take the utmost care not to endanger the public "peace" and "safety." Not a word should be said, in public or in private, that might render the slaves displeased with their masters or dissatisfied with their station. Nothing could be more clear than the instructions given John Smith when he left England for Demerara:

> You are not sent to relieve them from their servile condition, but to afford them the consolation of religion, and to enforce upon them the necessity of being "subject, not only for wrath, but also for conscience sake." Rom. 13.5, 1 Pet. 2.19. The holy gospel you preach, will render the slaves who receive it the most diligent, faithful, patient, and useful servants; will render severe discipline unnecessary, and make them the most valuable slaves on the estates: and thus, you will recommend yourself and your ministry, even to those gentlemen who may have been averse to the religious instruction of the negroes.[46]

The other missionaries the London Missionary Society sent to Demerara—John Davies, Richard Elliot, and John Wray—received analogous orders.[47]

They were warned against meddling with secular disputes. They were told that they should try to obtain the protection of local governments, and through their good behavior and respect to all who were in authority secure the enjoyment of liberty necessary to instruct and promote the salvation of the slaves. They were cautioned not to interfere in public or private with the slaves' "civil condition," and to instill in them notions of obedience and respect toward their masters. They were reminded that their only mission was to save the slaves' souls. Missionaries should not in any case engage in any of the civil disputes or local politics of the colony, either verbally or by correspondence with any person at home or in the colonies.[48] The Wesleyans also delivered very similar instructions to their missionaries, particularly those going to the West Indies. Such instructions, of course, were easier to give than to follow.

Most colonists in Demerara did perceive a profound contradiction between Christianity (particularly in its nonconformist evangelical version) and slavery. And, not surprisingly, most missionaries—if they had not learned it from the speeches of men like Thomas Clarkson and Wilberforce—soon arrived at the same conclusion. The difference was that the colonists wanted to get rid of the missionaries, and the missionaries wanted to get rid of slavery. There was no intrinsic contradiction, however, between Christianity and slavery. As the experience of the Portuguese and Spanish countries had shown, Christianity, in its Catholic version, could easily accommodate slavery. This had also been true in Protestant colonies where the Moravians had some success. But in those cases Christianity was still associated with a traditional view of the world, with a hierarchical concept of class relations that emphasized reciprocal obligations and sanctified social inequalities.[49]

It was only when the notions of personal freedom and individual rights that had merged with Christianity to produce what came to be called the Protestant Ethic were infused with a democratic view of the world, that people began to think that there was a fundamental contradiction between Christianity and slavery.[50] Specifically, it was only when people started questioning traditional institutions and power structures, and when they challenged a social and political system based on rank, monopolies, and privileges that they established a connection between spiritual bondage and physical bondage, slavery and sin, personal redemption and the emancipation of blacks. Once these ideas were linked—and only then—supporting emancipation became a task of militant Protestants.[51] Only then did they find in the same Bible that had been used to justify slavery equally compelling arguments to support an antislavery stance.[52] It was precisely in this new ethic that John Smith had been raised.[53]

The London Missionary Society's goal was certainly to rescue souls, not bodies. What remained to be seen was whether one was possible without the other. The society had been founded in 1795 with the purpose of saving the souls of the world's millions of "heathen." The idea was not new. Other missionary societies had been founded before, and some denominational groups—Presbyterians, Moravians, Wesleyans, Baptists—already had missions in different parts of the world.[54] What was new was the attempt to create a non-sectarian organization which would gather missionaries from different denominations—dissenters as well as men from the Established Church—in a universal crusade. In some way, this idea had been born as a response to the French Revolution. As the earliest historian of the London Missionary Society put it, "Christianity itself had been challenged, the new missionary policy was a bold and trumpet-toned acceptance of that challenge."[55]

The language of the Enlightenment, however, had a particular appeal to dissenters who were still struggling for the removal of restrictions imposed upon them in the past.[56] On the day of the inauguration of the London Missionary Society, the Reverend Mr. Bogue of Gosport celebrated the "funeral of bigotry," which he hoped would be buried so deep never to rise again.[57] The sermons preached on the occasion contrasted—in ways reminiscent of the Enlightenment—the times of "ignorance" and the new "luminous" times.[58] The preachers spoke of times of error, superstition, and persecution, and of the "fullness of time," the "refulgence of social truth." They rejoiced at the idea that "the gloom of superstition, error and sin" was going to be "forever banished from the face of the earth," and expressed the hope that dissension would be overcome and "the inhabitants of different climes, customs, colours, habits, and pursuits" would "be united in one large society, under the genial influence of Gospel and Grace." Their message was prophetic and ecumenical. The London Missionary Society defined as its goal "to promote the happiness of man and the honour of God."[59] And most important, it addressed all social classes without distinction.

In a few years the LMS created missions to the South Seas, South Africa, India, and the West Indies. It worked in close cooperation with other missionary societies and with the Religious Tract Society and the British and Foreign Bible Society. And, since Baptists, Wesleyans, Moravians, and Presbyterians already had their own missions, the London Missionary Society—which had started with the intention of bringing together ministers of different persuasions—ended by relying more and more on Congregationalists.

An article published in the French *Journal des Débats* on English missionary societies commented that their main purpose was not so much to extend the Kingdom of Christ as to consolidate the "Empire of the British Leopard," who sought to unite all his moral and physical powers, the better to hold in his grasp all his distant conquests.[60] Indeed, by sending British missionaries with their culture and values to different parts of the world to teach natives to live like Englishmen, the missionary societies did serve that purpose. But whatever other functions they may have performed as an arm of the empire, the LMS and most missionaries it sent to distant and often dangerous lands had only one conscious mission: to save the souls of millions who were perishing in sin. Yet neither the purposes of the imperial Leopard nor a devout determination to save the heathen exhausted the meanings and purposes of the missionary experience. For some of the young men who went among the heathen, what may have mattered as much as anything else was their own determination to have a career that would lift them out of obscurity and near poverty. All these different purposes and meanings were not incompatible, which may help explain the success of the LMS in recruiting missionaries.

The *Evangelical Magazine,* which was founded two years before the LMS with the explicit purpose of counteracting "the pernicious influences of erroneous doctrines" (like the ideas of Tom Paine and the French Revolution), became after 1795 the voice of the London Missionary Society. It published missionaries' biographies, memoirs, diaries, ecclesiastical history, book reviews, "authentic anecdotes," "striking providences," the last words of dying Christians—all to impress readers with examples of God's grace and to instruct them in the fundamental doctrines of the Gospel. It was in the pages of this magazine that young men and young women like John Smith and his wife Jane found guidance and inspiration. It was here that their letters about their mission, like the letters of other missionaries around the world, were published. Here they learned of the glory of missionary work, of their duty to devote the Lord's Day to preaching and self-examination, of their sinful nature and their dependence on God's grace.

Like many other religious journals of the period, the *Evangelical Magazine* aimed at reaching the growing number of literate men and women among the poor. Literacy was becoming so widespread in England that, in the optimistic view of the magazine's editor, in a few years it would be difficult "to find a beggar . . . who has not been taught to read."[61] The *Evangelical Magazine*'s explicit goal was to supply information fitted to "everyone's capacity and suited to every one's time and circumstances."

An egalitarian principle inspired the London Missionary Society's many calls for new missionaries. The directors stressed that none should be deterred from applying "by a mistaken opinion that learned men alone were qualified for such an employment." They argued that "successful missions, to those heathen countries which had made little or no progress in arts and sciences" should be chiefly composed of "serious mechanics." "Blacksmiths, whitesmiths, carpenters, gardeners, rope-makers, boat-builders, persons skilled in pottery and earthenware, and such as understand the smelting or fusing iron, might, therefore, provided they have the gift of communicating religious knowledge by their good conversation, be eminently useful."[62] The society welcomed men of good "natural abilities . . . well acquainted with divine truth and of good experience," even though they might not be learned, and promised to take charge of their education—opening the missionary career to men of modest origins.[63]

The recruitment policy was successful. Only a year after the foundation of the LMS, the *Evangelical Magazine* proudly announced that at Coventry, a minister, a buckle-and-harness maker, a weaver and his wife, and a gardener and his wife, had all been recommended to the directors by the subcommittee of examination responsible for choosing candidates for missionary lives. All had been accepted. Soon the society was sending such artisans-turned-missionaries to various parts of the world.[64] They were the kinds of people who were glad to devote themselves to a missionary career in distant and inhospitable countries. Upper- and middle-class students who graduated from Oxford or Cambridge had better prospects at home and were not likely to feel the call to preach to the heathen abroad. In 1814 the *Evangelical Magazine* noticed that "few students in our academies have offered themselves for missionaries"; in its first eighteen years the London Missionary Society received applications from only three or four academically trained students.[65]

So the salvation of "thousands of perishing sinners in the heathen land," particularly among the "Hottentotes and Negroes and a multitude of other rude tribes of mankind,"[66] was entrusted to men of modest origins, who were likely to see the world from their particular class perspective. Their social origins and their experience in England in a period of intense political debate, social polarization, and repression, at a time when evangelicalism was making progress among the working classes, may help to explain the abolitionist proclivities of some of them, their sympathy toward slaves, and their hostility toward masters. In some way, for them—as for millions of working-class people who signed petitions first in favor of the abolition of the slave trade, and later for total emancipation—the critique of slavery functioned as a metaphor.[67]

The debate over the abolition of the slave trade was still fresh when John Wray, the first LMS missionary to go to Demerara, left England. Even the *Evangelical Magazine*—usually indifferent to political matters unless they involved dissenters' freedom—had broken its silence from time to time to condemn the slave trade. In 1805 an article on the slave trade reminded Christians that their Master had come into the world to give life and to save it, "to proclaim deliverance to the oppressed." Another suggested that Protestant dissenters and ministers of the Established Church lead a petition campaign in support of the abolition of the slave trade. Still another observed that the "abolition of the slave trade is so desirable to men who truly love and fear God that it would be a matter of wonder to me it has not been accomplished if I did not know how exceedingly strong and powerful the argument of self-love and self-interest is with the majority of mankind."[68] For many evangelicals, slavery and the slave trade had become associated with sin.[69] The opposition the missionaries would meet in the colonies could only reinforce their conviction that the colonists' "self-interest" was an obstacle in the way of men who truly loved and feared God. And it is not surprising that when the missionaries from the London Missionary Society arrived in Demerara the colonists would see them as enemies.

Colonists and the Mother Country

History seemed to be running against the Demerara elites. Like slaveowners everywhere, they were caught in a process that seemed to condemn slavery and the system of values and sanctions associated with it to oblivion. To them, the missionaries represented new, powerful, and threatening historical trends that were undermining their ways of living. For it was not only slavery that was coming into question, it was the colonists' sense of status, their notions of discipline and punishment, their ways of conceiving relations between masters and slaves, blacks and whites, rich and poor, colony and mother country. It was not only their right to property that was being challenged, it was also the monopolies and privileges they had always enjoyed in the mother country. Debates in Parliament and in the British press were increasingly ominous. And the colonists were ready to resist.

Historical change seldom comes suddenly to people. And most of the time it is difficult, if not impossible, to say precisely when things start changing. But there are some historical periods when people do suddenly become aware that the world is not what it used to be. That was the feeling many colonists in Demerara had during the years between the arrival in 1808 of the first LMS missionary and Smith's trial in 1823. And the colonists responded as people usually do in such circumstances, with suspicion, fear, and anger.

Demerara originally had been a Dutch colony and although its definitive incorporation into the British empire occurred relatively late (1803), most plantations were in the hands of British citizens. By 1802, seven of every eight plantations in Demerara were British-owned. The increase in the colony's British population was in part related to the period of war and political strife which began with the American Revolution and ended only after Napoleon's final defeat in 1815.[70] During those years the colonists had been dragged into the wars of the mother country. They had suffered from naval blockades, occupations, and various sorts of retaliation. Between 1780 and 1803, Demerara changed hands six times. Initially, after American independence, Holland had tried to remain neutral. But the colonists, although prohibited from trading with the North American colonies, continued to receive American shipping. In 1780, England declared war on Holland, and Admiral George Rodney seized the colony. This British occupation lasted only one year. In 1782 the colony fell under French control, and two years later Demerara was given back to the Dutch.

Since the first British invasion, trade had been relatively free and in spite of political turmoil the area had continued to develop. So when the Dutch company attempted to recover its power and control over the colony the colonists resisted, forcing the company to back down. In 1791 the company lost its charter, and from that point on, the colony was directly subordinated to the Dutch States General. Then, in 1794–95, France invaded Holland and the Prince of Orange fled to England. When the Batavian Republic was inaugurated the colonists were divided. Royalists and republicans marched in the streets of Stabroek, the royalists wearing orange cockades and hailing the Prince of Orange, the republicans carrying tricolor cockades and shouting liberty, equality, and freedom—oblivious to the contradiction between their slogans and the reality of slavery that surrounded them. Soon, however, the republicans lost their enthusiasm. Orders came from Holland that the ports should be closed to all nations except Holland and France. Support for the republic vanished entirely when whites heard that Victor Hughes—the French "Negro Commissioner" for the West Indies—was talking about arming slaves and promising them emancipation. The specter of Haiti continued to haunt the colonists, confirming their fears and cooling their republican fervor.

Facing the prospect of losing not only the freedom to trade with England but also their slave property, some colonists decided to seek British intervention. These efforts coincided with similar pressures in England from supporters of the Prince of Orange. It is also possible that part of the pressure came from British merchants and textile groups interested in the colony's cotton. Whatever the reason, the British occupied the area again in 1796 and stayed until the Treaty of Amiens in 1802. These were years of great prosperity for the colonies of Essequibo, Demerara, and Berbice. Integration into the British market, combined with the disorganization of the economy in Haiti and other French Caribbean colonies, created particularly favorable conditions. There was an increasing flow of British capital and of colonists, some of whom came from Barbados and other British colonies in the Caribbean, attracted by the fertility of Demerara's soil and its freedom from the hurricanes that so often plagued the islands. Much Dutch property was sold to Englishmen,[71] and their manners, customs, and language were adopted.[72]

The importance of the ties between the colonists and Britain became obvious in 1802 when British rule was again interrupted and Essequibo and Demerara were returned to the Batavian Republic. Once again the Dutch government prohibited the shipping of goods to England. The new governor made the mistake of decreeing the banishment of British citizens who would not swear allegiance to the Dutch government. Nothing could have been more damaging and irritating to the majority of the colonists, and it did not take long for them to start conspiring for the return of British rule. They were supported by merchants in London, Liverpool, and Glasgow who feared for the capital they had lent to the planters. In 1803 the Dutch were displaced by the British—this time for good.[73]

As a result of all the changes during the last two decades of the eighteenth century, the colonists had acquired relative autonomy in administering their own affairs. So, although the return of the British was welcomed and even desired by many settlers, they tried to keep as much independence as they could when they signed the formal terms of surrender.[74] They demanded that the traditional laws and usages of the colony remain in force, the mode of taxation not be altered, and that their religion be respected. They also demanded that public officers (except for the governor) continue in their offices and that all inhabitants be protected in their persons and properties. They refused to take up arms against external enemies, and stipulated that the costs of building new barracks, erecting batteries, and provisioning soldiers and civil officers should be paid from the Sovereign's or Government Chest. They also insisted that no slaves be required for any "Black Regiment." All these conditions were accepted by the British as part of the colony's Capitulation Act. But such things were easier to insist on than to maintain in practice.

As time passed, the colonists found themselves struggling without success against British authorities encroaching on what the colonists thought were their rights. When the governor proposed to build a custom house, the colonists opposed—arguing that according to the Capitulation Act no new establishment was authorized. And when the government tried to levy a 4.5 percent export duty, it met with similar resistance. In the end—fearing that British vessels would not be authorized to enter Demerara for lack of a custom house—merchants and planters backed down. The struggle between colonists and royal authorities was still continuing when John Wray, the first LMS missionary sent to Demerara, arrived. Against this background of conflicts, it would not have been difficult to predict that the colonists' resentment of British interference in colonial affairs could easily be transferred to Wray or any other British missionary, particularly if he showed any sympathy for the cause of emancipation.

The year before Wray's arrival, local planters had protested loudly against the abolition of the slave trade. The *Royal Gazette* reproduced a petition signed by West Indian planters and merchants which had appeared in London newspapers.[75] The petitioners expressed their regret and alarm and asserted the legitimacy of the slave trade. They argued that agriculture in the colonies was impossible without African labor and talked about the bad consequences that would result from the abolition of the trade, not only for the planters but for the empire. The end of the slave trade would destroy the great capital invested in the West Indies. It would cut down on a commerce which paid almost three million pounds in annual duties to Great Britain, employed more than 16,000 seamen, and accounted for a third of British exports and imports. If the slave trade were abolished whites would leave the colonies, creating a dangerous situation. The petitioners concluded by saying that the proposal to end the slave trade violated laws of property, the well being of families, and the security of creditors. They cautioned: "Politicians must view, with peculiar alarm, a renewed discussion of that question, at a period when the existence of a black power [Haiti] in the neighborhood of the most important British island in the West Indies" afforded a "memorable and dreadful lesson." In spite of their protests, however, Parliament approved the feared legislation.[76]

Aside from the tensions between the colonists and the imperial government, Demerara was also torn by many internal conflicts. The basic struggle, of course, was that between masters and slaves (which will be examined in the next chapter). But, there were others born of the colonists' differences in nationality, class, ethnicity, and religion. Although Dutch and English colonists tended to agree on the whole on issues related to economic policies, they were often at odds with each other. Governor Bentinck's decision that no petition should be written in Dutch unless accompanied by an English translation brought these tensions to light. In 1812, English was substituted for Dutch in local pleadings, and as part of the new policy of anglicizing the colony, the city of Stabroek was renamed Georgetown by the next governor—Lyle Carmichael—in honor of George III. Step by step, the British government extended its prerogatives at the expense of the old rulers.

Trying to restrain the power of some important planters who sat in the College of Kiezers—an administrative body elected for life—Governor Carmichael extended the right to vote to all persons who paid income tax on 10,000 guilders. (Previously this right had been granted only to those who owned twenty-five or more slaves.) He also merged the College of Kiezers with the Financial Representatives, who were appointed every two years. Since the Financial Representatives also sat in the Court of Policy, the other important administrative body in the colony, Carmichael managed through these expedients to extend British influence over local government. All these administrative changes provoked protest, particularly from the Dutch, who considered them breaches of the Capitulation Act.[77] Although three-quarters of the property of Demerara was in the hands of British planters or managers, the Dutch still retained some power in the Court of Policy and the College of Kiezers. They could not but resent Carmichael's measures, which would increase the influence of wealthy merchants (mostly English) and a few middle-class professionals and bureaucrats to the detriment of Dutch planters.

When General John Murray, then acting as governor of Berbice, was appointed in 1812 to replace Carmichael, the Secretary for the Colonies warned him that the animosities among the colonists might make his situation difficult, but that the crown relied on his continuing to exercise "that spirit of forbearance and that firmness of character" which had rendered his administration of Berbice so popular.

Aside from tensions between the old Dutch elite and the English, there were others among the British colonists themselves. English, Scots, and Irish were often at odds. Their factional disputes were sometimes translated into religious bigotries. The English attended services at St. George's, the Scots at the Scottish Presbyterian church. There were also the Methodists and the Catholics. And almost everyone—Scot, Irish, or English—looked with suspicion on evangelical missionaries. But such conflicts among the British could be overcome by their perception of common interest and their growing sense of sharing a common identity. An 1806 notice in the *Royal Gazette* described the festivities of St. Patrick's Day, when "the greatest community prevailed. All national prejudices and party distinctions were forgotten. The Rose, the Thistle, and the Shamrock were happily entwined."[78]

In the colony, Europeans felt like exiles. And, like exiles everywhere, they tended to idealize the world they had left behind, and sometimes to take its ideological representations as adequate descriptions of everyday life.[79] They surrounded themselves with European things, symbols of their culture, marks of affiliation: pieces of mahogany furniture, billiard and card tables made in London, decanters, tumblers, wine glasses, shades, China tea and coffee sets and dinner services, silver knives and forks, ivory-handled carving knives and forks, mirrors, clocks, pianos, and bookcases. They hung on their walls views of Edinburgh, London, Greenwich, Dublin. They collected Sir Walter Scott's books, poems of Byron or Milton, books on the history of England, works on science and nature. They avidly read (40 to 60 days late) the news from England. They eagerly followed debates in Parliament, the intrigues of the court, the trials of political radicals. Against this background, life in the colony seemed uneventful. And when a London editor complained once that the colonial newspapers did not have local news but only reproduced extracts from English journals, the editor of the *Royal Gazette* answered apologetically that "in such small communities as colonies are in general, domestic occurrences worth mentioning, scarcely take place."[80]

The colonists followed with anxiety the ups and downs in the prices of sugar, cotton, and coffee in the London market. They delighted themselves with the lavish descriptions of the latest London fashions. They imported all sorts of food and drink: Madeira wine, port, claret, champagne, cheeses, hams, and even such small things as candles, soaps, boots, shoes, parasols, clothes, stockings, stationery—everything that reminded them of their origins. Living away from the day-to-day conflicts at "home," they idealized British society and British mores, and longed for what they could not get. Even the weather in Britain seemed to have healing powers. So, when they could afford it, they traveled to Britain to take the baths—of civilization or to recover from their physical ailments.

Most of all, the colonists wanted to be treated by the mother country as equals. In recent years, however, they had been constantly under attack, particularly from British abolitionists, who insisted on portraying them as brutish and retrograde. The English colonists were particularly vulnerable to the attacks. Although they depended on slave labor, they saw themselves as the inheritors of a libertarian ideological tradition with deep roots in their past history. Since the beginnings of their imperial history the English had contrasted the "freedom" of English rule with the "tyranny" of their enemies. In Guyana they were no different. They liked to stress their superiority over the Dutch in the management of plantations. They boasted about their entrepreneurship and diligence and their skills at making profits, and they often contrasted the "fair" and "benign" management of the slaves by the British with the "savage" and "brutal" way of the Dutch. They claimed they had put an end to tortures and had abolished the use of the wheel, replacing it with more "humane" methods of punishment.[81]

The British colonists had a divided consciousness. In spite of their complaints against the government in London they shared the imperial faith worshiped by British colonists throughout the empire and reiterated constantly in Parliament and in the press. The best evidence of the pervasiveness of this faith was that Demerara colonists shared it with one of their most egregious enemies, the emancipationist Thomas Babington Macaulay. Although they vigorously repudiated his condemnation of slavery, they subscribed to his words celebrating England and the empire.[82] Few documents could be more expressive of this ideology than a speech Macaulay made in 1824 to the Society for the Mitigation and Gradual Abolition of Slavery Throughout the British Dominion. Macaulay argued that England could not tolerate slavery anymore without renouncing her claim to her highest and most peculiar distinction. She had indeed much in which to glory, she could boast of her ancient laws, her magnificent literature, her long list of maritime and military triumphs, the extent and security of her empire, but she had a still higher praise. "It is her peculiar glory," he said, "not that she has ruled so widely,—not that she has conquered so splendidly, but that she has ruled only to bless, and conquered only to spare!"

> Her mightiest empire is that of her morals, her language and her laws;—her proudest victories, those she has achieved over ferocity and ignorance;—her most durable trophies, those she has erected in the heart of civilized and liberated nations. The strong moral feeling of the English people—their hatred of injustice—their disposition to make every sacrifice rather than participate in crime; these have long been their glory, their strength, their safety. I trust that they will long be so. I trust that Englishmen will feel on this occasion, as on so many other occasions they have felt, that the policy which justice and mercy recommend, is that which can alone secure the happiness of nations and the stability of thrones.[83]

The applause with which the speech was received testified to the appeal of this imperial faith, which neither slavery in the colonies nor all the atrocities committed by British troops and all the protests of the people who were forcibly incorporated into the British empire all over the world would shake for many years to come. This faith was inextricably related to another: the belief in the superiority of British institutions, and of British practices of civic and political liberty.[84] What explains their pervasiveness and resilience is that such notions had been (and continued to be) consciously or unconsciously manipulated by different groups: by the ruling classes in their struggles for power and in their attempts to put limits on the crown on one hand, and on "riotous people," on the other; by the crown to consolidate its legitimacy; and by the people to protect themselves from the arbitrariness of their rulers. Abroad, the imperial ideology served as a weapon of the empire. It was shared by Britons all over the world, for whom it helped to mark boundaries and to stress their superiority over others. But it was also used by subjects of the empire, who struggled against discrimination and exclusion and claimed a place in the sun in the name of British liberty, justice, and law.

The eighteenth-century imperial wars, the American Independence and then the French Revolution and the war with France, the economic changes, social dislocations, and political turmoil of those years had only strengthened this ideology. The ideology was ritualized and monumentalized, it created an image of the past and projected it into the future. It was celebrated in schoolbooks, in verse and prose, in novels and history books, in sermons and political speeches, in Parliament and in the press. It was celebrated in countless popular songs, which could be heard in the most remote corners of the world. The imperial ideology had a compelling power. It became something that had to be contended with, a banner around which elites in Great Britain or in the colonies could rally and that even oppressed people could invoke in their struggles against the abuses of imperial domination.[85]

The imperial ideology gave the colonists a sense of identity but did not blind them to the growing conflict of interest that separated them from the mother country. In time of economic recession and declining prices, when planters could not pay their mortgages, their relations with British merchants and royal authorities went sour. The system of political patronage also gave rise to personal enmities, rivalries, and conflicts. As a consequence of his practice of favoring friends and persecuting enemies, Governor Murray was admonished several times by the British government. There were several appeals to the King-in-Council making serious accusations against Murray's administration and casting doubt upon his character.[86]

Inter- and intra-class conflicts in Demerara were complicated by racial issues. There were conflicts between whites and mulattos, some of whom—like John Hopkinson, the owner of *John* and *Cove,* and the Rogers family, proprietors of several plantations, including *Bachelor's Adventure* and *Enterprise*—had managed to become relatively well to do.[87] Although the power white men had over black men and women tended to facilitate affairs between white males and black or mulatto females, and although in Georgetown there was no significant residential segregation,[88] the color line was jealously kept in public settings. Racial discrimination was conspicuously displayed in the social space, and color separated people into different groups, with different privileges. The burial ground of the English church was divided into three areas, one for the whites, one for the "free colored," and one for the slaves.[89] In the local theater free people of color had to sit in the back rows.[90] And even among blacks there were subtle forms of discrimination. In the church, mulattos often refused to sit next to blacks.

In a highly fluid society where fortunes were made and lost at a hectic pace and elite boundaries were often trespassed by newcomers, a society where plantation owners were hobnobbing with attorneys and managers, and where a few mulattos had managed to become plantation owners, a society where the protocols of race were often disregarded by white men living with black women—class and racial tensions often took the form of status anxiety, to which racial overtones were added. As a result, people constantly felt the need to mark boundaries by aggressively displaying their authority in boastful gestures and by clinging to traditional symbols of prestige. In spite of its humorous tone, an 1807 letter to the *Royal Gazette* betrayed these concerns. The author, under the pseudonym "Cow Skin," commented on the indiscriminate use of "Esquire," a term he thought was "misapplied and prostituted" in the colony. He complained that anyone with a few bunches of plantain felt entitled to call himself Esquire.

> I was a few evenings ago at my door when a boy handed me a letter. I looked at the address, and found it was my overseer B. W., Esquire. I should not be at all surprised to see one of these days, a letter addressed to my driver "Quaco, Esq." or in his absence, Nelson, Esq., Second Driver.

Cow Skin asked the printer whether he quoted his advertisements verbatim as they were sent to his office, or added Esquire, "by way of compliment."[91] So many were the conflicts in the colony that no one would have had difficulty endorsing a remark made in a letter sent to the *Royal Gazette* in August, 1822: "There is not a country on the face of the earth, where classes are more numerous and party spirit more firmly rooted than in this."[92] In a society with so many divisions and protocols it would be impossible for the missionaries not to blunder.

The conflicts and general malaise in the colony were exacerbated by the colonial press. In spite of the small size of the reading public, several newspapers circulated in Demerara—the *Colonist*, the *Royal Gazette*, the *Guiana Chronicle*. All were subject to censorship. Attacks against the British government were prohibited, but the local press always managed to insinuate some critical comments about colonial policies being debated in England. The *Royal Gazette* was more restrained than the *Guiana Chronicle*, which seemed always ready to cater to its public's taste for scandal and gossip. This itself was a sign of the times, for government patronage was being replaced by the patronage of the public. This trend was lampooned in a letter criticizing the *Chronicle* sent to the *Gazette* in 1819:

> It may be said that the Press, like the Stage, to be successful when the treasury is empty, must in a great measure, be what the town requires! So that if buffoonery on the one, and scandal in the other, please in the first respect and enriches in the second—all's well that ends well![93]

Three years later the *Guiana Chronicle* was again accused of factionalism and Whigism, of being mercenary, of courting scandal and detraction for the gratification of private ends, and of showing a disposition "to sacrifice the harmony of society and feelings of individuals to selfish sordidness and the maintenance of a guilty popularity." In its defense, the *Chronicle* argued that it was open to all parties, without taking sides with any, and that the practice of publishing addresses or communications "in the manner of advertisements" was as old as the trade. All journals in England, "ministerial and opposition," made regular charges for the insertion even of articles which contained "important matters of public intelligence." It concluded by saying that there was no need to defend itself at a time when it had "gained to itself a circulation which nothing but its possessing some claims to public patronage could have procured for it."[94] In fact, the aggressively sensationalist policy of the *Guiana Chronicle* would add virulence to the conflicts and problems undermining Demerara society.

Like other colonies, Demerara was affected not only by the wars of empire, but by the wild price fluctuations in the international market. When prices of cotton, coffee, or sugar went up, plantations expanded, planters found easy credit, and fortunes were made. When such prosperity lasted long enough, those who accumulated capital often returned to the mother country, leaving their plantations in the hands of managers and attorneys. In time of crisis, when prices collapsed, many planters could not pay mortgages and lost their properties to merchants to whom they were indebted. Most of these creditors lived in England. As a result, for one reason or the other, Demerara, like many other Caribbean colonies, did not have a large resident planter class.

The late eighteenth century was a period of extraordinary prosperity. The number of plantations increased rapidly, and a large number of slaves was brought into the colony.[95] Between 1789 and 1802, exports of sugar rose by 433 percent, coffee by 233 percent, and cotton by 862 percent.[96] Fabulous fortunes were made in a short period.[97] A cotton planter, it was said in 1799, could make a profit of 6,000 pounds sterling on one crop of 60,000 pounds. For a short period, the colonies of Demerara, Essequibo, and Berbice were the greatest cotton producers in the world. But soon American competition was felt, and by the time John Wray arrived in the colony in 1808 the bonanza had passed. Not only had cotton production declined but coffee had also suffered. The Napoleonic blockade and the refusal of the British government to give West Indian planters access to the United States narrowed the market. So planters started turning to sugar. But sugar, too, was affected by oscillations in the international market; and sugar prices, like those of cotton and coffee, declined after 1816–17, when the European market began to be flooded with sugar from Brazil, the East Indies, and other colonies in the Caribbean. Prices reached their lowest point in 1822–23.[98] While the prices of commodities exported by the colony went down, prices of the products sold in the colony went up. Everything became very expensive.

Wray complained of the high prices he had to pay for everything he bought and commented on the problems overwhelming the planters.[99] The high prices and the scarcity of food made the lives of the slaves increasingly difficult. Because of the war with America, there was no saltfish to be given to the slaves and they had to content themselves with plantain. "We cannot give five or six sterling for half a barrel of salt beef and fresh beef is out of the question. Mrs. Wray and I have had it once since we have been married. . . . A piece of mutton costs as much as a sheep in the East."[100]

Planters were in serious trouble. During the years of bonanza they had borrowed large sums of money in England, but when prices declined they found themselves unable to pay their debts. Planters in Essequibo expressed their distress in a petition addressed to Governor Bentinck in 1811:

> Your petitioners, have for some time past laboured under great inconvenience, from the distressed state of the European market for all kinds of West India produce, which has progressively increased to such a degree as to render it impossible for many of even the most respectable planters to pay their debts. . . . Many estates have been thus sold. . . . If suits are suffered to be continued . . . a great majority of the planters will be under execution and sequestration. . . . Many planters have produce sufficient in their buildings to pay twice as much as they owe, if it bore its usual price, but under the present circumstances, were they to be sued for even a thousand guilders, they could not pay it in cash, and for such comparatively trifling sums as these, they would be obliged to pledge their estates or go to jail.[101]

This petition was soon followed by another, signed jointly by the planters and inhabitants of Demerara and Essequibo. They said again that although their "logies" (warehouses) were filled with produce, creditors were not willing to take it in payment except at prices far below the standard by which a planter could subsist. Those who had already shipped their produce and drawn bills of exchange—as was generally customary—had not been able to sell it and their bills had been returned protested, adding more to their debts. Their experience had already shown that "some creditors, more craving than humane," would proceed against them.[102]

Governor Bentinck understood the plight of the planters and forwarded their petitions to the Earl of Liverpool, the Secretary for War and Colonies. The governor explained that the situation had become so distressing that he thought it necessary to submit their petition to the Prince Regent so that redress could be granted to them. He suggested that all execution sales of estates be suspended—provided that the governor and the Court of Justice took care that in the meanwhile the property did not deteriorate. Something had to be done quickly, he said, since in a few months several estates would be brought to the hammer, and unless some relief was granted ruin would come not only to those whose property was sold but also to their creditors.

The colonists were limited in their options by their creditors, who resided either in England or the Netherlands. The creditors determined not only what should be grown, but where it should be marketed. This constraint on the development of the colony was made clear in a letter the governor sent two years later to Earl Bathurst, who was now the Secretary for War and Colonies. Demerara could supply any quantity of cassava flour and maize, he said, "if sufficient inducements could be offered." Rice also grew extremely well, but the demand was not sufficient to establish the necessary threshing mills, barns, and other needed equipment. All the borrowed capital, "the interest of which now presses on the planter," had been secured by mortgages on slaves, land, and buildings, where they had to remain till the debt was either satisfied or the property changed owners. As a result, shifts of capital and labor into new channels of industry were difficult and could only take place if creditors changed their perception. But that, the governor recognized, was not easy. The creditors in the colony, he said, would soon be convinced, but European creditors might offer serious objections. And he explained that the "mercantile creditor stipulated that the crops are to be remitted to him for sale on which he gains a commission in addition to his interests; the present produce of coffee, sugar, and cotton is therefore advantageous to the European mercantile creditor, but the sale of timber, rice and corn, finding a market at the door or near at hand, would be of no advantage to him, beyond the mere interest of the capital he has at stake."[103]

Trapped in the logic of mercantile capitalism, the colonists were indeed in a desperate situation.[104] They were not free to trade with other nations, the prices of their staples were falling in the English market, their debts growing and their profits shrinking. Reviewing the state of the colony in 1812, one planter told the Court of Policy that in 1799–1800 on three estates under his charge the returns had amounted to 40,000 pounds, while expenses had been only two-thirds of what they were in 1812. In the past three years, however, the returns had been just enough to cover expenses. The proprietors were not even getting interest on their capital.[105]

The planters were caught in a terrible contradiction. The same historical process that in England was leading to abolitionism and free trade had opened new opportunities for investment in Great Britain, and was also turning investors toward the East Indies and other parts of the world, making them increasingly indifferent to the fate of the colony. Meanwhile, the Demerara planters, while struggling to increase their trade with other countries, could only try to defend their privileges in the British market and hold onto slavery. The economic crisis made the Demerara planters particularly hostile to those who first had abolished the slave trade and were now talking about legislation to improve the situation of the slaves. Adding insult to injury, those who campaigned in Britain in favor of the slaves drew an ugly picture of the planters, and even seemed to side with the blacks against their own countrymen. Even worse, after having abolished the slave trade in the British colonies the government continued to allow sugar to be imported from countries like Brazil, where the slave trade was still active.

Like colonists everywhere, the Demerara planters were at the mercy of the arbitrary policies of the home government. Their profits depended on the political support of metropolitan groups. From the mother country they received capital and imported most of what they needed. To the mother country they exported their products. From the beginning, the colonists had benefited from a privileged position in the metropolitan market, but to their distress many people in England were talking about free trade. During the 1820s Parliament was overwhelmed by a number of petitions from different interest groups favoring free trade. The West India lobby was having difficulty in defending its colonial privileges from the reformers' attacks, which were coming not only from East Indian interests but also from merchants involved in international trade, and from manufacturers and consumer groups.[106] The privileged position the colony enjoyed in the British market was threatened. The decline of sugar, cotton, and coffee prices irritated the colonists and put them even more on the defensive. They started complaining again of their low returns, attributing it to the "burdensome duties," "oppressive regulations," and to the advantages enjoyed by their competitors "unfettered by the restrictions of the British colonial system." And they anxiously followed the often bitter debates in Parliament and in the British press.

In 1820 the *Royal Gazette* transcribed from London newspapers several articles for and against free trade. One was a free-trade petition from the merchants of the City of London to the House of Commons. The petitioners condemned the restrictive and protective policies followed by the government and suggested that they all operated as a heavy tax on the community at large. They asked Commons to adopt "such measures as may be calculated to give greater freedom to foreign commerce, and thereby to increase the resources of the State."[107] A few days later the *Gazette* published a petition from shipowners against free trade. They warned that if restrictions in the corn laws and taxes on foreign wool were lifted, fields would go uncultivated, laborers would be unemployed, and national distress would follow. Ships would lie rotting in the harbors, sailors would go into foreign service, and "no nursery would remain to supply the fleets in the time of war." The petitioners claimed that the interest of those whose capital had been invested in "traditional trades" should be protected.[108]

The debate over free trade would occupy the attention of the colonists for years to come. Joseph Marryat, the indefatigable spokesman for the West Indies, brilliantly advocated their point of view, stressing the common interests of the colony and the mother country. In one of his speeches (transcribed in the *Gazette*) he said that the British colonist had to draw all his supplies from Great Britain only:

> Everything about him and belonging to him is British, his woolens, lines and leathers, the axe. . . . He roasts his meat at a British grate, on a British spit,

> or boils it in a British pot, eats it off British plates and dishes, with British knives and forks, drinks out of British mugs or glasses, and spreads his meal upon a British table-cloth. All his surplus means are spent on British manufacturers and produce, and this expenditure gives life and animation to British industry.

He claimed that all this benefited British manufacturers, landholders, and workers, and that in the end Britain and the colonies benefited from each other. In a parliamentary debate over timber duties, Marryat argued as a matter of principle against taking off the restrictive duties. "Principles," he said, "are immutable in their nature, and cannot be taken up and laid down at pleasure, adopted in one instance, and abandoned in another. If we abolish all restrictions on the importation of foreign timber, how can we refuse to abolish those on the importation of foreign corn?" Marryat feared that one measure would lead to another and sooner or later all protections would disappear, including those which shielded the West Indian planters.[109]

Like Marryat, the Demerara colonists were concerned with maintaining the protection they had always enjoyed in the British market. They sympathized with their West Indian neighbors, who also saw their profits shrinking. In 1821, the *Gazette* reproduced a petition from the Jamaica House of Assembly representing the state of "extreme distress" to which Jamaica and the British West Indian colonies were reduced by the inadequate returns their staple commodities obtained because of the burdensome duties, oppressive regulations, and the advantages which rival colonies and possessions enjoyed. A month later the newspapers published another petition from Jamaica. This time the petitioners went right to the point. They asked for relief, arguing that the price of sugar had so diminished since 1799, and the costs of production had risen so far, that the value of their crops had become barely equal to the cost of production, "leaving no rent for the value of the land, and no interest for the large capital employed upon it."

Soon after, Demerara newspapers were reproducing debates over additional duties on East Indian sugar to protect the West Indians. Inevitably, the issue of slavery was raised. In the House of Commons, a member who opposed such duties used the opportunity to attack the slave trade. He said that there was no reason why the English consumer should be obliged to pay one shilling more for sugar from the West than from the East Indies. "It was intolerable that the people of England should be called upon to pay high prices to enrich persons who had chosen to employ their capital in that trade of human flesh." Those who spoke in defense of the West Indian planters argued that they were obliged to bring their produce in British ships to a British market. If such restrictions were lifted they would be glad to send their sugar where they could get a better price for it.[110]

The colonists wanted it both ways: to be free to trade with any country and at the same time continue to have certain monopoly privileges in the British market. They had often disregarded restrictions on trade. They had sent their ships to Caribbean islands and traded with the Spanish and American colonies whenever they needed. Colonial policies had always been the result of a complex negotiation among the metropolitan government, various metropolitan groups of interests, and the colonists. But since the abolition of the slave trade the colonists had been feeling that they were losing their grip.

The elimination of the protective tariffs that guaranteed the colonies' preferential position in the British market was a gradual process that ultimately was to dismantle the entire mercantilist system of Great Britain. So was slave emancipation. But the prospect that such things might come to pass placed the Demerara planters on the defensive. Ironically, the Demerara colonists had been confirmed as members of the British empire just at the time when the debates about free trade and abolition were becoming more frequent, and the abolitionist movement in Britain was gaining momentum. In 1815, discussions in Parliament of the "Registry Bill" triggered strong protests both in England and in the West Indies. The Society of Planters and Merchants representing the West Indies complained that slave registration imposed by act of Parliament would infringe upon the constitutional rights and interests of the colonial legislatures and private individuals and would be a blow to their property. Defending the proposal, James Stephen published a series of pamphlets illustrating the evils of slavery in the Caribbean. The large number of pamphlets produced by both sides exacerbated their antagonism.[111] West Indian newspapers joined in against the proposed measure. The *Royal Gazette* expressed the point of view of the colonists in a series of angry editorials published between March and July 1816, attacking the British government and the African Institution for a measure they saw as an undue interference in the affairs of the colony.[112] The proposal was finally defeated and the decision to implement slave registration was left to the colonies. But abolitionist pressures continued.

The colonists were watching the movement toward emancipation with growing apprehension. They also followed attentively debates over the colonial trade. From 1821 to 1823 no month passed without the Demerara newspapers discussing such portentous questions. The year 1823 did not start very auspiciously for the colonists. The prices of their export commodities were at their lowest. Abolitionists intensified their campaign in Britain. In January an impressive group of notables founded in London the Society for the Mitigation and Gradual Abolition of Slavery Throughout the British Dominion. In March, Wilberforce introduced in the House of Commons a Quaker petition for the abolition of slavery. And two months later he published *An Appeal to the Religion, Justice and Humanity of the Inhabitants of the British Empire in Behalf of the Negro Slaves in the West Indies,* which had a profound repercussion on both sides of the Atlantic. To make things worse, a group of stockholders of the East India Company petitioned for the equalization of sugar duties, challenging the preferential treatment given to the West Indies.[113] Under such circumstances it is not surprising that Parliament's discussions of measures intended to ameliorate the slaves' conditions of living would throw the colonists into a rage.

The planters' point of view was made crystal clear in a book published by Alexander McDonnell in 1824, *Considerations on Negro Slavery, with Authentic Reports Ilustrative of the Actual Conditions of the Negroes in Demerara.*[114] The curious thing about this book is that in its defense of slavery and of the traditional colonial system, the author adopted many of the ideas that were used by people who argued against them. Both defense and critique belonged to the same ideological universe. Both were committed to industry and self-discipline. Both had the same faith in the redeeming qualities of education. Both believed in "progress" and "civilization." Both were confident of the power of ideas to change the world. Both shared a deep respect for human reason. Finally, both claimed to defend the interests of the British empire.

In spite of all these similarities, however, the planters whose views McDonnell typified and the critics of slavery and/or advocates of free trade had opposite views of slavery and of the relations between colonies and the mother country. The Demerara planters' position was full of ambiguities. Although they might have been willing to concede that free labor might be superior to slave labor in principle, they defended slavery not only because at that time they did not see any other viable alternative to their problem of labor, but also because emancipation was a direct attack on their property. And while they wanted to secure the privileged position they had in the metropolitan market they also wanted to be free to trade with other nations. The metropolitan elites were not more consistent, as McDonnell pointed out. They opposed slavery, but treated their "free laborers" worse than slaves. And while they were ready to challenge the right to property the planters claimed they had in slaves, the metropolitan elites who supported emancipation were jealously defending their own right to property.

The contending parties had serious differences. But it is probably because they shared so many beliefs that their struggle was characterized by so much hostility and resentment. It was because of their commitment to property rights that the planters reacted with such bitterness against those in Parliament who argued in favor of emancipation. It was because they shared with them the same concern with profit that the planters repudiated their policy of free trade, and it was because they had in common with the metropolitan elites a fear of the subaltern classes that the planters denounced with such eloquence the risks of abolitionist rhetoric and pointed to the hypocrisy of men who did not hesitate to preach the abolition of slavery while they exploited mercilessly their own labor force.

During the battle between the abolitionists and the West Indian elites, their similarities tended to be buried under layers of violent rhetoric on both sides, a rhetoric which emphasized only differences. The image that emerged was of two contrasting elites: in the colonies, a planter class, backward, arbitrary, and violent, almost feudal, holding onto traditional habits, defending slavery and the traditional social order; at home, a progressive, liberal-minded, reformist, legalistic, modernizing elite, fighting for emancipation and free trade. This dichotomic view, born out of the struggles of the nineteenth century—and perpetuated by historians—obscured both the divisions within the British elite and the complex reality of the planter class, a class divided between those who lived in the colonies and those who lived in the mother country, and whose opportunities for investment were constantly enlarging and becoming diversified but whose profits still continued in great part to depend on slavery and trade privileges at a time when an increasing number of people in Great Britain were ready to support measures in favor of emancipation and free trade.[115]

The changes taking place in Great Britain were deepening the gulf that separated West Indian planters and merchants who lived in Britain from those who lived in the colonies. The former—gathered in the Society of West India Planters and Merchants—were a wealthy and powerful group with an effective representation in Parliament, where they constituted a strong lobby for the defense of the West Indies.[116] But while the colonists continued to depend exclusively on slavery and the colonial trade, the interests of those who integrated the West India lobby were expanding and turning to other activities, such as insurance, banks, urban development, manufactures, and international trade.[117]

McDonnell's book revealed the Demerara planters' predicament and ambiguities. He argued that slavery was a legitimate institution, sanctioned by law and history. Slave property should be treated like any other property in the mother country. On these grounds he denied that Parliament had the right to take away from the planters "without indemnification, the privilege of obtaining from their slaves six days labour in the week."

Although he criticized mercantilist notions and accepted some of Ricardo's most advanced theories, McDonnell tried to demonstrate in quite traditional ways the advantages the colonies afforded to the mother country.[118] The colonies, he argued, provided opportunities for capital investment that yielded more profit than any foreign trade. And it was erroneous to consider British dealings with colonies as equal to their dealings with other countries, because, in fact, contacts with the colonies were much more extensive and frequent, since emigrants carried with them British customs, manners, and feelings. Even more important, in the case of the West Indies, trade benefited both the mother country and the colonies, for the proprietors either resided in England or ultimately returned to England, taking all their wealth with them. "There can be no difference whatever," he wrote, "in the encouragement given to the various artificers, between a gentleman of Yorkshire who resides and spends his income in London, and a West India proprietor who also lives there and spends an equal amount." He predicted that distress would befall the artisans in England if the West Indies were abandoned.

McDonnell argued that the party that was the loudest in denouncing the West Indians was made up of those engaged in the East Indian trade.[119] The outcry in favor of East Indian sugar had been, in his opinion, "solely to delude the credulity of the population at large by making them imagine they paid more for their sugar, than they would do if the duties were assimilated." But, in fact, the price of sugar in England was entirely regulated by the international market. Thus the public in England did not pay a fraction more for their sugar than if no protecting duty existed.

In McDonnell's opinion, the continuation of the slave trade in other parts of the world was benefiting foreign countries because it was much cheaper to import slaves than to rear them. The abolitionists' expectation that after blacks became acquainted with the precepts of the Christian religion, they could be emancipated and converted into a free and happy "peasantry" was simply wrong. In the West Indies nature was bountiful; in one month a person could raise food for a year.[120] So people lacked the driving wants that characterized a more industrious and civilized community. "What makes a man work in Europe?" he asked. "A much sterner task-master than any to be found in the West Indies—the dread of starvation." Without that compulsion men—whether black or white—would not work. Left to their own devices the ex-slaves would "sink to the conditions of the savage," and spend their hours "lounging in listless apathy under a plantain tree." Yet, he believed, slavery was doomed to die.[121] Sooner or later the cost of maintaining a slave would become equal to the value of his labor, and his master would find no advantage in keeping him, particularly since a free man worked better than a slave.

McDonnell also examined the effect produced on the slaves by the debates over emancipation in Parliament and the abolitionists' critique of slavery and attacks on the planters. The slaves, he said, believed in the power of the King to intervene for them, and were familiar with Wilberforce's opinions. Comparing the situation in the colony with that in England, McDonnell, who—like other nineteenth-century men—had very clear notions of class and class struggle, wrote:

> Say for example the weavers of Spitalfields were taunted with the crime of working from morning to night; if their hard fare and innumerable hardships were derided; if the sumptuous living and luxurious ease of the rich were called to mind; if they were told that the Christian religion authorized an equality of ranks; if an assembly of men were sitting discussing their claims; and if in that assembly they had zealous friends, clamorous in their support, and eager that they should divide the possessions of the wealthy—in such a case, would they toil on at the loom as hitherto? No! They would soon arouse themselves. If defeated in debate, they would speedily endeavour to obtain by force, what they would conceive they were denied by injustice. . . . Stir up the working orders against the rich, and in any community you will have disturbances.[122]

And in his opinion that was exactly what abolitionists, the African Institution, and the missionaries were doing. What kind of charity was it that led people, under "a vague, indefinite universality of feeling they affect," to regard the most remote inhabitants of the globe with the same degree of affection they felt for their nearest kindred? This attitude, he predicted, would lead to the ruin of the British empire.

What McDonnell could not see was that the empire was taking new directions: newly independent nations in Latin America and new colonies in India and Africa were becoming more important concerns than the few colonies in the Caribbean. And, however important the West India lobby still may have been, it could not stop the new emancipationist tide that seemed to have conquered the hearts and minds of the British people. It also could not (or would not) stop people like John Wray and John Smith from coming to the colony to preach to the slaves. Behind the missionaries were respectable and powerful people. Even the Bishop of London had sent a circular letter to the clergy and proprietors in support of teaching the slaves how to read.[123] And among those who were behind the African Institution were members of the high nobility. Moreover, members of the opposition were agitating in Parliament the issues of emancipation and free trade. And when the British government, rather than conceding to radical demands tried to compromise, as it did in 1823, its half-measures were already too much for most colonists.

CHAPTER TWO

Contradictory Worlds: Masters and Slaves

> The devil is in the Englishman. He makes everything work. He makes the Negro work, the horse work, the ass work, the wood work, the water work, and the wind work.[1]

The new trends that so preoccupied the colonists had contradictory effects on the slaves: they led simultaneously to increasing oppression and growing hopes for emancipation. This contradiction aggravated the tensions that had always existed between masters and slaves and created an explosive situation.

From the time Demerara was integrated into the British empire, the slaves' conditions of living and perceptions had been changing in significant ways—and would change even more as emancipation approached. Massive capital investment transformed the landscape and altered both the nature of plantation life and the slave experience. With British capital came new machinery, a more intense pace of work, new ideas, and a new style of living. Abolitionism, the increasing intervention of the British government, and the presence of evangelical missionaries in the colony altered the balance of power and redefined the terms of the relations between masters and slaves. From the struggles that in Europe and in the New World were eroding the institutions and ideologies of the *ancien régime* emerged notions about citizenship, social control, law, and judicial procedures, notions that undermined the ideological framework that supported slavery. The powerful statement made by the slaves in Haiti echoed on both sides of the Atlantic. The slaves in Demerara, like slaves elsewhere, were participants in this larger process. In their struggle against oppression they pushed the slave system to its limits.

The integration of Demerara into the British empire opened new market opportunities, but this process was full of contradictions. Massive capital investment and the expansion of production, at a time when the supply of slaves was dwindling as a consequence of the interruption of the slave trade, led masters to intensify labor exploitation and curtail many of the slaves' "privileges." This was reinforced by the shift from coffee and cotton to sugar on many plantations and by the decline of the prices of all these commodities in the international market. The shift to sugar and the introduction of steam mills forced plantation owners to borrow large sums of money, making them particularly vulnerable to the downward trend in sugar prices after the bonanza of 1815–16. Unable to pay their mortgages, many planters were forced to sell their plantations, or lost them to British merchants. As a consequence there was a gradual concentration of land and labor in the hands of British merchants, who had capital enough to cope with periodic crises and wait for better times. For the slaves, all these changes meant longer hours of work, a faster pace of labor, less time to cultivate their own gardens and provision grounds or to go to the church and the market, diminishing supplies of food and clothing, more rigorous supervision and punishment, and more frequent separations from family and kin. Slaves responded to these pressures with increasing rebelliousness.

No one captured better the beginnings of these changes than Henry Bolingbroke, an Englishman who lived in Demerara from 1799 to 1805, working as a clerk for an important merchant house, and who, after his return to England, published a book about the colony.[2] For Bolingbroke what historians later came to describe as a process of transition from merchant capital to industrial capital,[3] from a system of monopolies and privileges to a world organized according to the principle of free trade, from slavery to free labor, from colonialism to imperialism, was mainly a question of "national character." In his descriptions of plantation life in Demerara, Bolingbroke, with characteristic British pride, contrasted an almost seigneurial system of running plantations—which he attributed to the Dutch—with the entrepreneurial style of the British. "There is a wonderful dissimilarity between the Dutch and the English colonists," he noticed. "They naturally both go out with a view of making money, but the one with an intention of ending his days abroad, and the other of returning to his native country to live in ease and independence on the fruit of his industry."[4]

Bolingbroke praised the taste with which the Dutch laid out their plantations, their "general neatness and formal regularity, their handsome and comfortable houses and their beautiful gardens." He commented on their cult of leisure, their largesse, and their fondness for horses and boats. He relished their hospitality, their good food and wine. With grace and color, and perhaps a bit of irony, he portrayed the Dutch farmer sitting after dinner in front of his house, smoking his pipe, while the slaves came one after another to thank him for their daily allowance of rum.

For Bolingbroke the Dutch were planters of the "old school." Nothing could, he said, "divert their attention from the traditional manners in which they settled their estates." They seemed to "aspire only to a competency not to a fortune." By contrast, the system the English had introduced in the colony seemed to Bolingbroke much more profitable. It insured "as much cultivation in one year as a Hollander would accomplish in four. . . . The one dashes on and prepares a hundred acres to plant while the other is content with twenty-five." The Dutchman's greatest ambition, he said, "is to make his plantation look like a garden, while that of the Englishman is to get the greatest quantity of cotton under cultivation possible, as it has been found by the experience of a series of years that the quantity and not the quality constitutes the profit of the crop."[5]

Bolingbroke described the slaves on Dutch plantations as if they were medieval serfs. Although he opposed the interruption of the slave trade and considered slavery a "civilizing" institution, he dreamed that one day, with the help of the British, the slaves could be brought "on a level with the English peasantry" of his time.[6] He was appalled by the subservience of overseers and slaves toward their masters on Dutch plantations, and contrasted them with the "more independent" slaves of the British.

> The negroes belonging to the Dutch estates, copy the overseers' humble politeness, and are considerably more respectful to whites than those belonging to the English plantations. A certain erect carriage in John Bull imperceptibly introduces itself into the incult address of the English negroes. Or it may arise from their not being kept so strictly, nor considered in so degraded condition as other negroes are.[7]

The slaves' increasing rebelliousness, Bolingbroke thought, was explained by the "liberal" conduct of the English, which he contrasted with the "severity" of the Dutch. Such notions had great currency among British colonists and were repeated by almost every English traveler who visited Demerara during this period, by colonial administrators both in the colony and in the mother country, and by British politicians of different persuasions.[8]

The contrast Bolingbroke drew between a Dutch manorial and a British capitalist style of management of a plantation was too neat. He himself met Dutch planters who, like one Mr. Voss, started with nothing and through perseverance, industry, and frugality—qualities Bolingbroke would attribute to the British, not the Dutch—built such a fortune that he could give his natural daughter (whose mother was an Indian) £20,000 a year. And there certainly were Englishmen no less extravagant than the Dutch described by Bolingbroke, Englishmen who indulged themselves in frolics, gambling, hunting, cockfighting, and horseracing, vainly trying to enact in the colony an idealized version of the life of the English country gentry. And, although it is true that particularly barbarous forms of punishment used by the Dutch—such as breaking a criminal on the wheel—were abandoned under British rule, there is no evidence (aside from Bolingbroke's own testimony) that the Dutch planters he met were more brutal than the British in day-to-day dealings with the slaves.[9]

The picture of passive slaves under Dutch rule is hardly consistent with the abundant evidence of slaves' rebellious behavior during the Dutch period, particularly the bloody uprising of 1762–63 in Berbice, when the slaves kept the colonists at bay for almost a year.[10] But if Bolingbroke may have exaggerated the contrast between the Dutch and the English, his overall picture contrasting two different styles—or, in truth, two different moments in the history of Demerara—is confirmed by other sources.[11] There is enough evidence that the changes in the conditions of production and the debates over the abolition of the slave trade did indeed alter the management of plantations and create new motives for dissatisfaction among the slaves and new opportunities for resistance.[12] We may dismiss Bolingbroke's national character theories, and discount the bigotry that probably made him overestimate the entrepreneurial qualities of the British. But there is no doubt that things had changed from the time the Dutch ruled the colony.

It is possible that some of the old Dutch planters continued to run their plantations in "traditional" ways.[13] But in a period of rapid change, planters who did not adapt to the new requirements of production, invest more capital, introduce new machinery, and exact more labor from their slaves would soon be out of business. A few years after Bolingbroke wrote his *Voyage to Demerary,* D. S. Van Gravesande, who boasted of being the grandson of two Dutch governors, lamented the changes that had taken place in Demerara and remembered the "happy days" when owners and slaves plowed together, when slaves had Saturdays to cultivate their gardens and to sell their produce, "by which the Sabbath was kept holy." For slaves, the changes that Gravesande characterized with nostalgia, and Bolingbroke celebrated with enthusiasm, meant growing exploitation and an encroachment upon what they defined as their customary "rights."[14]

The integration of Demerara into an expanding capitalist world gave slaves not only new motives for protest, but also new notions of rights and new opportunities for resistance. More important, it raised expectations that they soon might be free. Debates both in England and in Demerara over the abolition of the slave trade in 1807, the "Registry Bill" in 1815–16, and the amelioration laws in 1823 redefined the parameters of the slaves' struggle. From irate remarks about abolitionists and the British government made by managers and masters at dinner tables, from comments made by frustrated missionaries in their interminable confrontations with managers and local authorities over the right to preach to slaves, and from newspaper articles some slaves could read in British and local newspapers, they came to believe that they had powerful allies in England—men like Wilberforce—who favored emancipation. This perception encouraged them to become bolder.

The impact of such world-wide economic and ideological changes on people's lives can only be assessed in the context of the particular conditions that prevailed in Demerara.[15] The colony was not a blank slate on which history was being written from the outside. It had an ecology and a history of its own. From its past it had inherited a system of land use, a peculiar pattern of settlement, a body of law, political and administrative institutions, means of social control, and codes of behavior, all of which defined the framework within which missionaries, masters, and slaves had to contend with each other in the new and changing world. Both masters and slaves used the image of the past (as they constructed it) to assess the changes that were taking place in their lives. But their images of past and present were also measured against a vision of the future, which at this point in history seemed to promise emancipation for the slaves and bankruptcy for the masters.[16]

Compared with other Caribbean plantation societies, Demerara was late in developing. The European settlement began in the neighboring colonies of Berbice and Essequibo in the middle of the seventeenth century, after the Dutch West India Company failed to establish colonies in Brazil. But until the middle of the eighteenth century the Demerara region remained practically unoccupied. The first land concessions in this area were granted in 1746 along the Demerara River, at a distance from the sea. The choice of locating the first settlements upriver reflected not only the West India Company's preoccupation with finding a place relatively protected from pirates and smugglers, but also the difficulties in occupying the coastal lowlands, which in Demerara were below sea level and periodically flooded. By the end of the 1760s all land along the west bank of the river had been granted and there were 130 plantations, a third of them already in the hands of English settlers. As in other plantation societies, African slaves constituted the main labor force. The company realized early on the advantages of having the natives on its side, so it outlawed the enslavement of Indians, and brought in an increasing number of Africans.

When all land along the river had been granted, the company distributed new grants along the coast, east and west of the mouth of the Demerara, in the areas that came to be known as the East Coast and the West Coast. A limit of 1,000 acres was established for sugar and 500 for coffee plantations. Before the area could be settled a complicated system of canals, dams, and sluices had to be built to improve drainage—a task the Dutch were particularly qualified to accomplish.[17] Plantations were laid out next to each other, with frontages of 100 roods (one rood = 12 feet) and depths of 750.[18] Planters had the right to acquire a second grant of more land reaching into the interior. Although with time the original pattern was slightly modified, with some plantations having less and others having more than the number of acres originally stipulated, the overall scheme was maintained. This peculiar pattern of settlement led to a dense concentration of slaves in a relatively small area,[19] and it made contact among slaves easy, particularly since plantations were connected by canals and a good road leading to Georgetown. The work required to keep canals, trenches, and sluices in working order added to the tasks plantation slaves normally had to perform in other ecological settings, where such problems did not exist. The exceptional fertility of the soil, the favorable climate, and the variety of crops grown had the same effect, since there was no slack period and slaves were kept constantly busy.[20]

In 1772 the Dutch government, concerned with growing slave unrest, established rules governing relations between masters and slaves. The *Rule on the Treatment of Servants and Slaves* defined their rights and obligations.[21] It tried to curtail masters' violence and neglect and to enforce slave discipline. It prohibited slaves from selling—and colonists and seamen from buying from slaves—any staple such as sugar, coffee, cacao, rum, indigo. But it authorized the purchase of cattle, greens, and vegetables "out of their fields." Slaves were to be paid in money, clothing, or knick-knacks, but "by no means in Guns or any other kind of Fire arms nor in Gun powder or lead." Slaves who violated this prohibition were condemned to be severely whipped; free persons trading with them illegally were to be fined. The regulations further stipulated that no plantations should be left at night or on Sundays and holidays without at least one white person in attendance, and it prescribed a series of measures relating to ways of dealing with runaway slaves, acts of thievery committed by slaves, and punishment. Slaves were to be kept under close surveillance. They were prohibited from moving about outside their plantations without a pass, and forbidden to use boats on rivers or canals at night, or to walk on any dams or public paths after seven at night, unless they had a proper pass from their masters and carried a lantern or torch. Slaves could not carry firearms or sharp weapons while walking on any dam or public road, unless they had their owners' written permission. But slaveowners were forbidden to give arms to slaves, except for two designated huntsmen on each plantation. They were also supposed periodically to order slaves' houses to be searched for firearms, gunpowder, or shot. And recognizing the symbolic and revolutionary power of songs, all inhabitants were ordered to warn their slaves not to sing on board any vessel unless a white person was present.[22] The *Rule* limited the punishments masters could inflict on their slaves to twenty-five lashes.[23] For harsher punishments, masters were to send their slaves to the fortress.

Although the regulations tried to restrain communication among slaves from different plantations by requiring that they obtain permission from masters to move about, it did authorize slaveowners to sanction "the reveling of their negroes once in a month," aside from the customary holidays. The *Rule* also obliged proprietors to set aside provision grounds, about one acre for each five slaves; to give them a "reasonable weekly allowance agreeable to the custom of the colony"; and to supply them with cloth. Masters and managers were strictly forbidden to force their "servants" to work on Sundays and holidays, except in an emergency such as the breaking of a dam or other "urgent" work. In such cases masters were to secure official permission.[24] Both masters and slaves who disregarded the instructions set by the *Rule* were to be punished.

There was a great similarity between these regulations and others found in slave societies throughout the New World. And, as elsewhere, the Demerara regulations were often disregarded by both masters and slaves. Slaves wandered around without their masters' permission. They walked at night along roads and dams without passes. They continued to sing when they went up and down the river on boats. Masters and managers often forced slaves to work on Sundays, ignored the stipulations concerning food and clothing allowances, eluded their obligations to set aside provision grounds, and continued to inflict much harsher punishments on slaves than those accepted by law. In this respect, Demerara was not very different from other slave societies. But one feature set Demerara apart: in Demerara, it was the responsibility of an official known as the "fiscal" to make sure that the *Rule on the Treatment of Servants and Slaves* was respected; and slaves could appeal to him for redress. So, many years before the English created the office of protector of slaves, a similar institution already existed in Demerara.[25]

The Dutch apparently had inherited this practice from Spain and introduced it in their colonies.[26] When Demerara was taken over by the English this institution was kept. The fiscal was in charge of investigating cases of conflict of interest and maintaining order in the colony.[27] He was to hear slaves' complaints, and see that they were treated according to the law, that they were adequately clothed and fed, and that planters and managers did not punish them unfairly or excessively. He was also supposed to hear masters' complaints, though, not surprisingly, these were rare.[28]

Demerara slaves were quick to learn about their "legal" rights. They learned that they could bring their complaints to the fiscal, and they did; although in the early period their complaints often fell on deaf ears.[29] When they went to the fiscal with their grievances, they were more likely to be punished than to get redress. For things to change, slaves had to wait until the time when, under abolitionist pressure and the impact of the slave rebellion of 1823, the British government decided to make this institution more effective.

The colonial status of Demerara made people's lives particularly vulnerable to political and economic forces that transcended the narrow boundaries of the society in which they lived. Like other colonial plantation societies in the New World, Demerara, as we have seen, was trapped in a system that imposed restrictions on trade with other nations, and its economy was extremely sensitive to fluctuations in the international market. It was perhaps even more vulnerable because Demerara had a relatively small internal market and its economy was mostly geared toward the outside.[30] Shifts from one staple to another were exclusively determined by conditions in the international market, and fortunes were rapidly made and lost as a result of the market's ups and downs. Since English merchants provided credit, many plantations passed into their hands when planters failed to meet mortgage payments. This explains, in part, why an increasing number of owners lived abroad.[31] Plantations were left in the hands of attorneys and managers. This situation was not likely to give rise to the kinds of "paternalistic" practices common to slave societies with a large resident planter class. Demerara functioned almost like a factory, in which a tiny minority of whites—soldiers, merchants, clerks, doctors, attorneys, managers, and other plantation employees, amounting at the time of the rebellion only to about 4 percent of the total population—and an equally small number of free blacks lived surrounded by an overwhelming slave majority.

With the incorporation of Demerara into the British empire the number of slaves increased dramatically. The slave population doubled between 1792 and 1802, and from 1803 to 1805, the United Colonies of Demerara and Essequibo imported an additional 20,000 slaves from Africa—almost a third of its total slave population at the time.[32] As a consequence, in the years that followed, Africans constituted a large percentage of the slave population and there was also a clear imbalance between males and females. In 1817, some

55 percent of the slaves were Africans—of whom 63 percent were male and 37 percent female. When the rebellion broke out six years later, the number of Africans had declined, but they still constituted about 46 percent of the total slave population. Judging by the more complete records for the neighboring colony of Berbice, most African slaves imported during this period came from Central Africa, the Bight of Benin, the Gold Coast, the Bight of Biafra, and Senegambia. Kongo, Coromantee, Papa, Igbo, and Mandingo were the largest groups.[33] The highest concentrations of Africans were on plantations with more than 400 slaves, and these were often sugar plantations.[34]

Planters had initially devoted themselves to the production of coffee and cotton, but in the beginning of the nineteenth century sugar was becoming the main source of wealth. On the East Coast—the area between the Demerara River and Berbice, where the rebellion was centered—the shift to sugar came late. When George Pinckard visited the East Coast in 1796 as a doctor accompanying the British Expeditionary Force, there were 116 estates. All the plantations were still growing cotton, except for *Kitty,* which had just been planted with cane. Pinckard calculated that on average cotton production required one prime slave for two acres, while two slaves were needed for every three acres on a coffee estate, and one for every acre on a sugar plantation.[35] This meant that sugar planters had to invest more capital in slaves. Sugar also required greater investments in buildings and machinery. Perhaps for this reason East Coast planters were initially reluctant to invest in sugar. Almost ten years later—when Bolingbroke left the colony—coffee and cotton were still the main staples. Judging by the tax lists published in the *Royal Gazette* in 1813, most plantations on the East Coast were still devoted mainly to cotton, although several were growing coffee, and a few produced both cotton and sugar.[36] In other parts of Demerara, it was common to find plantations producing simultaneously sugar, rum, cotton, and coffee.[37] By then, approximately 8 percent of the plantations had more than 300 slaves, 40 percent had between 200 and 300, and another 46 percent had between 100 and 200. The remaining plantations had less than 100.[38] Plantation *Le Resouvenir* had 396 slaves, *Good Hope,* 433, and *Dochfour,* 376.[39]

The decline of cotton and coffee prices and the extraordinary rise in sugar prices in 1814 and 1815 led several East Coast planters to shift to sugar.[40] This trend continued in the following years.[41] By the time of the rebellion in 1823, out of 71 plantations on the East Coast (including those located on the East Bank of the Demerara River), about half were producing sugar, but only eleven were devoted exclusively to it. Twenty-five produced only cotton; fifteen produced sugar and coffee; nine, coffee and cotton; five, cotton and sugar; two, sugar, cotton, and coffee; and four only coffee. Plantation *Triumph,* which in 1813 had only cotton, had started producing sugar. *Good*

Table 1. African and Creole Slaves in the Demarara-Essequibo Population, 1817-1829

	1817		1820		1823		1826		1829	
Africans										
Males	26,725	35%	24,858	32%	21,768	29%	18,898	27%	16,362	23%
Females	15,499	20%	14,471	19%	13,005	17%	11,592	16%	10,329	15%
All	42,224	55%	39,329	51%	34,773	46%	30,490	43%	26,691	38%
Creoles										
Males	17,056	22%	18,569	24%	19,457	26%	19,860	28%	20,730	30%
Females	17,893	23%	19,678	25%	20,748	28%	21,032	29%	21,947	32%
All	34,949	45%	38,247	49%	40,205	54%	40,892	57%	42,677	62%

SOURCE: Minutes of the Evidence before Select Committee on the State of the West Indies Colonies, PRO ZMCI.

Hope, Mon Repos, Lusignan, Annandale, Enmore, Bachelor's Adventure, and several others had done the same. *Chateau Margo* had doubled its sugar production in few years.

The shift to sugar brought increasing concentration of land and labor. By 1830, *Triumph* had merged with *Ann's Grove* and *Two Friends* and had 383 slaves; *Bachelor's Adventure* had merged with *Elizabeth Hall* and *Enterprise* and together they had 694 slaves.[42] *Success,* which in 1813 had 186 slaves and cultivated only cotton, had shifted to sugar and had 481 slaves. It belonged to John Gladstone, a Liverpool merchant and father of the future prime minister.[43] (At the time of emancipation Gladstone owned several other plantations in Demerara and about 2,000 slaves, for which he received a very large compensation. By then, most East Coast plantations had been converted to sugar, several plantations had merged, many had passed into the hands of British corporations, and steam mills had replaced traditional wind- and cattle-driven mills.)[44]

Between 1807 and 1832, sugar production in Demerara as a whole showed a remarkable growth.[45] Even when prices started going down after the brief bonanza of 1815 and 1816, and stayed down for the next ten years, plantation owners continued to grow more cane and produce more sugar in a desperate attempt to keep their profits high. Sugar production almost tripled, while coffee and cotton production was reduced by about half. (See Figs. 1 and 2.) During the same period, the costs of production increased, the total slave population declined, and prices of slaves rose. Although slave productivity was one of the highest in the British colonies (an annual average of 10 and ¾ cwt. per slave engaged in sugar cultivation—a figure surpassed only by Trinidad and Saint Vincent)[46]—plantation returns diminished and planters started complaining again that they could hardly realize any profit at all.

When Peter Rose,[47] who had resided in Demerara since 1801 (except for

Table 2. Male and Female Slaves in the Demarara-Essequibo Population, 1817-1829

	1817		1820		1823		1826		1829	
Males										
African	26,725	35%	24,658	32%	21,768	29%	18,898	27%	16,362	23%
Creole	17,056	22%	18,569	24%	19,457	26%	19,860	28%	20,730	30%
Total	43,781		43,227		41,225		38,758		37,092	
Females										
African	15,499	20%	14,471	19%	13,005	17%	11,592	16%	10,329	15%
Creole	17,893	23%	19,678	25%	20,748	28%	21,032	29%	21,947	32%
Total	33,392		34,149		33,753		32,624		32,276	
Sex Ratio, African	1.72		1.70		1.67		1.63		1.58	
Sex Ratio, Creole	0.95		0.94		0.94		0.94		0.94	
Sex Ratio, Slaves	1.31		1.27		1.22		1.19		1.15	

SOURCE: Minutes of the Evidence before Select Committee on the State of the West India Colonies, PRO ZMCI.

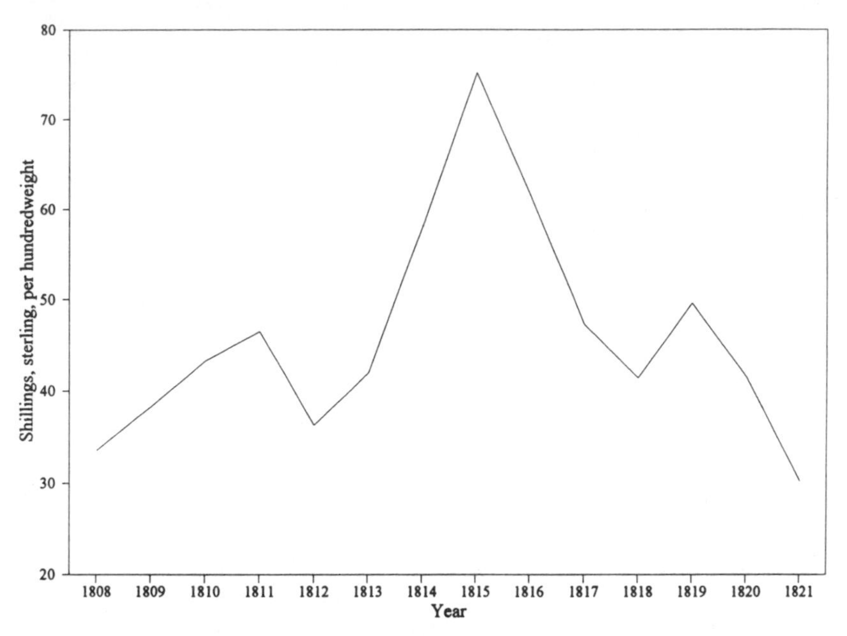

Figure 1. Sugar Prices in the British Market, 1808–1821
Source: *Royal Gazette*, December 13, 1821

a period of six years), appeared before a select committee of the House of Commons in 1832 to testify on the state of the colony, he said that when he had arrived in Demerara two-thirds of the population were employed in the cultivation of cotton and coffee, but probably no more than one-fifth still devoted themselves to such crops.[48] Most people had shifted to sugar and great improvements had been introduced in the manufacturing of sugar and rum. He calculated that with the introduction of new machinery, planters had managed to decrease labor by one-third. But the fixed costs of production had increased.[49] Rose acknowledged that although the outlay of capital was greater, planters not only had saved labor, but could also make the same quantity of sugar in less time and with less loss of cane. He estimated that an estate located in the best part of the colony with a gang of 500 slaves could produce an annual average of 10,769 cwt. of sugar (twice the average for the colony) and 58,354 gallons of rum. After calculating a long list of necessary expenses—slaves' food and clothing, drainage, doctors, salaries, sugar hogsheads, nails, timber, lime, tar, pitch, cordage, cane punts, machinery implements, the cost of repairing buildings, etc.—he demonstrated that the cost of production was such that at the current sugar prices planters made little profit. Estates were yielding at the most 2.5 percent on the capital invested, at a time when the legal interest in the colony was 6 percent.

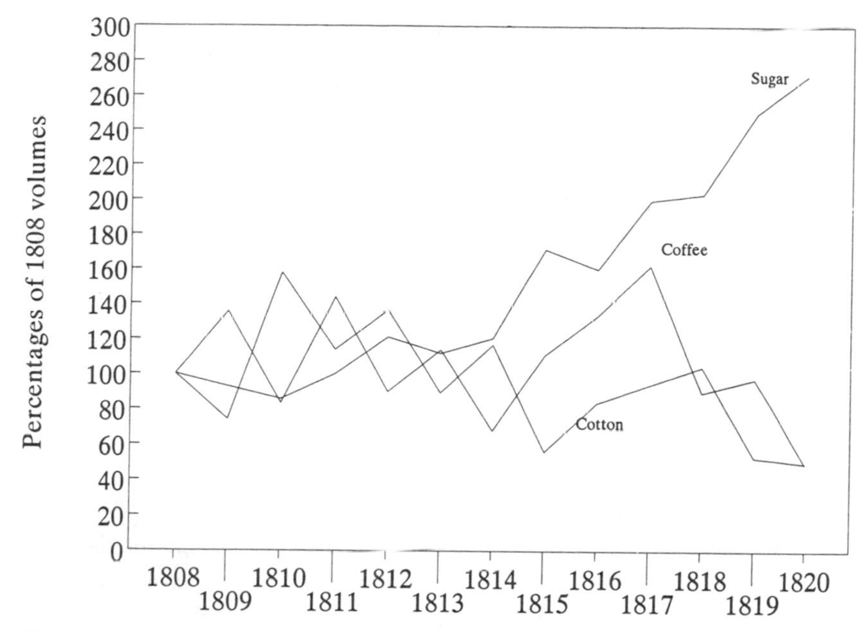

Figure 2. Exports from Demerara and Essequibo, 1808–1820
Source: The Local Guide, Conducting to Whatever is Worthy of Notice in the Colonies of Demerary and Essequebo, for 1821 (Georgetown, 1821)

Within the preceding nine months there had been a depreciation in property value of about 30 percent.

In his testimony, Rose also touched on the problem of labor supply. It had always been difficult to get enough slaves, but the abolition of the slave trade had sharply intensified the problem. He was convinced that the Demerara planters would not be able to compete with planters from countries like Brazil or Cuba that continued to import slaves from Africa.[50] As a consequence of the interruption of the slave trade, the relative number of females increased, the number of slaves of productive age declined, while the percentage of old people grew considerably. These changes affected the estates' productivity. The problem was not that females did not perform as well as males, for most of the time they did. But there were circumstances in which they could not match the sheer physical strength of the males. There was no difference, for example, between male and female labor in light weeding. But in heavy weeding, thirty females were needed to do the work twenty-five males could do. And women of reproductive age were unable to meet normal work standards because of child-bearing and child-rearing obligations. Managers usually divided the slaves into two or three gangs according to age and sex. Women typically worked in the "second gang," children, the old, and the disabled were in the third, which was given light tasks.[51] A plantation with many children under ten and many people over fifty could not be very efficient. Rose calculated that the really effective labor force in Demerara was about one-third of the slave population, while during the slave trade it had been two-thirds.[52]

The population figures for this period show that Rose indeed had reasons to be concerned. The number of children younger than ten did not increase—as he thought—but actually declined slightly from 1817 to 1829. But the proportion of women did in fact grow from 43 to 47 percent.[53] Even more important, the percentage of slaves between twenty and forty (the so-called prime negroes) declined from 50 percent in 1817 to 29 percent in 1829, while those over forty increased from 14 to 33 percent during the same period. (See Table 3 and Fig. 3.) But the more ominous trend was the gradual decline in the total slave population. In 1817 the total number of slaves living in Demerara and Essequibo was 77,163. In 1829 the slave population had fallen to 69,386, a reduction of about 10 percent in twelve years.[54] (See Table 2.) As a consequence of the end of the slave trade and decline of the slave population of productive age, the price of a prime slave rose from about £50 at the beginning of the century to around £150 in the 1820s.[55]

Between 1808 and 1821, planters managed to bypass the British government's restrictions and to transport—with or without license, and under a variety of pretexts—about 8,000 slaves from neighboring areas in the Caribbean.[56] Most came from Berbice, while Dominica and the Bahamas provided the second and third largest groups. A smaller number came from Barbados, Saint Vincent, Saint Christopher, Grenada, Trinidad, Antigua, Suriname, Martinique, and Tortola.[57] There were even cases of free blacks brought to Demerara and sold as slaves.[58]

If these imports are taken into account, then the actual decline in the slave population in little more than a decade was closer to 20 than to 10 percent. High slave mortality combined with a low birth rate was responsible for this trend. The high mortality was in part due to the unhealthy conditions in the colony.[59] Many diseases afflicted the slaves, particularly dysentery, typhus, smallpox, yaws, tetanus, syphilis, leprosy, and a variety of lung and bone diseases and verminosis. But disease was aggravated by the intense rhythm of labor, unhealthy working conditions in the mills and in the fields, the precarious nature of medical assistance,[60] and inadequate diet.[61]

Adult slaves received a weekly allowance of a pound and a half to two pounds of saltfish and a bunch of plantains. Children were given half that quantity. Slaves complemented their diet with the products of their own gardens.[62] Every traveler who visited Demerara in the early period marveled at the slaves' gardens, where they grew yams, corn, and a variety of squashes. They also raised chickens, ducks, goats, and turkeys and (more rarely) pigs. In addition to the small gardens near their houses, slaves also had access to

Table 3. Age Distribution of the Slave Population of Demerara and Essequibo, 1817-1829

	1817		1820		1823		1826		1829	
Under 5	9,814	12.72%	8,617	11.14%	7,721	10.23%	7,052	9.88%	7,607	10.96%
5–10	7,412	9.61%	7,723	9.98%	7,729	10.24%	5,736	8.04%	5,245	7.56%
10–20	10,080	13.06%	11,197	14.47%	12,831	17.00%	13,677	19.16%	13,033	18.78%
Under 20	27,306	35.39%	27,537	35.59%	28,281	37.47%	26,465	37.08%	25,885	37.30%
20–30	19,044	24.68%	12,403	16.03%	8,824	11.69%	8,792	12.32%	9,498	13.69%
30–40	19,998	25.92%	21,169	27.36%	17,872	23.68%	15,524	21.75%	10,818	15.59%
20–40	39,042	50.60%	33,572	43.39%	26,696	35.37%	24,316	34.06%	20,316	29.28%
40–50	7,414	9.61%	11,185	14.46%	14,074	18.65%	14,623	20.49%	14,836	21.38%
50–60	2,470	3.20%	3,553	4.59%	4,640	6.15%	4,505	6.31%	6,228	8.98%
60–70	714	0.93%	1,191	1.54%	1,409	1.87%	1,193	1.67%	1,609	2.32%
70–80	111	0.14%	234	0.30%	299	0.40%	218	0.31%	426	0.61%
80–90	17	0.02%	44	0.06%	44	0.06%	31	0.04%	36	0.05%
Over 90	11	0.01%	16	0.02%	7	0.01%	7	0.01%	9	0.01%
Over 40	10,737	13.91%	16,223	20.97%	20,473	27.12%	20,577	28.83%	23,144	33.35%
Unknown	78		44		27		24		41	
Totals	77,163		77,376		75,477		71,382		69,386	

SOURCE: Minutes of Evidence before the Select Committee on the State of the West India Colonies, PRO ZMCI.

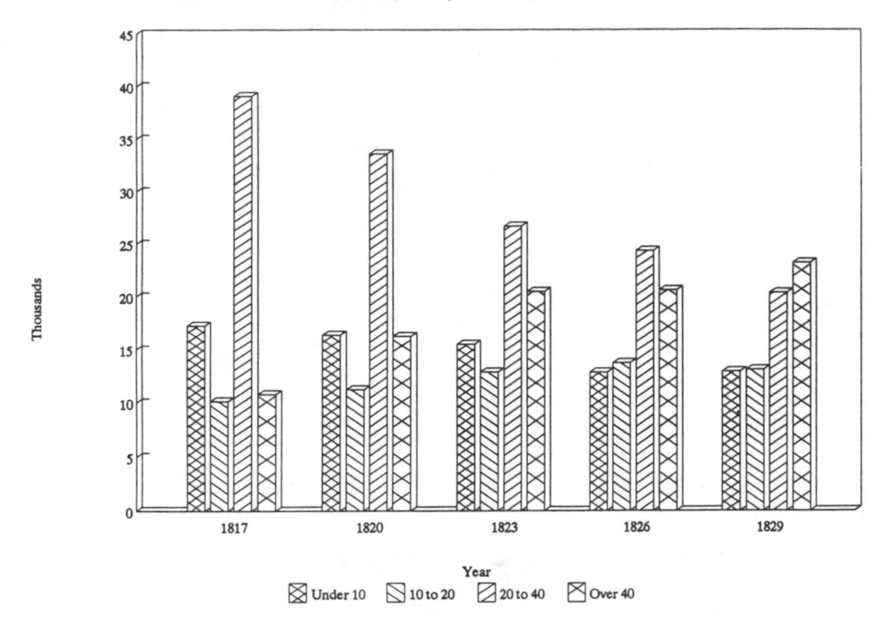

Figure 3. Distribution of Slave Population by Age, 1817–1829
Source: Minutes of Evidence before the Select Committee on the State of the West India Colonies, PRO ZMCI

provision grounds: a parcel of land given to each family to grow what they needed.[63] They worked on their gardens and provision grounds during their "free" time. Slaves sold their surplus to each other, or to free blacks and whites in the neighborhood. On Sundays they took their produce to the market in Mahaica or Georgetown.[64] The markets were more than a place of commercial exchange. They were a place for gathering and socializing, for meeting with friends, for gambling and drinking and participating in other forms of entertainment. The sale of the slaves' surplus gave them an opportunity to participate in the market economy and to accumulate some cash, introducing in their lives a small space of freedom and autonomy.[65] Bolingbroke mentioned an old woman on a sugar estate in Essequibo who when she died left nearly 300 pounds sterling which she had acquired "merely from raising feathered stock."[66] Slaves also obtained cash from hiring themselves out to perform different tasks on Sundays. But after the abolition of the slave trade slaveowners put so many demands on the slaves that they were left with little time to devote to their gardens and provision grounds. As a consequence, slaves became more dependent on their weekly allowances, and the colony more and more dependent on food imports.

The situation turned critical in 1812, when war between England and the United States interrupted the flow of trade. Deprived of supplies, whites and blacks risked starvation. The colonial administration was forced to intervene. On September 18, 1813, the *Royal Gazette* published a notice from the fiscal announcing that he intended to inspect the provision grounds, "which should be maintained on every estate." He threatened to prosecute "to the utmost sejour of the law" those who had not complied with the regulation. This measure, however, apparently did not work, since in 1821 Governor Murray felt compelled to issue a proclamation reminding planters and managers that they were to keep sufficient provision grounds "well and properly stocked with provisions, to be cultivated at the ratio of one acre to every five slaves, under penalty of *f*90 [guilders] for each acre of provision ground cultivated short of the number so required." The governor also ordered the respective burgher captains to inspect those grounds every year in January, accompanied by the manager or overseer of each estate.[67] The situation did not improve. When the Reverend Wiltshire Staunton Austin, who had lived in Demerara for many years, testified before a House of Commons committee in 1832, he said that there were no provision grounds in Demerara as there were in other colonies. His testimony was confirmed by others.[68] Even though such statements may have exaggerated the disappearance of provision grounds, there is no doubt that, with the increasing demand for labor in the export sector, slaves were left little time to devote to their gardens and provision grounds. Not surprisingly, slaves were constantly complaining of a lack of food.

Equally serious were the problems arising from inadequate medical assistance. Although every plantation had a "sick-house" where slaves were supposed to be treated, and there were always a couple of slave nurses or midwives and a visiting or resident medical doctor in charge of keeping an eye on the sick, the conditions were extremely precarious. Medicine was often primitive, most doctors were careless and ill-trained, and the "sick-houses" looked more like prisons than hospitals, since it was there that slaves were usually locked up in stocks. To make the situation even worse, managers, who always suspected that slaves were malingering, dismissed their complaints and forced them to work when they were sick.

All this helps to explain why death rates were so high. Predictably, mortality was much higher among men than women (by fully 20 to 25 percent). But it was even higher among children. A doctor who was in charge of about a thousand slaves on the West Coast reported in 1824 that of the sixty-seven children born on one estate, twenty-nine died in the first two years, which he attributed to tetanus, worms, and other infant diseases.[69]

The effect of high mortality was compounded by a low birth rate.[70] The imbalance of males and females, and the relatively small number of women of reproductive age, limited the number of births. After the interruption of the slave trade, the imbalance tended to diminish with the growth of the

Table 4. Births and Deaths Among Slaves, Demarara and Essequibo, 1817-1829 (by triennium)

	1820	1823	1826	1829
Births	4,868	4,512	4,494	4,679
Deaths	7,140	7,188	7,634	5,724
Net loss	2,272	2,676	3,140	1,045

SOURCE: Minutes of Evidence before the Select Committee on the State of the West India Colonies, PRO ZMCI.

creole population, but the marked predominance of males over females among African-born slaves persisted. (See Tables 1 and 2.) In 1817 there were about 27,000 African-born males and only 15,000 females. And although among the creole slaves the figures showed an almost even balance, many of the creole females had not yet reached reproductive age. Frequent miscarriages and the practice of abortion also reduced the birth rate.[71]

The triennial registration of May 1820 revealed a wearisome trend (Table 4). During a period of three years, about 7,000 slaves died in the United Colony of Demerara and Essequibo, while only 4,800 were born. The next two triennial reports, for 1823 and 1826, did not indicate any improvement.[72] Even before such figures were made public, planters had been worrying about the trend—so much so that the Essequibo Agricultural Society decided to offer a gold medal valued at ten guineas to any person in charge of an estate who raised the greatest number of children in proportion to births and to females of reproductive age.[73]

Caught in the vise of having to increase production with a shrinking labor force and probably fearing the growth of the free black population,[74] the colonial government had in 1815 imposed restrictions on manumission, complicating the procedure and imposing severe economic burdens on slaves who wished either to buy their own freedom or the freedom of their relatives and friends.[75] The restrictions also applied to masters wishing to grant manumission to slaves. The new regulations reserved to the Court of Policy the power to grant manumissions, limited personal arrangements between masters and slaves, and required that anyone filing a petition for manumission pay from 250 to 1,500 guilders, plus post a bond of up to 500 guilders. The regulations also made it clear that slaves would have no rights to freedom until they actually had in hand their letters of manumission granted by the Court of Policy.[76] The intention was to discourage manumission altogether.[77]

Manumissions had never been very numerous but they had always been a source of hope for the slaves. "I have known many instances of negroes, who paid their owners a proportion of the purchase money and were allowed after emancipation to work out the balance," wrote Bolingbroke in 1805.[78]

But the number of manumissions diminished after the abolition of the slave trade. Between 1808 and 1821, only 477 manumissions were officially granted in Demerara and Essequibo, 142 to males and 335 to females.[79] The chance an adult male had to be emancipated was slim. Only one out of ten males manumitted during this period was an adult. The others were boys, often manumitted with their mothers. Many manumissions were bought by the slaves themselves or by free blacks rather than granted by whites.[80] Apparently, more women were able to buy their freedom and that of their children because they had more opportunities to earn some cash, either by selling produce in the market (since huckstering was mainly a female activity) or by doing "favors" for whites. It is also possible that some white men, ashamed of publicly recognizing their liaisons with black women, preferred to give them money to buy their freedom and the freedom of their children. Slave men may also have chosen to use the little cash they earned by performing small jobs on Sundays to free their wives and children.

Although manumissions were few, the free black population continued to grow in Demerara. In 1810, freedmen constituted 3.5 percent of the total population; twenty years later they were a little over 8 percent.[81] This growth can be explained in part by natural growth, and in part by the influx of freedmen and free blacks (mainly women) from other colonies,[82] but also by an increase of manumissions after 1823[83] as a consequence of a change in British policies.[84]

Most free blacks lived in Georgetown, where they either performed a variety of services for the white elite or worked as independent artisans and small shopkeepers. Many kept stalls in the market or lived as hucksters, going from plantation to plantation selling goods. Those who had managed to accumulate enough capital to buy a slave or two sent them to sell goods on the streets, or rented them out.[85] Free black artisans often hired slaves to work for them in their shops. A few worked as wood cutters in the Mahaicony, an area where there was a great abundance of mahogany—a wood then in great demand, which made this activity quite profitable. Many freed blacks continued to live on plantations working as independent artisans or maintaining small shops that often became the gathering point for slaves from the neighboring area. Whites were constantly complaining against "grog shops," which they suspected of being receptacles for stolen goods.[86] Because of the variety of activities they were involved in—some making them dependent on whites, others linking them to the slaves—free blacks' behavior was always unpredictable.[87]

The decline in the number of slaves was particularly serious in Demerara because it coincided with the expansion of the area under cultivation and the shift to sugar.[88] Although labor had been "saved" by the introduction of new machinery in the processing of cotton, coffee, and sugar, the system of labor in the fields had not changed fundamentally and more slaves were now working on sugar plantations, where the cultivation of cane required more workers per acre than cotton or coffee. The process of growing cane continued to be extremely laborious. Everyone agreed that if cultivating, harvesting, and processing cotton and coffee were arduous tasks, growing and cutting cane and manufacturing sugar were even worse. To start a new cane field, the land had first to be cleared and empoldered[89]—a task often given to hired task-gangs. Cane cuttings were planted between ridges. During the next twelve to sixteen months, while the cane was growing, the fields had to be weeded, two or three times. When cane reached maturity it had to be harvested and transported by boat to the mills to be processed. The work in the mill required a large number of slaves with different skills.

Bolingbroke described in detail the various steps in the production of sugar. After the cane was ground the "liquor" that had been collected into a cistern was sent through spouts to a "boiling house." Once boiled and skimmed and treated with lime, the syrup was reduced and clarified in a succession of copper vessels. It was then cooled in a wooden gutter and put into hogsheads in the curing-house. Molasses drained for a fortnight out of the hogsheads, and was channeled into a separate cistern.[90]

For the slaves, the complex process of making sugar involved many risks, but increased the opportunities for effective sabotage. If the furnaces were not kept hot enough the cane juice would ferment. The pulp left from the grinding of cane ("megass") had to be removed quickly, otherwise it would accumulate causing serious hazards that could force the mill to stop. Coopers in charge of making the puncheons and hogsheads might slow down, producing fewer than were necessary for the day. Slaves responsible for providing the coopers with staves and hoops could withhold them. Men or women in charge of firewood could damp the fire by choosing green or wet instead of dry wood. Cane feeders could put coal, chisels, and other things into the mill, forcing it to stop. Any distraction in the boiling house might cause a fire. Thus sugar production required great vigilance and rigor from managers and overseers, and discipline, coordination, and skill from slaves. And when steam mills were introduced (something that happened early in Demerara) the tyranny of the machine, which imposed its own pace, was added to the discipline of the drivers and overseers.

Since a manager was given a commission of 2 to 10 percent of an estate's production, managers were relentless in their demands and severe in their punishments. And they certainly were less concerned with the slaves' health or welfare than with their productivity. As long as plantations were producing profits, managers could count on the support of their distant employers.

If they abused their authority by making unreasonable demands or punishing the slaves excessively, they only had to respond to the attorneys and local authorities, who for their part—at least until the rebellion—tended to dismiss the slaves' complaints. Only when the situation got out of hand did managers lose their jobs. So with the help of slave drivers and overseers—who usually were indentured poor whites who had migrated to the colony in the hope of improving their economic situation—managers drove the slaves as hard as they could.

The division of labor depended on managerial decisions and probably reflected plantation requirements as much as English social practices, although it is possible that the slaves had some say in the matter.[91] The monopoly of women, both slave and free, over huckstering (also found in other places in the Caribbean) suggests some continuity with African traditions, since women tended to control this activity in most African societies. The situation of women was aggravated by their double tasks of production and reproduction. Many confrontations between female slaves and managers originated in contradictions between women's roles as mothers and as workers—which were intensified by growing labor exploitation after the abolition of the slave trade.[92] Managerial decisions about the type and the amount of labor to be performed and the free time allowed to slaves were a permanent source of resentment and conflict.

Most slaves of productive age who lived on plantations were field laborers.[93] About 7 percent worked as domestics, 5 percent as tradesmen, and another 5 percent as drivers or gang leaders.[94] Although both males and females worked in the fields, there was a clear division of labor along gender lines. This was even more obvious among skilled workers. Carpenters, coopers, and masons—the largest groups among the tradesmen on plantations—were all males. The same was true of sawyers, wood cutters, blacksmiths, coppersmiths, boat builders, tailors, fishermen, watchmen, transport workers, and even basket makers. By contrast, seamstresses, laundresses, weavers, coffee cleaners, midwives, children's nurses, and slave hucksters were all females. Other activities, such as cooking, domestic labor, jobbing, and nursing, were performed by both males and females. And although most drivers were men, a few were women.[95] In sugar mills, most tasks were assigned to males. The only mill job that was conventionally given to women was the burdensome work of removing the "megass." The division of labor created hierarchies within the slave community, with artisans and drivers at the top and field laborers at the bottom. The relative autonomy that drivers and tradesmen enjoyed made them even more likely to resent slavery and entertain dreams of emancipation. Slaves who performed domestic labor enjoyed a "privileged" position because they had direct access to managers and masters, with whom they often established personal relations, and from whom they could expect special favors, but they were also watched more closely, and could easily lose their position. This explains why although they tended to side with their masters, they often betrayed their trust.

Whatever their activities, slaves were constantly occupied. As soon as they were finished with one job, they had to start another.[96] Planters who grew only cotton or coffee and temporarily had no work for some of their slaves rented them out, something that would have been more difficult in plantation societies where monoculture was the norm and everyone needed labor at the same time. The variety of crops cultivated in Demerara may explain why the system of hiring slave gangs was so common. The newspapers always carried advertisements from people wanting to hire task-gangs. Although the practice was used throughout the year, the demand increased at harvest time. A number of people who had no plantations made a living exclusively by renting their slaves.[97] That is why the slave population registered in the city of Georgetown tended to increase rather than diminish after the abolition of the slave trade, even during a period when the total slave population was declining, and the demand for slave labor on plantations was increasing.[98] In 1815 there were 62,411 slaves listed as attached to estates and 10,103 as "belonging to individuals" in Demerara and Essequibo. Some of these "individuals" were little more than labor contractors who took the responsibility for certain tasks and moved from one plantation to the other with their gangs. Others derived their main source of income from sending slaves to sell things on the streets. This practice was common even among free blacks or mulattos who had managed to acquire a couple of slaves.[99]

Task-gang slaves were employed in a great variety of activities: harvesting, construction, digging or cleaning trenches, empoldering land, opening canals, cutting staves or shingles, timber or firewood, growing plantain, or doing other equally heavy work.[100] These slaves were among the worst treated. People who hired them did not give them the "privileges" they gave their own slaves, and they tried to exact the maximum possible amount of labor in the shortest period of time. Task-gang slaves were usually deprived of access to gardens and the other small benefits enjoyed by slaves attached to plantations. They constantly complained of lack of food and clothing, excessive labor, and unfair punishment.

The situation of slaves who were rented out without any specification of the tasks they were to perform was even worse. Typical was a case reported by the fiscal of Berbice in 1822. Five slaves complained that they had been hired as part of a large task-gang to work on an estate for a period of twelve months. They did not have the privileges of the slaves of the estate but had all the obligations. (Apparently slaves from the estate had received clothing but the complainants had not.) So, when they were ordered to bring grass and fuel they refused. Other members of the task-gang joined in. As a consequence, twelve men and one woman were flogged and one of the slaves, Pompey, was put in the stocks. The slaves were then told that if they did not bring grass they would be flogged again, so five slaves had gone to complain to the fiscal. When called by the fiscal, the manager explained that the slaves had been hired for the year (and were not part of a regular task-gang). They had refused to do their job and when he repeated the order, Pompey had stepped out and with "an impertinent tone and awkward gesture" stated that he would not bring grass at night as it was not part of his duty. To punish his defiance, which could have influenced others, the manager put Pompey in the stocks.[101] This case and many others like it show that whether they were rented out for a period of time or worked on specialized and supervised task-gangs, these slaves were worse off than those who belonged to plantations. So universally acknowledged was their bad situation that managers threatened unruly plantation slaves with renting them out or sending them to work in a task-gang. And sometimes this was indeed used as punishment. On the other hand, by moving from one plantation to another, task-gang slaves managed to establish networks and links of friendship both with plantation slaves and free colored people. These bonds could become instrumental in a rebellion.

The amount of work to be done on plantations was set by custom.[102] The regular work-day was supposed to be twelve hours. Slaves would rise at five and work from five-thirty or six in the morning until sundown. At midday they would have an hour or an hour and a half to rest and eat. But in practice it was not unusual for them to be kept working until late hours, sometimes until two in the morning, and those who worked in the mills frequently had to get up at two or three a.m. They scarcely had any time to have an adequate meal. And although slaves were supposed to have Sundays free, managers often forced them to work, at least for part of the day. Pregnant women were, in principle, entitled to certain privileges. They were to work on the weak gang until the sixth or seventh month and then be in charge of light tasks until the child was born. They were to return to work after five or six weeks, but for the next one or two years, managers were expected to grant them an extra half-hour mornings and afternoons and some free time during the day for breastfeeding. Managers complained that women used this "privilege" as a pretext for taking longer breaks, while women often complained that their "rights" had been violated.

Experienced and competent managers had a general notion of how much work to expect from each slave, but tasks had to be adjusted to a variety of circumstances and there was a great deal of arbitrariness in the assignment of tasks to individual slaves. Slaves often contested managers' decisions, and the amount of work had to be constantly negotiated between them. The customary practices that resulted from these confrontations and implicit negotiations came to be seen by the slaves as rules they could invoke whenever managers stepped beyond such limits. Since circumstances often changed—slaves accustomed to one activity were assigned another, and crop, soil, and weather conditions varied—the allotment of tasks was always a potentially controversial issue.

One manager who was asked how much work was given to slaves on his plantation said that an "able man" working to dig a trench was "to throw out" 450 to 500 cubic feet of earth a day. Another said that he expected the slave to dig ten feet by twelve of a four-foot-deep trench (480 cubic feet). Others set the quota at 600 hundred cubic feet. In cane-holing, a slave was supposed to dig from one-twentieth to one-twenty-fifth of an acre per day. In weeding and moulding young canes—which was considered a light but dangerous and tedious task usually assigned to women—a slave was expected to go over an eighth or a ninth of an acre a day. In 1824, a manager on a plantation in Leguan calculated that for each heavy hogshead of sugar, twelve to fifteen slaves were needed to cut the canes and carry them to the punts. Two more were needed to feed the mill, three to carry canes and "clean the vessels about the mill," plus one to every copper vat, two firemen to each set of coppers, and one fireman to the steam engine. The number of slaves required to carry "megass" depended on the distance they had to walk from the place were the cane was ground to the storage building, or "logie."[103]

On a cotton estate, weeding requirements varied during the season from six prime slaves an acre when the crop was young, to three or four late in the season when it was time for the third weeding. Managers expected a male slave to be able to gin fifty pounds of cotton and a woman thirty, but this operation depended on the dexterity of the individual worker. Weeding was simpler on coffee than on cotton plantations, because the coffee trees were usually shaded with plantains and the grass did not grow so fast. Still, four men were put to weed an acre, trim the plantains, and pull suckers off the coffee trees. A male slave was supposed to pick an average of thirty to thirty-six pounds of coffee beans a day. Picking quotas varied, however, not only according to the slaves' skills but also the richness of the crop. If beans were plentiful, one slave could pick three baskets a day, but when the crop was thin they might be able to gather only one.[104]

All this meant that labor requirements had to be adjusted to the nature of the soil, the stage of growth, and weather conditions. The amount of work expected from a slave in charge of holing land depended on whether the soil was dry or wet, soft or hard. If the task was weeding, the amount of work varied according to the heaviness of the grass and the foulness of the plants, and this in turn depended on the age of the plants and the state of the weather. The amount of cane a slave was suppose to carry varied according to the quality of the cane. The work assigned to a slave digging trenches depended not only on the nature of the soil but also on the size and depth of the trenches, and so on.

When managers made their decisions about how much work a slave or a gang was expected to perform they also had to consider the slave's ability and strength. A prime slave could do more work than an average slave (one manager calculated that an average slave did a fourth less than a prime slave). An old man obviously could not perform as well as a young one, or a sick man as well as a healthy one. Individual slaves were better at certain tasks than others. Managers always found such decisions difficult because slaves always seemed to be feigning sickness or pretending they were not able to accomplish their tasks. They were convinced that slaves could—if they wished—do much more work than they did. Typical was a story told by one manager: that he had seen two slaves, who had bet on who could work faster, do twice as much ginning as was usually assigned to them.[105]

Customary norms and uses generated expectations on both sides. Managers used them to assess the slaves' performance, the slaves to assess the managers' fairness. But all the different circumstances affecting labor performance introduced an element of unpredictability and gave managers a great amount of discretionary power. In spite of all the confrontations and negotiation that went on between slaves and managers, the final decision after all was theirs. As one manager aptly put it, decisions about how much work a slave should or could do depended more on the manager's "judgment" than on any particular rule. And managers, out of inexperience, spite, or carelessness, sometimes broke the rules, triggering slave resistance and punishment.

Another common source of contention and confrontation between managers and slaves was food and cloth allowances. Aside from weekly allowances of saltfish and plantain, each man was supposed to receive every year a blue jacket and trousers, a hat, four ells of osnaburg, four ells of checks, and a cap. Women got a hat, a blue-cloth wrapper, a blue petticoat, three to six ells of checks, six ells of osnaburg, a handkerchief each, sometimes needle and thread, and a pot or two. On holidays, slaves could expect to be given a little salt, some sugar, and tobacco or rum. Occasionally they received a blanket. Everything else they were supposed to provide for themselves, except for housing and "medical assistance."

When norms for labor performance, food and cloth allowances, and medical assistance were violated, slaves protested. From law and custom they derived notions of "rights." And it was in the name of these "rights" that, individually or in groups, slaves went to the fiscal to complain, rejected their allowances (when these were insufficient or spoiled),[106] did not perform their tasks when they thought the assignment was "unreasonable," and from time to time even resorted to strikes, collectively refusing to do any work until their demands were met.[107]

An experienced manager who lived in Demerara for almost twenty-five years gave in 1824 eloquent—though somewhat biased—testimony about the slaves' attachment to what they perceived as their rights:

> No class of people are more alive to their own rights than the slave population of this country; and when these happen at any time, or in any way, to be the least infringed upon, they do not hesitate to seek for redress; if oppressed or wronged by the managers, they apply to their owner, if present, or if absent, to his attorney, who on all occasions is most ready to enquiry [*sic*] into the cause of their complaints; sometimes they are of a very trivial nature, being hardly worth noticing, and are only brought forward by some of the ignorant of the slaves, who have been urged on by the more artful and vicious, for the purpose of forwarding their own private designs; but when a complaint is better founded, it does not unfrequently happen that the manager is discharged, the overseer or whomsoever the complaint may be against. In case the slave considers himself aggrieved, and thinks he is not redressed as he ought to be, he does not hesitate to apply to the fiscal, whom I have never known to neglect an application of this kind, or omit making the necessary enquiry into the cause of such complaint. . . . If slaves are curtailed in the least, either by mistake or design, of the time allowed them for breakfast and dinner, or made to work at improper hours, or punished on trivial occasions, they do not fail to make complaint of it, and are redressed accordingly.[108]

The picture was somewhat utopian. Most slave complaints were dismissed, and slaves were often punished when they went to complain. It was only after the rebellion of 1823—when the British government decided to intervene more directly in the management of the slave population—that fiscals became somewhat more responsive to the slaves and started fining a few managers.[109] Until 1824, when Sir Benjamin D'Urban replaced Governor Murray, it had not been an ordinary practice to record slave complaints and the proceedings in the fiscal's office.[110] The new governor, following the guidelines of the British government, ordered the fiscal to keep records and made it clear that all slave complaints should be patiently heard, carefully investigated, and even-handed justice observed, "as well to the masters, as the slaves." He recommended that unfounded and malicious complaints be duly checked, but that well-grounded ones be "redressed to the utmost extent of the Law," and according to the circumstances of the respective cases. He also took measures to inform the slave population that he would at all times personally hear and attend to the complaints.[111] Governor D'Urban did in fact take a personal interest in controversial cases. When Darby from

Bel Air was punished unfairly by the manager for having complained to the attorney, the governor wrote the fiscal a letter extremely revealing of the new notions of social control that inspired his policy, notions that stood in direct contrast to the position adopted by his predecessor.

> If this be tolerated, there is an end to all justice: a slave is wronged, he complains to his legitimate protector, who interferes in his behalf, and what is the consequence of his interference? why, the manager punishes the slave again for having dared to complain, so that all the advantage he derives from an effort to get redress is to get two punishments instead of one. This practice must of course put an end to complaints altogether, but then it must also put an end to hope; and, if it be at all general, cannot but end in universal desperation. If such was the case among the managers of the Eastern District, which I trust it was not, there would be no need to seek further for sufficient cause of the insurrection which burst forth there last year. Such causes must, in the nature of things, produce such effects; and cannot be too deeply deprecated.[112]

The governor, however, did not receive much cooperation from the fiscals. While the governor was responding to the new guidelines that originated in the Colonial Office, fiscals judged the situation from the perspective of slaves' proprietors, who felt that discipline would be ruined if anyone was allowed to interfere with their authority. This became clear in a letter Fiscal Charles Herbert sent Governor D'Urban in 1825.

> The servants of the Crown, although in foreign climes, are actuated by a sincere desire to ameliorate the slave population, and to carry into effect the wishes of His Majesty's Government. But, your Excellency is well aware that the feelings for one part must not obviate a just consideration for the other, and that the tranquility of a valuable colony will be effectually destroyed if acts of insubordination and misconduct on the part of the slave, and particularly in the unsettled state of 1824, are tolerated under the name of complaint.[113]

So, in spite of the new governor's efforts, Charles Herbert and his successors continued for the most part to dismiss slaves' complaints, preferring to side with their masters. But that did not dissuade the slaves from complaining. In 1829, an old and sick man named Charles complained that he could not do the work he was told to do. Recognizing the man's frailty, the fiscal called his master. But when the master answered that "every care and attention" had been paid to Charles and that the work was light—he only had to "look for a few cattle" and tie them up with the calves at night—the case was dismissed. The same thing happened when Lewis, who belonged to a gang sent by the manager of *Porter's Hope* in the East Coast to cut cane in Mahaica, complained that the overseer did not give the slaves time to eat.

The manager informed the fiscal that Lewis did not perform as well as he should and had a regular amount of time for his meal. He explained that the "negroes" were dissatisfied because of the change from cotton to sugar. The fiscal dismissed Lewis's complaint. A similar fate befell Goodluck, who had been cutting firewood for some months and went to complain that he had received no allowance of clothing from his owner and only got a bunch of plantain every two weeks and a small "bit of piece of fish." Goodluck also reported that the managers had given him 150 lashes with the "long whip" because he did not cut six cords of wood a week, and when he requested a pass to go to the protector of slaves, the pass had been denied. When the manager was called to testify he declared that Goodluck had been given an allowance of one good bunch of plantains and two pounds of saltfish every week and in twenty-four days had carried only ten cords of firewood, when the "usual" was to carry one cord per day. The manager excused himself for not giving clothing, arguing that he did not have means to purchase it. After one witness who lived on the estate corroborated the manager's testimony, the protector dismissed all of Goodluck's complaints but one: he told the owner to provide Goodluck with clothing "with little delay as possible."[114]

Women's complaints fared no better.[115] In fact there was a generalized opinion among managers that women were more difficult to manage, more "refractory" than men, and always ready to instigate the men to insubordination.[116] So managers were as severe with the women as they were with men. Women's recalcitrance was in part a consequence of their double exploitation. As workers, they were subjected to the same abuses as their male counterparts. But women had to cope with additional problems: rape was one, separation from their suckling children another.[117] Not every case of rape was reported, but some extraordinary cases involving children ended up in the office of the protector of slaves, and even in the Court of Criminal Justice. Such was the case of a girl of about ten who was allegedly raped by her manager. After the incident the girl became sick and was taken to the "sick-house." The "nurse" called a doctor, who examined the girl and discovered she had been raped. Pressed by the nurse and then by her stepmother, the girl first said she had been raped by a seventeen-year-old slave. But, under further pressure, she confessed that the manager had raped her. When the manager heard the girl had accused him, he confronted her just to hear her again confirming the story. When he asked her who had done it she answered, "It was you, master." The girl died a few weeks later. Her father took the case to the protector of slaves. The inquiry was inconclusive, but an examination of the body confirmed that the girl had indeed been violently raped.[118]

More typical of the kinds of abuse women endured and of the ways they resisted was the case of Rosey from *Grove,* who went to the protector of slaves to complain that while she was at work in the field she had had a pain in her bowels and had stopped working. The manager ordered her to go on with her work and struck her with a small stick and then with his fist, knocking her down. Then he had her hands tied behind her back and sent her home, where she was put in the stocks for three days and nights. During this time her suckling child was kept from her. Her breasts swelled. She left the estate without a pass and went to the protector of slaves. When the manager came to testify he said that Rosey was sitting in the field instead of working and did not reply to his questions. So he had slapped her face, "in consequence of her great impertinence." He added that Rosey had not been confined in the stocks but in one of the rooms of the hospital. Her sixteen-month-old child had been in the "yaws-house" and he thought the child was "fit to be weaned." The manager produced a statement from the overseer confirming his deposition and saying that the woman had resisted orders and used "abusive" language. Rosey was said to have gone to the "yaws-house" and taken her child away by force, beating the woman in charge. According to the overseer, she apparently had been instigated by her husband to complain against the manager for not allowing one of her older children to stop at the "yaws-house" to take care of her sick child, "there being a nurse there to take care of the yaws people and every attention paid to them in respect to food." Rosey's complaints were dismissed. This and other similar cases reveal not only indifference on the part of the slaves' "protector" and brutality on the part of managers, but also different conceptions of nursing, weaning, and child care. Rosey and other women were probably following African practices of nursing babies for several years, something managers did not tolerate because it intruded on work. Conflicts around such practices were common throughout the Caribbean.

Women also complained about unfair work assignments and punishment. Typical was a case brought to the attention of the protector of slaves in 1829. Jacuba, Julia, Dorothea, Una, and Effa, five female slaves belonging to *Le Repentir* but working on *La Pénitence,* went to complain against the manager. They said he had ordered them to carry "megass" from the mill on Thursday and again on Friday, and told them to get all the "megass" away by nine o'clock at night. They worked from three in the morning until nine at night without stopping, and had no time to eat until after work, although "boiled plantains were brought to them." On Saturday they were again ordered to carry "megass," but refused to do so. They told the manager that other carriers should be assigned to do the job, "as it was the custom that those who carried megass should, after doing that duty for two days running," be sent to do some other work, because carrying "megass" from the mill was the hardest work on the estate.[119] They complained to the manager that he was being unfair, giving light work to some people while killing others with heavy work. On Saturday, instead of obeying the manager, they went to the fields to cut cane. The manager sent for them and confined them in the stocks, both hand and foot, for several hours, then locked them in a room until Sunday morning, when he put them in the stocks again. They were kept there until three in the afternoon. And once again they were confined in a room until the next day.

Interrogated by the protector of slaves, the manager claimed that he had assigned more women than were needed to do the job. But still they had not taken away the "megass" as fast as they should. As a result, it had accumulated "to the great danger of the machinery." The next day they did even less, he said, so he had ordered them to the same duty on Saturday. He explained that if they had carried away the "megass" on the first day as they were supposed to, he would have appointed another working team. He was convinced that the women could have performed their task with great ease, since the mill always stopped three hours every day. (The manager did admit that while they were confined the women had received only water and plantains.)

After hearing the manager, the protector of slaves called the overseer, who confirmed the manager's report. The next to testify was the engineer. By and large, his testimony corroborated the manager's and the overseer's, but he made a new and crucial point. When asked whether the women had carried the "megass" away as fast as they should, he said that they had not. They had allowed the mill to choke. But when the protector asked whether he thought the women were *able* to keep the mill clear, he answered no, "the megass comes from the mill too fast; it is a large mill; while I looked after them and hurried them, they worked as fast as they could; but I do not know if they did so when I turned away." The protector decided that the manager had acted in contravention of the "14th Article of the Slave Ordinance, and First of the Amended Act," and fined him 200 guilders for each of the slaves (the lowest penalty stipulated by the regulations).

Another case typical of conflicts that emerged between female slaves and managers involved Beckey and Lydia, two women working in a weak gang. The two complained that for the past three weeks they had been confined every night in a "dark room" because they were unable to weed two rows of young canes. Beckey stated that she was locked in as soon as she came from the fields and had no time to have supper. Her child—a three-month-old baby—was taken away from her every night. Lydia had similar complaints, but she added that her two-month-old child "gets no suck; that the milk sours in her breasts, and that when she gives the child suck next morning it purges [vomits] it."[120]

When the protector of slaves called the manager, he argued that the women had been given only as much work as was usually allotted to others—two rows of twenty-four beds. He also claimed they usually finished only half their work and were always late in the mornings. He explained that he had assigned a "nurse" to take care of the children, and two other young women to suckle them, to prevent the mothers from harming the children out of spite, adding that this sometimes happened because the women wanted to injure the manager, who was responsible for the welfare of the children. After much consideration the protector told the manager to discontinue this practice.[121]

There were many other motives for conflict between slaves and managers. The "right" that slaves had to the products of their gardens and to the animals they raised often led to confrontations that ended in the fiscal's or the protector of slaves' office.[122] The story Thomas told the protector of slaves is very revealing. Thomas belonged to a carpenter in Georgetown. On a Sunday, when he was leaving his work with two chickens he was taking to his wife, his master took the fowls from him, cut off their heads, and threw them in the public road. Not satisfied, the master had also taken two bits' worth of yams, cut them up and thrown them away. He kicked Thomas "in his private parts"—which indeed appeared to the fiscal to be swollen—and then gave him a "dose of salts" and locked him up in the stocks and "otherwise ill-used him by tearing off his hair, which he, Thomas, produced" as evidence. Thomas said that several slaves and the wife of his owner had witnessed the episode. The protector of slaves sent Thomas to the jail, with directions to the medical attendant to examine him, and summoned his owner, a Mr. Milne, who denied ever having touched the slave, although he did admit having taken the fowls belonging to Thomas's wife. Milne said he had repeatedly ordered her away from his premises, "in consequence of her trafficking by day and by night in his yard." Everyone who was interrogated denied knowing anything, and the medical attendant stated that the swelling had been of long standing and not produced by any kick. The protector reprimanded Thomas and sent him away. But in the following months several slaves complained against Milne. This suggests either that he was a particularly abusive master, or that the slaves out of solidarity with Thomas had decided to take revenge on Milne.

The "right" to own fowls was also the subject of a complaint against John Quarles. Several of his slaves went to complain that not only did they have to work nights and Sundays, but their manager ate their fowls. The fiscal went to the plantation and found that the slaves were right. He "directed" Quarles to pay them the full value of any of their ducks or fowls he had ordered to be killed, and warned him that if there were any more complaints, he would be prosecuted criminally. A similar case was reported in Berbice, although its outcome is not known. Two slaves, Philip and Leander, complained that the manager had ordered the overseer and the driver to kill their hogs, and when Philip asked permission to sell their hogs in town the permission was denied and he was "licked up" and put in the stocks. A story equally suggestive of the ways slaves understood their "rights" was reported by several slaves who went to complain that the manager, after hiring them on a Sunday to gin cotton for himself, had failed to pay them.[123]

Another source of conflict was the transfer of slaves from one activity to another, since slaves lacking the necessary skills were punished for underperformance.[124] Typical was the case of Quamina, a slave from Berbice who appeared at the fiscal's office complaining that he had been sold as a cooper and carpenter but instead had been sent to pick cotton. He had not been able to pick as much as the others and had been flogged, his back washed with brine and rubbed with salt. A similar complaint was filed by Azor, who had been put to pick over coffee in the logie—a delicate task usually assigned to women, and work he had never done before. He was flogged and put in the stocks because he had not filled his basket. Perhaps the most pathetic story of this type was that of a slave who said that although he did not know to make baskets he had been ordered to replace the basketmaker who had died. He failed to produce as many baskets as expected and was punished. Called by the fiscal, the manager took refuge in stereotypes. He argued that the slave must be faking, since *all* blacks knew how to make baskets.[125]

Managers often claimed that the slaves were just pretending they could not accomplish their tasks and accused the slaves of deliberately slowing down and being careless. Slaves complained that masters made unreasonable demands and were unfair in their punishment. That was the case of George, who complained that—although he had never been guilty of "impertinence," never neglected his work, or ran away, and when he had a pass on Sunday never failed to return on Monday—he had been punished for failing to make an adequate fire in the mill. He argued that there was no one to split wood and that the available wood was green, so that he could not make fire. When the fiscal called the manager, the man argued that the previous year he had been forced to change the slaves who fed the mill because coal, chisels, and other things had been put in the mill to prevent it from working. Since then he had had no problems. George had been one of the feeders dismissed and lately had been employed to make fire for the engine. The manager also reported that he had been told that before he had come to the estate George had been detected stealing with others "upward to 300 pounds of sugar."[126] A similar case was filed, involving a slave from *Friends*. The slave complained of unfair punishment. He had been told by the manager to make fire under the engine at midnight, but there was no firewood at hand and when he went to fetch wood, rain was falling, the wood was wet, and the fire would not catch. The complaint was considered unfounded and the slave, instead of getting redress, was punished.

Sometimes it is indeed difficult to know whether the slaves were really involved in sabotage, whether they were making up stories and trying to use the fiscal against their managers and masters. But there are cases in which obviously guilty masters got off with a mere warning from the fiscal. Only rarely did fiscals impose fines on managers or masters, although this practice seems to have become more common after 1823. In 1825, for example, the fiscal fined a certain Pollard 900 guilders for having inflicted fifty-seven lashes on a slave. Pollard was well known for cruelty toward the slaves. He is often mentioned in John Smith's diary as a manager who persecuted slaves who came to the chapel, and there were also several complaints against him on the part of slaves. G. Stroek was also punished. Several complaints had been made against him too. One morning the "schout" reported to the fiscal that Stroek had put a "negress" in irons that were "so small as to occasion her considerable pain." The schout had heard the woman scream and after releasing her brought her to the fiscal's office. Stroek was fined 900 guilders. The manager of *Friendship* was also fined 150 guilders after an old slave complained to the fiscal that he was not given enough food. Also fined two shillings for every "negro" was the manager of *Land of Canaan,* when twenty-seven slaves from the estate complained of "a want of clothing, lodging, food, comforts in sickness &c." In his report to the governor, the fiscal explained that he had visited the estate and had found that the proprietor was about to abandon it and remove his people to a more fertile location.

> I went through the negro houses and buildings if they may be so called. Everything was very miserable, and the negroes wanted many comforts which I afterwards procured for them. I spoke to the people told them to behave well, and that I would always attend to them and take care of them.[127]

When a manager's behavior generated slave unrest that could compromise the functioning of a plantation, the fiscal sometimes forced his dismissal. But even then, slaves risked being punished for having brought the complaint. Such was the case of a slave who reported that the manager had "connexion" with his wife and with the wives of ten other "negroes." The fiscal visited the plantation, determined the validity of the accusation, and recommended the manager's dismissal. At the same time he ordered the slave to be punished for having neglected his work.[128]

Occasionally, the penalty imposed on masters was more severe. When a woman complained that her mistress was going to send her to Suriname, the fiscal apprehended a Dr. Ferguson, who was tried, found guilty, and sentenced to three years' imprisonment and hard labor. Ferguson, allegedly, had illegally transported many slaves to Suriname. Serious cases of managers' violence resulting in slaves' deaths were referred to the Court of Criminal Justice, but even then punishments were often minor. When the manager Angus MacIntosh tied the slave London so tight that he died, MacIntosh was initially suspended from his functions and considered incapable of ever again exerting managerial functions. But when several people testified that he was a "very human man," he was acquitted.[129]

Even though cases in which managers were punished because of slaves' complaints were rare and slaves' complaints were usually dismissed, the slaves continued to go to the fiscal and file their grievances.[130] The fiscals' and protectors of slaves' records registered hundreds of such complaints involving a great variety of issues. Slaves complained that they were forced to work when they were sick, and punished if they did not meet the manager's expectation; that their children were mistreated and overworked[131] or did not get adequate medical assistance when they were sick;[132] that they had been separated from their families and sent away to work; that they were told to do tasks they were not familiar with and then punished because they did not perform well;[133] that they did not have enough food and clothing; that their work was excessive; that they were kept until late hours and had very little time for themselves; that they had not been paid for the work they had done on their "free time" or for fowls they had sold to the managers; that their punishment had been in violation of the law; and that they were forced to work on Sundays and were not allowed to go to the chapel. In addition, women complained that they had no time to nurse their children, and men that managers were sleeping with their wives. There were also blacks who claimed that they had been given manumission by their masters and then reenslaved, either by an heir or attorney or by someone else. Even though the slaves who went to the fiscal to complain represented a small percentage of the total slave population and rarely got any redress, one cannot but be impressed by their persistence.[134] Their stories of helplessness, deceit, and abuse constitute one of the greatest indictments against slavery ever written. But they also testify to a people's astounding resilience. Every one of the stories reveals not only the slaves' strong attachment to what they perceived as their "rights," and their disposition to fight for them with all the means they had, but also the importance they attributed to a grievance procedure that, whatever its limitations, at least allowed them occasionally to be dealt with as people rather than as things.[135]

Grievances always presuppose a notion of entitlement, thus a close reading of the slaves' complaints uncovers their notions of "rights." Conversely, the complaints also reveal the world they wished to create within the limits imposed on them by their masters. But it is important to remember that such complaints involved negotiations with masters and public authorities, and thus a search for common ground, a sort of compromise. The plaintiffs invoked norms that they thought whites in positions of authority might find acceptable. It is thus a "public transcript" that we find in the records of the fiscals and the protectors of slaves.

From these records it becomes clear that while masters dreamt of total power and blind obedience, slaves perceived slavery as a system of reciprocal obligations. They assumed that between masters and slaves there was an unspoken contract, an invisible text that defined rules and obligations, a text they used to assess any violation of their "rights." Slaves expected to perform a "reasonable amount of work," to be defined according to customary rules and adjusted to the strength and competence of individual workers. In exchange, they felt entitled to receive an allowance of food and clothing according to custom, to be given time enough to have their meals, to have access to land and "free" time to cultivate their gardens and provision grounds, to go to the market and to the chapel, and to visit relatives and friends. They felt they were entitled to the produce of their gardens and provision grounds, and that they should be paid for services they rendered on their "free" time. They expected to be relieved from work, to receive some kind of assistance when sick, and to be given food and clothing allowances in their old age. They also believed they should not be punished if they accomplished their tasks and behaved according to the rules, and that punishment itself should not go beyond the limits of the "acceptable." In addition, women felt entitled to nurse their babies according to their habitual practices, and to have some control over their children. The Demerara slaves' "public transcript" could be summed up in a few words: all slaves should perform according to their abilities, and all should be provided according to their needs.[136] And whenever this norm was violated and the implicit "contract" broken, they felt entitled to protest.[137]

Less visible but equally compelling was their commitment to "rights" that they did not claim publicly but that remained inscribed in a "hidden transcript," grievances that did not reach the fiscal's ears, but still fed anger, generated forms of behavior that masters considered devious, and eventually triggered rebellion. Among these were the right to freedom, which included the right to the fruit of their labor, the right to constitute and maintain a family according to their own criteria of propriety, the right never to be separated from family and kin against their wish, the right to move about without constraints, to celebrate their rituals, play their drums—in short, the right to live according to their own rules of decency and respect. The "hidden transcript" can only be guessed at through slaves' behavior. When they ran away, when they performed rituals secretively in the middle of the night, but most of all when they rebelled, they were asserting rights they did not dare to assert publicly.[138]

The boundaries between "hidden" and "public transcript," however, were not fixed once and for all. Rights that belonged to the "hidden transcript" would become public as soon as slaves perceived any chance of having them acknowledged. This happened whenever a shift in the balance of power favored the slaves, as when abolitionist pressure increased and evangelical missionaries started arriving in the colony, bringing new protocols and prompting new desires—and new motives for conflict between masters and slaves. The protocols of Christianity that prohibited work on Sundays and required attendance at religious services became a source of contention immediately. Slaves started publicly claiming their "right" to attend religious services and not to work on Sundays. The same thing happened when the British government, in spite of the opposition of planters and managers, prohibited the flogging of women. Soon slaves were reporting cases of flogging to the fiscal. By contrast, the desire to learn how to read stimulated by the evangelical missionaries remained in the hidden transcript for a long time because it was opposed by most planters and local authorities. No slaves went to the fiscal's office to complain that their masters did not allow them to learn how to read. But they continued to teach each other in secret.

The conflict between managers and slaves was not simply about work or material needs. It was a conflict over different notions of propriety: of right and wrong, proper and improper, fair and unfair. As we have seen, these notions derived from the written law and from custom. But sometimes slaves seemed to be assessing the situation from a script that fiscals and managers ignored, a script that slaves had brought from Africa and that was being rewritten under slavery. The evidence, however, is scanty, elliptical, and very difficult to interpret because the documents were all recorded in the language of the fiscals and masters. The voice of the slave that reaches us in inadequate translations, through layers of biases and misperceptions, is barely audible. If it is easy to reconstitute what slaves were complaining about, it is much more difficult to know the meanings they attached to their complaints—a problem any historian will have to confront who tries to understand the meanings that subordinate people who left no records assigned to their actions.

Slaves were not "Africans," except in the eyes of missionaries and other Europeans, who were inventing Africa. When they were brought to the New World, slaves did not share a single culture. They came from different places, spoke different languages, belonged to different social groups. The original "web of meanings" was torn. Threads unraveled in many directions to be rewoven again in different ways. Angolans, Igbos, Mandingos (did they ever see themselves in these terms, or did they identify themselves with a particular kin, a particular place?), whoever they once had been, however they had once defined themselves, all found themselves lumped together with others who had come from other places, other kins, other villages, under such generic categories as "negroes" and "slaves." If the documents are to be believed, they too had come to designate themselves and their peers as "negroes" and "slaves"—new identities perhaps that would serve as powerful ideological weapons for forming new solidarities and waging new wars—new scripts written upon old scripts, new boundaries, new meanings, and a new language. But for the first generation of those who had been uprooted, it would have been impossible to forget the worlds that were left behind: moral sanctions; notions of right and wrong; of what was desirable, what was proper; the mutual obligations that bound wives and husbands, family and kin; rituals of initiation; the things one should teach children; the ways the young should address the old; ways to celebrate life and mourn death; the ways humans related to nature, history, and gods; the boundaries between the living and the dead; and the taboos and rituals that had been such an integral part of their lives. They also would remember their ways of sowing and harvesting, or waging war, their dances and songs, their food and their dress, their tools, their houses, their villages, their medicine plants, an infinite number of ways and things that could not be easily forgotten or given up, but could never be the same again. In the process of creation of a new culture, the memory of the past would become dimmer and dimmer, and a new culture was created.[139]

The signs of a past denied every day by new experiences, and, like palimpsists, written over again and again, are difficult to detect and decipher. Sometimes one has a glimpse of them and suspects there is still a hidden history of slave resistance waiting to be discovered, a history that will go even deeper than historians have gone so far. Historians of slavery have often stressed that such a history is not only a history of men and women whose bodies were exploited and whose minds missionaries tried to conquer, but of men and women who created a world for themselves, imperceptible to the eyes of the outsider, not made in the image of the white men, a world invested with meanings that were not a mere reflex of the slave system, but a creative synthesis of past and present. But with few exceptions, most historians have continued to neglect this culture when they study slave resistance.[140]

Read with such ideas in mind, both the fiscals' and the protectors of slaves' records seem to suggest that much more than meets the eye was at stake when slaves registered their complaints. This possibility was implicit in a tragic case that ended in the Court of Criminal Justice in Berbice. The family of the slave Christian had apparently committed collective suicide by jumping into the River Canje. They all drowned, except for Christian, who was found hiding in the bushes, and one boy—Christian's "adopted" son—who was rescued. Christian was brought to trial.[141] He argued that his wife and children had committed suicide because they had been separated from each other. The manager testified that Christian had wanted him to buy not only this wife, but two other young wives and their brothers, and his request had been denied.[142] The fiscal asked that Christian be hanged, but the court did not find sufficient legal proof to condemn him. Christian's drama seems to have been that of many others who silently endured the violation of their notions of family and kinship. He no doubt originally belonged to some African group for whom brother-sister relations were fundamental, a notion that even the most understanding manager would have found difficult to accept, and even more difficult to accommodate.[143] Equally moving are the tales of slaves who were brought to trial for being involved in practices that the colonists were too ready to label "obeah," and to condemn as "evil" and "dangerous" practices, which from the slaves' point of view were attempts to re-enact rituals designed to remove suffering and disease.[144]

The slaves' willingness to settle disputes through different types of mediation also finds resonance in many judicial practices typical of African societies, and the same can be said of their attachment to Sunday markets and to their provision grounds (although here, as in many other similar circumstances, it is impossible to separate what was a legacy of the past from what was born of the present, since slaves' attachment to markets and provision grounds and their uses of mediation often can be interpreted, without reference to their past, as simple strategies for survival). The conflict over the time allowed for nursing is a good example. It was common among African women from different regions to nurse their children until they were three and sometimes even older. But since this interfered with their work, managers and overseers often punished them. Here again, slave expectations born out of a different culture collided with managerial interests. But the attachment to this nursing practice could also be interpreted (as it was by the managers) as a strategy to avoid work. Funeral rites, which in most African societies were extremely important, also proved difficult to maintain, although there is evidence that slaves tried to keep them. They also tried to preserve the language of the drums, songs and dances, and many other traditions whose meanings escaped the perception of managers and missionaries.

All in all, the historian is left with more questions than answers. What happened to the cultural and social scripts the slaves brought with them? How quickly were they transformed? Did the traditional respect shown for the elderly crumble entirely under the impact of the slave experience? How long did slaves retain their original languages? Did they try to re-create the forms of association based on age groups or crafts common in many African societies? And what about other forms of kinship? Is it possible that the slaves' previous experience with secret societies was instrumental in the plotting of conspiracies? When we read in Oleudah Equiano's memoirs that the Igbo people in Africa used to wear a long piece of calico or muslin wrapped loosely around their bodies and ate plantains, yams, beans, and Indian corn, and we find similar practices in Demerara, shall we speak of "survivals"? And when masters gave slaves cloth as an allowance instead of skirts or blouses, were they accommodating to slaves' preferences, just trying to save money, or both?

It is true that no group can transfer "its way of life and the accompanying beliefs and values intact, from one locale to another," as Sidney Mintz and Richard Price have often stressed.[145] And slavery, of course, made it particularly difficult for blacks to maintain their traditions in the New World. But one could argue that precisely because they *were* enslaved, they tended to hold unto their traditions as long as they could—as a strategy of resistance and survival—and that in their day-to-day lives there was a constant, sometimes visible, sometimes invisible, struggle to keep traditions alive. This is far from saying that they would succeed in keeping them intact. It is also possible that some cultural traditions, however modified by the slaves' experience in the New World, were more resilient than others,[146] particularly those which did not interfere with the slave labor and discipline. Why would masters care about the tales, jokes, and stories told by slaves to each other at night when they returned from work? Why should masters care about the slaves' ways of fishing and hunting, of making their own furniture and utensils, or growing their own gardens, or preparing their food. And, at least until the missionaries arrived, why should anyone care about slaves' notions about life and death, about nature and the universe, their ideas of causation, or their cosmogony? Masters were concerned with the slaves' public behavior. As long as ideas and traditions did not interfere with life on the plantation in any way or with the social "order," slaves could keep them. They also could play their drums, dance, and sing at prescribed times and places. But who could control what happened secretively in the slave quarters? Who could control what happened in their minds?

Ironically, the very reason that made it possible for cultural practices to survive—their invisibility to white eyes—makes it difficult to identify them. Most of what can be documented for nineteenth-century Demerara has to do with aspects of slave culture that slaveowners were troubled by, and it was usually only when the slaves' beliefs or practices led to conflict that they found their way into the documentary record. That is why we know something about slaves' notions about family, nursing practices, and "obeah." But even in these matters, the information reaches us in distorted versions because managers, public officials, and missionaries simply had no understanding of the slaves' culture. It is even more difficult to trace its roots back to the place of origin, since slaves came from different places, and borrowed from each other's cultures.[147] Slaves also appropriated (within the limits imposed by slavery) symbols, values, and practices from the masters' culture. This complex process of "Euro-Afro-Creolization" is essential to an understanding of the rebellion of 1823.[148]

For us, what is important here is only to remember that the slaves' assessment of their masters, their reasons for protesting, and their notions of "rights" were rooted in a variety of experiences that included the written law (local or metropolitan), customary rights issuing from their day-to-day dealings with masters, public officials, and missionaries, and memories of the African past. Echoes from Africa and Europe resonated in a dissonant polyphony in Demerara. And when the rebels of 1823 spoke of their rights, they carried to its ultimate consequences a long history of struggles for justice.

The slaves' notion of "rights" took on a new dimension in the late eighteenth century, when revolutionary discourse gave universality, and thus new legitimacy, to the notion of rights and made it much more encompassing. Prisoners of juridical concepts and a legalistic rhetoric, governors, fiscals, masters, and managers were always talking about slaves' legal rights, even if in practice they tended to deny these same rights. But by doing so they could only enhance the slaves' commitment to those rights.

When, in 1816, news of a rebellion in Barbados reached Demerara, Governor Murray issued a proclamation saying, among other things, that the slaves in Barbados had been "misled" to believe that the king had ordered them to be free, but that "it was not in the nature of things" that such orders could have been sent. "Every history proves that slavery has existed since the world was made. . . . The Holy Bible commands slaves to be obedient to their masters and they must be so, it is no less their interest than their duty," Murray said. But after this strong statement legitimizing slavery, he went on to say that all that slaves could expect was to be made as happy as possible in their servitude. "I appeal to yourselves, whether your Master and the Government of the colony, have not manifested a strong desire to make you so? Whether your situation has not been better every year?" Murray concluded his proclamation by promising that he would be ever ready to give the slaves "the benefit of those laws, which protect you from oppression." Although he warned the slaves that in case of insurrection he would be "like an arrow from a bow, to execute an instant and terrible justice," he also represented himself as the slaves' protector, the man responsible for the implementation of laws that favored them. So it is not surprising that in 1823, when the slaves heard that "new laws" approved by the British Parliament had been withheld from them, they were prepared to appeal to the governor and to claim their "rights." By then the abolitionists had legitimized the slaves' own notions of "rights." The language of the universal rights of man conferred a new meaning on slaves' struggles, since it acknowledged the slaves' "humanity" and their right to be free.[149] This was reinforced by the missionaries' rhetoric.

As has been demonstrated time and again by historians of slavery, slaves were not the passive victims of oppression often portrayed by abolitionists and missionaries. They fought back in every way they could, always trying to gain more control over their own lives. Every time masters invented ways to keep slaves under control, slaves managed to turn things around, defeating (at least in part) the masters' intentions. But every time the slaves managed to get some advantages, masters were ready to take them away. So it was with the provision grounds. Originally it may have been a good idea for the masters to grant access to these plots of land, freeing slaveowners from having to worry much about feeding their slaves. But soon the slaves seemed more interested in working on their own grounds than on the plantation. But when slaveowners, in an attempt to compensate for declining prices and a shrinking labor force, decided to put so many demands on the slaves that they did not have much time to devote to their gardens and provision grounds, slaves started to complain and "free time" became an issue for contention.[150] Something similar happened with the task system (not to be confused with the system of task-gangs). Recognizing the difficulty of supervising slaves' labor, some managers adopted a task system, assigning a certain work to be done by a slave gang within a given time. Managers expected that this would encourage the slaves to work harder and faster in order to have some "free" time for themselves. Their expectations were initially fulfilled, but as soon as the slaves finished their tasks, managers started giving them more work to do, thus defeating the original intention and provoking protests.[151] Manumission, too, could have been an effective means of social control. Nothing could be more dear to the slaves than the idea of freedom, and if they thought they could obtain manumission by satisfying their masters' and managers' expectations, they might have worked harder and better. Yet if manumission were easily obtained, the slaveowners would soon have to come to grips with a growing free black community and a shrinking labor force. So masters felt the need to make manumission more difficult. If manumission became an impossible dream, however, it would lose its effectiveness as a way of inducing slaves to be submissive. And when managers, in the hope of increasing slaves' willingness to work, gave them permission to go to the market, to visit relatives, to sing and dance, slaves were late for work the next day, or used the opportunity to meet friends and plot rebellions. Every means devised by managers to extract the maximum surplus from the slaves, every means they used to put down resistance, could become the pretext for new forms of struggle and resistance.

There had been no major rebellion in the area since the famous uprising in the neighboring colony of Berbice in 1762–63, when plantation after plantation had fallen into the hands of rebels who pillaged and burned their masters' houses, set sugar and rum stores on fire, massacred a great number of whites, forced the remaining white population to seek refuge in boats, and finally assumed control over the colony for almost a year.[152] In that rebellion, which lasted one year, divisions among the slaves, and reinforcements sent from Holland and other colonies in the Caribbean, had finally enabled the Dutch to prevail. The repression had been merciless. For those who were living in 1823, however, 1763 was very much in the past. Long past too was the time when the Dutch had been forced to sign a treaty with the "Bush negroes" of Suriname. Rebellion in other areas, particularly in Haiti in 1791, Jamaica in 1807, and Barbados in 1816, had periodically reactivated the whites' fears. Small uprisings had also occurred from time to time in Demerara and Berbice—in 1772 on plantation *Dynemburg,* in 1794 among the Coromantee of the West Coast, and in 1808 and 1812 on the East Coast.[153]

For slaves, however, rebellion was risky and its lesson always ambiguous. It taught that rebellion was possible, but repression implacable. So most slaves normally preferred other forms of resistance. Running away was one of the strategies they used. The pattern in Demerara was the same as elsewhere. Slaves ran away alone or in groups. More men than women ran away (approximately one female for every eight males, judging by the number of captured runaways in the local jail).[154] And although many runaways were caught, some were able to elude their hunters for months, sometimes years. A few were never captured.

The proximity of wooded areas and the dense network of rivers created an ideal setting for runaways and maroons. At the same time, the heavy concentration of blacks on plantations and in the city made it difficult to identify runaways. They could also hide temporarily in the uncleared areas in the "backs" of plantations—where the thick brush made for perfect hideouts. There they could count on friends to supply them with food. Sometimes runaways went further, toward Mahaica Creek, or to the neighboring colonies of Essequibo and Berbice, where there were several bush communities in the Mahaicony area. A few were even caught on their way to the Orinoco.[155] Some runaway slaves managed to establish more or less permanent camps in the bush, where they would grow corn and other crops.[156] More often, however, they would hide in the slave quarters on plantations where they had friends or relatives, or they tried to pass as free and looked for jobs some place else. They also hid among the many hucksters who came to the town market on Sundays, or they offered their services to colonists who lacked capital to buy slaves and were most happy to find someone to work for small wages. Some, after a few days of being away, went back to their plantations, probably after they discovered that life alone in the woods could be even harsher than life in the slave quarters. The most daring and desperate runaways stole boats and took to the sea. But, sooner or later, most runaways found themselves enslaved again. Periodically, the Indians—the "Bucks," as they were called by both whites and blacks—came down the river to join whites on slave-hunting expeditions into the bush.[157] And they always returned with slaves who had been hiding there for long periods. Runaways also fell into the hands of free blacks, or even other slaves, who were lured by the rewards offered for their capture.

When runaways were caught they were severely flogged, and sometimes, sold away to other plantations far from friends and relatives. So most slaves preferred other strategies.[158] Sometimes as an act of vengeance slaves killed horses or poultry, broke tools and machines. In extreme cases they poisoned or attacked drivers, overseers, or managers. A few went as far as committing suicide or infanticide, but most of their resistance centered on work, and Demerara slaves sometimes resorted to strikes as a form of protest and as a means to force managers to meet their demands.[159]

Day-to-day forms of resistance have attracted great attention in recent years.[160] Some historians have correctly pointed out that there is a qualitative difference between resistance and rebellion, one intending to ameliorate the system, the other to overthrow it.[161] But these two forms of protest should not be seen as mutually exclusive. Although not every act of resistance leads to rebellion, without the daily and tenacious acts of defiance and sabotage, rebellions would have been difficult, if not impossible.[162] It was in daily resistance that slaves reinforced their commitment to their "rights" and tested the limits of their masters' power. It was in daily resistance that slaves' resentment grew, that bonds of solidarity were strengthened, that networks and leaders were formed, and individual acts of defiance were converted into collective protest.[163]

The reports of the fiscals and protectors of slaves are an ideal source for the study of the many strategies used by slaves to expand their control over their lives and labor. The reports registered not only slaves' complaints but also punishments and the alleged reasons for such punishments. And although, as we have seen, the early fiscal reports were very incomplete, they improved after the rebellion of 1823, particularly after the British government decided to intervene more directly in slaves' management. In the reports, slaves' offenses are put into several categories. These are somewhat imprecise and overlapping and may tell us as much about what masters and managers considered offenses as about what slaves perceived as such, and as much about the repressive nature of the system as about the slaves' willingness to fight back. But they still give a more precise image of the day-to-day strategies used by slaves than most sources. Also, since males and females are listed separately, the records help to identify differences in behavior.[164]

The "List of Offences Committed by Slaves in the Colony of Demerara and Essequibo," for the first semester of 1828, registered 10,504 offenses, and by the end of the year more than 20,000 were recorded—eloquent testimony to the intensity of conflict between slaves and masters. The offenses were divided into five categories.[165] The first, "Serious and Aggravated Offences," involved physical aggression ranging from murder and attempted murder, wounding others, sodomy, attempting to ravish, housebreaking and stealing, to attempting suicide, arson, cruelty to children and animals, and killing and destroying stock. During the first semester of 1828, only about 1 percent of the cases listed fell into this category. The second category involved mainly cases of theft and connivance to theft; these represented 5 percent of the total number of offenses. In the third category, "Insubordination Accompanied with Violence," were incidents of slaves striking or threatening to strike managers, overseers, or drivers, spitting in overseers' faces, breaking or attempting to break the mill, or destroying buildings. The number of slaves listed under this rubric was also negligible (less than 1 percent of the total). Typically, the drivers—usually slaves themselves—who were responsible for the direct control of field laborers were the most frequent target of the slaves' hostility.[166]

The overwhelming majority of cases (93 percent) were in the next two categories: "Insubordination Unaccompanied with Violence" and "Domestic Offences." A great variety of "offenses" was listed under these labels and for the purposes of simplification can be grouped into six distinct types: running away, work underperformance, insubordination and symbolic challenges to authority, destruction of property, minor cases of discipline, and causing problems within the slave community. Contrary to the common belief that running away was a frequent strategy used by slaves, it constituted a surprisingly small percentage of the total cases listed (only about 4 percent). By far the greatest number of offenses was related to refusal to work and underperformance: leaving work unfinished, laziness, neglect of duty, bad work, feigning sickness, getting to work late, and other things of the sort. These made up about 66 percent of the total number of offenses. But if we add those offenses classified simply as acts of disobedience (without specification), they would account for 74 percent of the total.

The largest single group included acts characterized as insubordination and challenge to authority. They referred to slaves who had refused to obey orders, defied managers, or held clandestine meetings. A smaller number of cases was listed as insolence or the use of abusive language toward a superior. Minor acts of indiscipline included fighting among themselves, rioting, causing disturbances, harboring runaways, lying, swearing, leaving an estate at night, dancing and carousing on a estate without leave, and drunkenness. Remarkably, in one semester, in a total of 10,054 slave punishments, only 209 had been caused by drunkenness. This contradicts the common opinion among whites that slaves were often drunk. Several other practices that whites thought of as common among slaves were also extremely rare. There was only one recorded attempt to commit infanticide, one attempt at suicide, one slave punished for practicing "obeah," and one woman punished for eating clay. One can argue, of course, that many more such cases might have gone unpunished because masters did not really care. But, although it might be true that masters were not much concerned with clay eaters, they certainly objected to drunkenness, suicide, and infanticide, and were firmly decided to curtail "obeah." Some of the cases, however, may have escaped the attention of masters because slaves would have kept the practices away from their eyes. In the case of "obeah" it is also possible that it had become less common in Demerara because of the influence of evangelical missionaries, whose flocks had grown steadily from 1808 to 1823, when their activities were restricted as a consequence of the rebellion.

Of little significance were cases of "destruction of property." These amount to less than 1 percent of the total. Some could be characterized as acts of sabotage. Others may have been simply a result of carelessness. Under this category appeared slaves who had allowed cattle to trespass on cultivated land, broken boats or carts, destroyed produce, or sold or run away with tools. Others were punished for "ill-using" horses or for unintentionally setting fire to the "megass logie." Slaves neglecting and concealing their own sores were also included within this category, since they were seen as injuring the masters' property.

Finally, there were the slaves who were punished for quarreling with each other, biting each other, mistreating or neglecting their children or parents, being unfaithful to their spouses, beating and maltreating other slaves, or for "fornication." Such matters were usually left to the slaves themselves, so the number of punishments was low. Only twenty slaves were punished for such familial offenses.

The record of offenses and punishments shows significant differences in the behavior of males and females. Of all the offenses reported, 63 percent had been committed by men,[167] who constituted 54 percent of the slave population. Many more men than women were punished for running away (387 in a total of 451) or for drunkenness (198 in a total of 209). Men were also more likely to commit crimes involving violence and theft. Of the 102 cases of aggravated offenses, only 14 involved women. And of the 509 cases of theft, only 53 were attributed to them. By contrast, a disproportionate number of women were punished for using "abusive language" (261 involving women as against 141 involving men). Women were also more likely to be punished for neglect of duty, and for coming to work late.[168]

Some of the reports are very detailed and include the names of the slaves, the plantations they belonged to, the nature of the offense, and the form of punishment. To the modern eye the punishments seem arbitrary, whimsical, and out of proportion.[169] By the time the records had been made, the British government had prohibited the flogging of women, so they were subjected either to confinement or were put on the treadmill. For example, Victoire, a female slave from *Le Retraite* accused of stealing money from the "negro" Friday belonging to the same plantation, was referred to the fiscal, who after an inquiry "ordered the said woman on the tread mill."[170] Annette, from *Goed Verwagting,* was also put on the treadmill for running away repeatedly. Picle and Alfred, who were accused of having run away for six months, stealing five sheep and three hogs from the manager, committing several depredations on neighboring estates during their absence, and "also endeavouring to entice more of the negroes to run away and join them," were condemned respectively to ninety and sixty-seven lashes. But Welcome and Geggy, two watchmen from *Mes Delices,* received seventy-five lashes each just for breaking open the estate's storehouse and stealing some "rum, pork, etc."

Two men and four women from *New Hope,* who had disobeyed orders and also instigated the gang to disobey and show contempt for the manager, received different punishments. The men were condemned to forty lashes and the women to forty-eight hours' "labour on the tread mill." Thirty lashes were given to Harry Quash for "coming into the yeard betwixt the hours at night with a horsewhip, and making a riot in the yeard with his wife, disturbing the neighbours."[171] Phillis, a female from *Retrieve* who had refused to go to work, supposedly "under pretence of having a stiff neck" (although the doctor said there was nothing wrong with her), was put in solitary confinement for twenty days. But Jessey, who was accused of riotous and insulting behavior, of having instigated the women gang "to shout and huzza" at the manager, and of then leaving the field, was condemned to solitary confinement for only four days. Three male slaves from *Retrieve* received between twenty-five and forty lashes for refusing to "turn out" the day after Easter holidays; while three others who had run away from *Hyde Park* received twenty-five lashes each.[172]

From these records, it becomes clear that formal punishment was an integral part of the system. But, as we have seen, intimidation was not enough. In order for the system to function at all, slaves had to have certain rights and privileges. The precarious balance between punishment and privilege was difficult to maintain. Too many privileges would lead to freedom, too many punishments to conflict and possibly rebellion. Masters saw slaves' "rights" as "privileges," as concessions that could be withdrawn at their own wish. Slaves clung to law and custom and saw "privileges" as "rights." Masters and slaves were engaged in a process of unending negotiation and confrontation. It was on this contested terrain that John Wray and John Smith had to fight their own battles. Not surprisingly they found themselves on the firing line, and precisely at the time the struggle was most fierce.

CHAPTER THREE 

The Fiery Furnace

The conflicts between missionaries and colonists that led to the tragic events of 1823 started fifteen years earlier, when the first missionaries of the London Missionary Society arrived in Demerara. They came deeply ignorant of the protocols and the unspoken rules of a slave society, and with their heads full of notions that were likely to provoke the colonists' outrage and to aggravate the tensions that pitted masters and managers against slaves, and colonists against the mother country. In their day-to-day dealings with slaves, masters, managers, and royal authorities, the missionaries generated irritation on the part of the authorities, loyalties among slaves, and hatred among masters. Increasingly threatened by the new economic and ideological trends in the mother country and fearful of losing control over their slaves, the colonists vented their anger on the missionaries. A close examination of their interaction not only sheds light on the process that led to Smith's indictment, but also helps to explain some of the circumstances that led to the rebellion.

Before the LMS sent John Wray to Demerara, no one had given "religious instruction" to the slaves. In 1794, British Methodists had applied to the government of the United Provinces for permission to send missionaries, but the Court of Policy refused.[1] Some time later, in 1805, when a Methodist missionary from Nevis approached the governor about settling in the colony to preach to slaves, he was told to leave on the first ship.[2] The colonists did not want evangelical missionaries in their midst for fear they would make the slaves "dissatisfied."[3] In 1808 there were still only two clergymen in Demerara, a Dutch reformed minister and an Anglican chaplain of the garrison.

With Wray, the situation was somewhat different. He had come to the colony at the request of a Dutch planter, Hermanus Hilbertus Post, who from the beginning had given him both material and moral support. Although there was nothing unusual about his career as a planter, Post was an unusual man in one important respect: unlike most of the colonists, he was pious. He had been born into a wealthy family in Utrecht in 1755. His father had a sugar refinery and was a member of the local senate, but business reverses had forced him to retire to a place in the countryside. The young Post had decided to try his luck in Demerara. He started as a manager, but after two years he bought his own land with the help of a friend. He began with thirteen slaves and, step by step, built a fortune during a period when market conditions were particularly favorable. This allowed him to travel to Holland for two years, and then to the United States, where he lived from 1791 to 1799 in New Rochelle, New York. In 1799, however, he had been forced to return to Demerara to take care of his plantation. On the brink of bankruptcy, he had to work hard to pay his debts and re-establish his fortune. He also seems to have undergone some form of religious conversion—probably when he was in New Rochelle—for he began to worry about giving religious instruction to his slaves. He first hired a free black schoolmaster to read scripture to his slaves on Sundays. Not happy with this solution, Post approached the directors of the LMS and convinced them to send a missionary to his plantation, promising to build a chapel and to support the mission with one hundred pounds a year.[4]

Post's plantation, *Le Resouvenir,* where Wray arrived in February 1808, was on the East Coast, about eight miles from Stabroek (later Georgetown). It was a large plantation with 375 slaves and about 700 acres, 225 of which were planted in cotton, 375 in coffee, and 100 in cocoa and provisions. The main building stood about a mile from the seaside, and behind it there was a canal leading to the back dam. Orange trees had been planted on each side of the canal. A green path shaded by rows of Mountain Cabbage trees led from the main house to the public road. Everything had been built or grown by slaves, under Post's supervision.

Many people in the colony criticized Post, saying that he would do better paying his debts than spending his money on missionaries and chapels. They considered him a fool and a madman, and charged him with introducing anarchy, disorder, and discontent. He was going to make Demerara a second Haiti, they said. The colonial authorities forbade Post "to hold any riotous meeting of slaves on the estate."[5] But in spite of the opposition Post kept his promises. In September 1808, after Wray had been in Demerara for just over six months, Bethel Chapel was inaugurated at *Le Resouvenir.* To Post's satisfaction a few whites and about 600 slaves attended services on the day the chapel opened. Post built a small house for the missionary. It all cost him about one thousand pounds (the equivalent of the cost of twelve slaves). To induce his own slaves to attend services, Post gave them permission to go to the market on Saturdays so they would have Sundays free. In January 1809, about a year after Wray had started his mission, Post boasted: "No drums are heard in this neighborhood, except where owners have prohibited the attendance of their slaves. Drunkards and fighters are changed into sober and peaceable people and endeavour to please those who are set over them."

Post was so enthusiastic about the mission that he decided to invite another missionary to run a school for the planters' children in Stabroek. In response to his request, the LMS sent John Davies, who reached Demerara in January of 1809. The society's director explained to Post that Davies had studied at a seminary in Gosport and originally had wanted to devote himself to the service of Christ among "the heathen," but, now that this opportunity had appeared, the directors of the society thought Davies could use his knowledge to impress the young colonists with "sentiments of humanity toward the Negroes, disposing them to promote their moral improvement." The directors expected that in his spare time, Davies would also be allowed to preach to the slaves.[6]

Wray was in his late twenties, and full of excitement and determination. But in spite of Post's optimism and support, things were not easy for Wray. At first, the slaves from *Le Resouvenir* did not show much willingness to attend services. When Post insisted that they go, they answered they had no clothes: "Massa, me no jacket, me no hat, no shirt to attend church." But when Post gave them clothes, they argued they had done nothing wrong, so they did not need to go to the chapel.[7] Most of those who did attend services belonged to neighboring plantations. Wray also met opposition from the Dutch minister—"a man of infamous character," Wray thought—who did not want him to baptize slaves, arguing that according to Dutch laws this would make them free (although he himself had baptized a few). "The Devil has made great use of the Dutch clergyman," Wray wrote to the LMS.[8] Most important, after Wray had been in the colony for just a few months he was already hearing rumors that the Court of Policy intended to expel him.

Wray reported his concerns to the LMS: "I have the unpleasant information to communicate that the cause of Christ meets with great opposition in this colony." The Court of Policy apparently had decided to get him out of the country, and the managers dared not permit the slaves to go to the chapel. Wray thought the Court of Policy was responding to pressure from colonists who wanted to make their slaves work on the Sabbath. And, he added, "It is the opinion of these gentlemen that the Gospel will ruin the colony; of the governor I had a better opinion, but he appears to be afraid of losing their esteem. Some of them think religion is very well for the white people, but it is best to keep the blacks in a state of ignorance. Others think it is an abomination to be in the House of God when negroes are there."[9] Wray feared that if "influence could not be used with the Government in England," the preaching of the Gospel would be entirely prohibited.

Conscious of this opposition, Wray asked people he knew to sign a petition on his behalf, and forwarded it to the LMS as evidence of the good results of his preaching. Sixteen people, mostly managers and overseers, and one landowner, Henry Van Cooten of *Vryheid's Lust,* a friend of the Post family, signed the document. It stated that Wray's labors had "inspired" the "coloured" and black people with reverence and obedience. The signers assured the LMS that after the singing of hymns or psalms and prayers for the good and prosperity of His Majesty, the government, and the inhabitants of the colony, Wray delivered "suitable discourses from the Holy Word of God," clearly pointing out "the duties of man and specially that servants should serve and obey their masters . . . as agreeable to God." The petitioners also certified that they had observed in the people under their "care" the good effect of Wray's preaching.[10]

When the LMS directors received Wray's letters and the petition, they immediately contacted the celebrated Wilberforce, asking him to forward the documents to Lord Castlereagh. They also asked for an interview with Henry Bentinck, the newly appointed governor of Demerara, who was still in England. Thanks to Wilberforce's efforts both requests were satisfied.[11] Thus from the beginning a pattern was settled. Whenever they encountered opposition from the colonists, missionaries appealed to the LMS, which in turn resorted to sympathetic people in government, who would then support the missionaries.

But in spite of such support, Wray's problems were far from over. In May, word went around that the slaves had made a plan to rebel and take the colony from the whites. Several slaves were arrested, among them two who attended services at Wray's chapel. Such incidents could only increase the colonists' opposition to missionary work.[12] Still, Wray felt confident. "The plot was laid about a year ago and was discovered [revealed] by one of Mr. Post's negroes, which is much in our favour," he wrote in June. He hoped that the innocence of those who had been arrested would be proven, and that the fact that one of Post's own slaves had exposed the conspiracy would make the colonists see his missionary work with more sympathy and less fear. A month later, things on the whole looked more promising "for the spread of the Gospel in this dark part of the world. . . . We have many enemies, but I hope more friends." The prejudices of the people had begun to subside—or at least so he thought—and the new governor seemed to be much in favor of preaching to the slaves.[13]

Wray still had one concern. He had left a certain Miss Ashford behind in England and had expected that (with the approval and support of the LMS), she would join him. But time was passing. After Wray had been in the colony for several months, Post wrote to the directors of the LMS that Wray was showing signs of irritation and was threatening to leave.[14] Wray himself freely expressed his bitterness to the secretary of the society in December and again in February. Mr. Post, he said, had all the inhabitants of the country against him. The cotton crop had been destroyed by a disease, and what was left was rotting on the plants because of heavy rains. Post was very much "cast down" on account of the bad prospects for cotton and coffee. "The state of country is at present gloomy," Wray wrote, "and unless a change takes place it certainly will be ruined." He also complained that his work as a missionary was very laborious. He sometimes had to make slaves repeat the catechism near a hundred times before they were able to remember it, particularly those who spoke Dutch. (Some time earlier he had praised the slaves' excellent memory and the rapid progress they made in learning.) "I have enough to do to fill ten or eleven hours, sometimes fourteen in a day, if I give that instruction to the people which I ought to." Finally, he mentioned the subject of the woman he wanted to marry, "a person," he said, "in every way qualified for a missionary's wife." With characteristic vigor he warned the directors that his engagements with her were as serious as those he had with the LMS, and he did not intend to break them. His "connection" with her had been too long formed "to be broken off by men, however good they may be," he wrote defiantly. Wray wanted his bride to be trained in midwifery, a skill he hoped would be very useful in the colony.[15] Post's letter and Wray's complaints were examined by the directors and some months later Miss Ashford and Miss Sanders (John Davies's bride-to-be) were on board the *Fortune,* bound for Demerara.[16]

Missionaries and their wives left behind the world in which they had been brought up. But they carried with them their dreams, their notions about the social order, about politics and religion, about family, class, gender, sex, and race, notions about protocol and etiquette, fairness and unfairness, right and wrong, of what was possible or impossible, notions about the ways men and women should live and die. They came to a new world in which some of their notions made no sense, and others were likely to be considered dangerous by the colonists. Theirs was a difficult mission. Not only did they have to gain the slaves' confidence and to compete with the slaves' traditional systems of beliefs—which they did with some success—they also had to overcome the colonists' unwillingness to have their slaves instructed in religion. And this would prove to be an almost insurmountable task.

Although they all had come from the same world, and had been exposed to the same basic notions, each of the missionaries arriving in Demerara would respond to the challenges according to the peculiarities of his social origins and personality and the specific circumstances of his work (whether he lived in town or on a plantation, preached to whites or to blacks, free or slaves).[17] Many years later, commenting on the situation of the missionaries who went to live on plantations, Wray wrote that slavery with all its evils was opened before their eyes. The whip reminded them every day that they were in a land of slaves, and they were looked upon with suspicion by planters. Missionaries who lived in towns could ignore the suffering of the slaves and were more likely to escape the animosity of planters.[18]

Wray and his wife lived at *Le Resouvenir.* Their main task was to save the slaves' souls. Davies and his wife settled in town. They were in charge of teaching free whites, although Davies was also expected to preach to whoever was in town—whether free blacks, mulattos, or slaves, including those from neighboring plantations who might be present. It was clear from the beginning that Wray took his work more seriously than Davies did. He was soon totally engaged in preaching, while his wife instructed slave women and children. Both were delighted at the slaves' progress and took obvious pleasure in talking to them. With the curiosity and skill of an amateur anthropologist and the biases of an English missionary, Wray took careful notes on the slaves' habits and beliefs.

John Davies was perhaps less religiously motivated than Wray. Success and comfort were more appealing to him than the "cause of Christ." Like Wray and others who had lived in England through a period of intense social and political tensions, Davies had developed a keen political instinct and was very aware of the power structures that governed the world in which he lived. But unlike Wray, he often seemed more concerned with using them for his own benefit than for the benefit of his religious mission, even when this meant that he had to bow to colonists' demands and defer to local authorities. A few years after Davies arrived in the colony, Wray commented in one of his letters to the LMS that Davies had been allowed 100 joes a year from the Court of Policy, but only on condition that he not teach slaves to read; the children that the governor had sent to him were put to work weeding the grass around the chapel, and Davies seemed to change his mind with every governor and "to become all things to all men." It seemed to Wray that Davies's purpose was not to win souls for Christ but to gain the colonists' favor and their joes. Wray was afraid that Davies's heart was "too much set on the riches of this world."[19]

Wray's accusations were probably not wide of the mark. Fourteen years later, when Davies advertised part of his furniture and library for sale, he listed several tables, chairs, sofas, bookcases, lamps, "a chaise and harness," nearly new, and an impressive collection of books, including "Newton 3 vols., Pearce 2 vols., Bossuet 19 vols., Massillon 15 vols., the poetical works of Milton, Walter Scott, Henry's History of England, 11 volumes, Hume and Smollets do. 15 vols., Colquhon's Police of the River Thames and Metropolis 2 vols., Buffon's Natural History, 20 vols., Biographical Dictionary 12 vols., Adam Clarke's Bibliographic Dictionary 8 vols., Dr. Franklin's works, Blair's Sermons, 5 vols., Boswell's Life of Dr. Johnson, 4 vols., Boswell's journal," and many other books of travels, chemistry, theology, history, and literature. Davies had managed to make a pretty comfortable life for himself and his family.[20]

Davies's willingness to accommodate and compromise, and his apparent lack of religious commitment and concern for the slaves, may have been a question of temperament and personality. But it is also possible that, living in town and depending on the colonists' patronage for the success of his school, he had become more sensitive to the colonists' point of view than to the slaves' predicament. Still, in spite of his willingness to accommodate, he found himself—like Wray—the target of criticism and harassment. A few months after he arrived, a group of whites came with rocks and bricks, threatening to stone him. He was only rescued by the interference of a "coloured man"—who the next day found himself in jail. Davies was then told by the fiscal (in a way that seemed to him brutal) that he could not hold any more meetings in town "on any pretence whatever," on pain of imprisonment. The problem was settled by the government on Davies's appeal, but this did not put an end to the colonists' harassments. The antagonism between colonists and missionaries went beyond personality and character. It was a matter of conflicting goals and different notions of social control.[21]

Wray's work as a missionary was riddled with contradictions. To be successful in his mission he needed to gain both the slaves' trust and their masters' support. But whatever he did to please slaves displeased their masters, and if he sided with masters he was sure to alienate the slaves. Thus whatever in his mission brought him pleasure was also sure to bring him pain.

Within a few months after he arrived at *Le Resouvenir,* slaves from neighboring plantations were walking for miles on Sundays to attend services at the chapel, and Wray was busy teaching children how to read. Except for a few white overseers and managers, most of his congregation were slaves. At *Le Resouvenir* many spoke only Dutch and had difficulty understanding him, but most of the others did speak English, and when he was in his optimistic mood, Wray sometimes thought the slaves understood him "as well as any congregation in England."[22]

Wray realized quickly that the slaves were extremely anxious to learn to read, and in a few months he could report with pleasure that several could already read the catechism.[23] It seemed to him that the slaves were teaching one another, so fast were they learning. Several times in his reports to the LMS, Wray wrote of the effort and perseverance blacks put into learning how to read, giving up their time for resting. "Those who think that blacks will only make any exertion under the whip should come here to see it," he once commented.[24] In his teaching, Wray used the Sunday School Union spelling book, several of "Lancaster's readings," and Watts's first catechism,[25] the same sorts of readings Wray would have used in any Sunday school in England. Through them the slaves were initiated in some of the mysteries of their masters' culture.

Like many children of the Enlightenment, Wray had an unshakable faith in the good effects of instruction, and would have had no difficulty subscribing to the words of another member of the LMS, the Reverend J.A. James from Birmingham, author of *The Sunday School Teacher's Guide,* for whom ignorance "was the prolific mother of crimes and miseries." The Sunday school teacher, James wrote in his *Guide,* should be acquainted "with the obligations of inferior to superior and of persons in dependent stations of life to those who are their supporters and their employers." His goal was "to tame the ferocity of their [the students'] insubordinate passions, to repress the excessive rudeness of their manners, to chasten the disgusting and demoralizing obscenity of their languages, to subdue the stubborn rebellion of their will, to render them honest, obedient, courteous, industrious, submissive, and orderly," though above all to save their souls. "Put the rod into the hand of conscience, and excite a trembling dread of the strokes which are inflicted by this internal censor. . . . Your efforts are to prevent crimes, instead of punishing them, and to prevent misery, instead of merely relieving it." All this was born out of a deep suspicion of the lower classes: "There is a sort of practical and vulgar infidelity, which weaves its toils in the dwellings of the poor," wrote James. Like James, Wray wanted the laboring classes restrained "within the bounds of subordination and order." For him too, reading was a powerful instrument for the progress of piety, virtue, and self-discipline. Instruction was a tool for self-improvement and a means of social control.[26]

The LMS missionaries, like a growing number of people in Great Britain, espoused a new concept of discipline and punishment, one in which literacy and religious instruction would replace the threat of the rod and the demoralizing promise of alms. They hoped that instead of being driven by coercion, people would be compelled by their own consciences. They would learn to behave "properly" out of conviction, not fear. And when men and women learned to be self-reliant, thrifty, hard-working, and self-disciplined, they would escape poverty, reject crime, and cease to be a burden to society. This approach to social problems was making great headway in Britain. Such ideas had inspired the creation of the Society for the Support and Encouragement of Sunday Schools in 1785. By the end of 1786 more than 250,000 children attended Sunday schools and the number continued to grow.[27] In 1803 the Sunday School Union was formed. The movement found great support among evangelical groups, and it is not surprising that having witnessed the multiplication of Sunday schools in England and being convinced of their success, Wray tried to introduce something similar in the colony.[28]

The new means of social control, however, were only compatible with a society of free laborers and could not work to the masters' satisfaction in a slave society where the illusions of the free market were lacking. What meaning (if not a desire to escape slavery) could self-discipline, self-reliance, thrift, or hard work have to men and women who had little to look for, and who, except for what they produced in their gardens and provision grounds, saw the product of their labor appropriated by their masters? What appeal could such notions have to men and women who could hardly expect to be made free through their own efforts, and who, together with their children and their children's children, were bound for life to their masters? As McDonnell and other colonists put it, in a society where workers were not driven by fear of unemployment and the threat of starvation (or the hope to improve their lives) only the fear of physical punishment could make them work for their masters. Demerara planters—like planters everywhere—opposed teaching slaves to read and considered the whip more effective than religious instruction. They saw education not as a means of social control, but as a profoundly subversive threat to the social order. They were convinced that if slaves began reading gazettes and abolitionist pamphlets, they soon would be plotting rebellions.[29] This was also the opinion of Governor Murray, who made it clear to Wray in a conversation they had in 1813.

The dialogue between the two men—as it was reported by Wray in his diary—exemplifies admirably two entirely different approaches to education, and it replicates in the colony debates taking place in England.[30] Murray told the missionary that he would not give him permission to teach the slaves to read. He feared slaves would be influenced by anti-slavery literature and cited Haiti as an example of the dangers of teaching slaves to read. Murray expressed his concern that if the Demerara slaves learned to read they might have communication with the "Spanish Main." Wray tried to convince the governor that teaching the slaves to read was not necessarily dangerous. It was the free people, not the slaves, who had risen in Haiti, he argued. And the slaves did not need to know how to read to communicate with each other:

they were known to walk twenty miles after working the whole day and could do a great deal of mischief if they were so inclined. It was his concern—he told Murray—to put only *good* books into their hands, but Murray answered that it would be impossible to keep them from getting *bad* ones, and that an ill-disposed person could do a great deal of mischief by distributing them among the slaves. Wray insisted that any "bad" books would soon be discovered and that the slaves could learn no more bad things from books than they knew already. They were well acquainted with everything respecting both slavery and its abolition. They got ample information from gentlemen's servants and others who had been in England, and from overhearing the conversations of the whites every day at their own tables. Whites spoke freely of these subjects and probably often magnified them, and their servants heard all they said. Besides, there were many people in the colony who could read books to the slaves if they wished.

The governor was not convinced by Wray's arguments. But Wray would not give up. In support of his arguments in favor of slaves' education, Wray invoked the examples of Saint Kitts, Antigua, Saint Croix, and Suriname, where the instruction of slaves by the Moravians had produced "the most happy effect." The horrors of the French Revolution and the Irish Rebellion, he continued, had been committed by people who were uneducated, stupid, and ignorant. Literacy could have the opposite effect. To give more weight to his opinion, he quoted the chaplain of Newgate, who contended that educating the poor would reduce crime in England dramatically. He also referred to the Bishop of London's address to the West Indies, which asserted that teaching people how to read was "the very means of preserving not only the negroes but all the common classes of people" from being corrupted by mischievous writings.[31] Wray agreed that the great bulk of the common people had indeed been at first a little staggered by those "bold licentious principles" which the partisans of the French Revolution—especially Thomas Paine and his disciples—propagated "with so much effrontery and so much indiscretion." But they had soon recovered from this "delirium," and the reason was precisely that in England the "higher orders" of the community could write and the inferior could read. More than 300,000 children of the poor had been religiously educated in the various charity schools, Sunday schools, and schools of industry, and they were capable of reading and comprehending "those admirable discourses, sermons and tracts of various kinds," which "the ablest and most virtuous persons, both among the laity and clergy" composed for the lowest classes. In Wray's opinion, instruction could neutralize sedition. But nothing he said seemed to move Governor Murray.[32]

As both a plantation owner and a career officer for the British crown, Major General John Murray, lieutenant governor and commander in chief of the United Colonies of Demerara and Essequibo, was in a difficult position. Being a colonial governor was like being a manager on a plantation belonging to an absentee owner. Like any manager, Murray had to please his masters in England and control a restless slave population, while making a living for himself. His powers were limited and his role was riddled with contradictions. As governor his mission was to implement the laws and regulations issued by the British government, to keep order in the colony, and to make sure that the colony would be not a burden but a profitable enterprise. At the same time he had to keep the colonists satisfied, so that they would not create trouble. If he succeeded in his difficult task, he could hope to have a brilliant career and perhaps one day retire as a wealthy man in England.

Like everyone in the colony—slaves, masters, managers, merchants, and missionaries—Murray was trapped in a historical contradiction over which he had no control. The constant struggle between masters and slaves was being redefined by a dangerous new trend. The interests of the colonists and those of the mother country seemed to be moving in opposite directions and it was becoming difficult to satisfy both. All this change was part of a larger historical process punctuated by struggles between those who upheld traditional notions about social hierarchy, forms of discipline and punishment, education and political rights, labor and trade, and those who repudiated these notions in the name of a new social order within which slavery would have no place. To make things worse, slaves were becoming more challenging. In this changing world, Governor Murray's role was increasingly difficult to play, particularly since Murray himself owned an estate on the Arabian Coast, between Demerara and Essequibo.

Governor Murray had lived many years in the West Indies before he had moved to Demerara, and unlike those planters who lived in England, Murray had had to contend day after day with the problems of a slave society. He knew it from the inside and scorned the "philanthropist" back home, who he thought knew nothing of the colonies. Murray was particularly irritated when, "supposedly" compelled by "generous" feelings and reformist impulses, the "philanthropists" seemed to side with slaves rather than with masters, or when they supported evangelical missionaries in their effort to preach the Gospel to the slaves. Not that he opposed "religious instruction" as such. What he could not tolerate were the missionaries' "democratic" manners, and their attempt to teach the slaves to read. He was convinced that reading would sow the spirit of rebellion among them.

In the debate over the good and bad effects of literacy, as in many other things, Murray was on the side of the past, Wray on the side of the future. The missionary—like many other evangelicals—was convinced that the elites could discipline the lower classes through education. Murray saw the education of the oppressed as a revolutionary step in itself, a step that could only lead to more harm than good. Education had always been a privilege of the upper classes, a badge of status, and Murray saw a risk in extending this privilege to other social groups. He was not alone in his thinking. His words echoed the writings of conservatives in England, who felt threatened by the new democratic trends, suspicious of the growth of the popular press, and fearful of popular mobilization and of the political challenge it represented to oligarchical power. One clear expression of such reservations came from the pen of "Cato" in a letter addressed to the Earl of Liverpool. Originally published in London on December 7, 1820, the letter was reproduced in the *Royal Gazette* (which always selected from the British papers articles that had a particular appeal to the colonists).

Cato contrasted an ideal past of harmony, order, and stability with a present of dissension, disorder, and chaos. England, he wrote, was on the brink of a precipice; "another step, and everything sacred and estimable in the country" would be "engulfed" to be seen no more. He attributed these circumstances to the "licentiousness of the Press," the circulation of "immoral, profane, libelous and treasonable publications," contaminating the lower classes. When the lower orders had not been "practised upon" by the apostles of faction and sedition, they were an "honour" to their country, and to human nature. They had felt no great degree of veneration for their superiors, but felt no hostility toward them. "Their King, their Country and their Religion, were the grand points round which all their honest enthusiasm and noble pride were concentrated." But now the situation had changed. Old feelings had vanished and had been replaced by others "diametrically opposite in their nature."

Cato argued that in the past the lower classes had had no opportunity to become entangled in politics; the circulation of the daily papers was limited and confined to the higher classes. The laboring classes had no one to form political opinions for them, and "they were utterly incapable of forming them for themselves." In recent years, though, a new class of writers had appeared, men who, in order to survive as journalists, had no scruples about flattering and courting common men and women. Their success could not surprise anyone who knew the teacher and the disciples. "The one was without principle, the other without judgment." As a consequence, one of the most fragile of all the bonds which held society together—that which "united" the poor to the rich—had been severed. The writers of sedition assured their readers that they were of the same class with themselves—that they were the only friends they possessed in the nation—that the rich oppressed them, the Government plundered them, the Church deceived them, in short that every man above them was a knave and that every institution in the country existed only for the purpose of robbing and cajoling them. They assured them that they (the people) were the source of all power, the depositaries of all knowledge, and the exclusive possessors of all virtue. Not surprisingly, the great body of the lower orders had become one mass of disaffection. The distinction between virtue and vice was daily vanishing, Cato complained. Moral and religious feelings were rapidly decaying, and on the Sabbath the laborer and mechanic, instead of reading their Bibles and "seeking the altar of their God," were occupied in studying the sheets of slander, sedition, and blasphemy. In Cato's opinion, the revolutionary press had to be destroyed, and that could only be done by the "strong arm of power." He ended by attacking the Whigs and appealing to the Earl of Liverpool to destroy the seeds of rebellion. "I demand of you to protect our houses from the flames, our property from the robber, and our lives from the assassin."[33]

Governor Murray might have been less eloquent but his fears were even stronger than Cato's. He was afraid that if the slaves received instruction they would start reading abolitionist literature and would eventually rise up. Wray, however, was convinced that there was no better antidote to rebellion than religious education, and that there could be no religious instruction without literacy.

After being in the colony for some months, Wray decided to write a catechism on the duties of servants and children.[34] In 1810 the board of directors of the LMS agreed to print one thousand copies of it.[35] No one who read Wray's catechism could have doubted that his intention was to teach the slaves to be obedient and respectful to their masters. To the question "What are the duties of servants and slaves to their masters, owners, and managers?" the answer was "Respect, faithfulness, obedience, and diligence." And the answer to "What is respect?" was "An acknowledgement of their superiority and authority and a respectful manner of speaking of them and to them, and a becoming behavior." Nothing could be more clear. Over and over, questions and answers stressed the obligation slaves had to their masters, even to "unfeeling masters." The catechism warned the slaves against the sins of theft, waste, and negligence, told them to be as careful of their master's property as if it were their own, condemned all spirit of "strife" and "idle talking and gossiping," recommended that they perform their tasks with cheerfulness, and reminded them that God had promised to reward servants who were attentive to their duties.[36]

After such a demonstration of loyalty to the social order, who could have complained against Wray's teaching the slaves? But the colonists were not reacting to Wray's personal intentions or words; they were reacting to everything he had come to symbolize. More particularly, they were reacting to the way he related to the slaves. What irritated the planters and made them suspicious was not only his insistence on teaching the slaves to read, but also his personal involvement with them, his recognition of their humanity and individuality, his attempts to mediate between managers and slaves whenever there was conflict, his sabbatarianism, his watchful eye, his concern with respect for the law, his connection with the LMS, and most of all the success of his preaching and the increasing number of slaves who attended his services. Planters feared that all this would eventually erode the principles of coercion upon which the whole system of slavery was built. With the passing of time their irritation and suspicion only grew. And so did Wray's commitment and determination.

For the slaves this struggle between missionaries and planters had a quite different meaning.[37] It offered them an opportunity to challenge their masters' authority and to extend the boundaries of their own freedom. Predictably, the more their masters opposed religious instruction, the more the slaves came to see it as a desirable privilege. In this struggle they were ready to side with those missionaries who sided with them.

As a missionary on a plantation Wray had little choice. His mission could succeed only if he gained the slaves' trust. As he once wrote, a missionary to the slaves had to be willing to converse with them "freely" on religious matters and to give them easy access to him, but at the same time know how to keep "the proper distance." He hoped that when slaves discovered that missionaries were their friends and were willing to intervene for them, the slaves would learn to trust them. He believed he could handle the slaves' complaints by using the Bible's many references to slaves, and was convinced that he could discourage the slaves from bringing "unjust" complaints. Conversely, he also expected that when masters and managers found that missionaries did not encourage indolence, impertinence, or bad behavior, they would trust them too. But, as he learned time and again, things were not as easy to accomplish as he had hoped.[38]

Living in the midst of slaves, Wray soon found himself involved in his black brothers' troubles and sorrows. One would come to complain that his wife had betrayed him with another man. Another would show him the blisters he had from being whipped and describe in detail the cruelties of the manager. Still another would lament that his master did not give his slaves permission to go to the chapel. Slaves sometimes complained that they could not attend services because they were kept by their managers until late on Sunday morning, performing chores or waiting for their weekly allowance of saltfish. Even when they did not have to work for their masters on Sundays, they sometimes went to the market or worked for themselves in their own gardens or for people willing to hire them. (As was customary in Demerara, whites, free blacks, and even slaves hired slaves on Sundays to do all sorts of small jobs. This practice annoyed Wray, but when he asked the slaves why they worked on Sundays, they told him that they needed money to supplement the food and clothes they received from their masters.)

Wray wished the managers would stop making the slaves work on Sundays and follow Post's practice of giving them an extra day or at least a half-day during the week to work in their gardens or go to the market, so that "the day of the Lord might be a day of prayer instead of a day of merchandise."[39] His hopes were not totally unrealistic. This practice was common in some Caribbean islands and had been followed in Demerara in the eighteenth century.[40] But under market pressures the situation had changed. Driven by the need to increase production at a time the labor force was not only shrinking but becoming increasingly expensive, planters were not willing to satisfy the missionary's demands. As a consequence, the debate over whether the slaves should work on Sundays became a permanent source of friction for missionaries, masters, and slaves.

There was nothing Wray could do to avoid getting embroiled in the slaves' conflicts with their masters, and these were many and constant. There was also a lot of bickering among the slaves themselves, or between free blacks and mulattos, blacks and whites, free blacks and slaves. Some of the conflicts were born out of the intricate patterns of race relations in the colony and the inevitable tensions generated by Wray's attempts to create a sort of "racial democracy" in the use of space in the chapel. Commenting on the racial protocols in the colony, he noted that "white men and colored women dance together, and sleep together, and live together in fornication, but they will not sit together in God's house, though they must both be saved by the same Savior."[41]

In Demerara every group was very conscious of what it thought were its rights, and rigid etiquettes marked the boundaries among them. Coming from a different world, with altogether different protocols and boundaries, the missionaries had difficulty understanding people's behavior in Demerara. The LMS missionaries' idea of a chapel in which all would be treated as equal was at odds with the complicated system which separated blacks from mulattos, mulattos from whites, and free from slaves, even in the graveyard.

If blacks were ready to accept the new "democracy" of color the LMS missionaries intended to create in their chapels, mulattos and whites were ready to oppose it. One day, Wray asked a free black man why he had given up coming to the chapel. The man replied that one evening he had sat among the white people and someone had told him that he was not to sit there, "as that place was for the white people only," so he would not come again, "as he thought there should be no difference in the house of God."[42] Another time, Wray asked a free mulatto woman why he never saw her at the chapel any more, and she replied that it was not right "to mix colors." She complained that a black woman had lately taken a seat in the pew in which she always sat.[43] "It is astonishing," Wray thought, "how proud some of these people are, who are a little above a common field negro. Some of the free coloured people left Mr. Davies' meeting a few Sabbaths ago because some of the poor slaves happened to sit down on their seat. Oh that the Gospel may humble them and bring them to Christ."[44] Such problems did not occur only among free people. Slaves too were punctilious in matters of etiquette, and they too had their notions of rank and propriety. One slave woman complained to Wray that another had offended her by refusing to address her daughter as "Miss."

Two of the most difficult problems Wray and other missionaries had to face were what they defined as "adultery" and "fornication." Slaves had many wives, and slave women often lived with white men without being married. And no one seemed to be willing to give up these practices. In his diary and his letters to the LMS, Wray complained time and again of managers who took female slaves as their mistresses and of black girls who went to live in "fornication" with whites. "The offer of a new coat . . . a little remission from labour and the honour of having a white man are temptations almost too strong to be resisted." Wray was told that on some estates the managers did not dare discipline some of the gangs for fear of being told "you have taken our wives from us."

In the towns the situation was even worse than on the plantations. Many women who had obtained their freedom thanks to relations with whites encouraged their daughters to do the same. Wray often told the young women how silly they were to enter into such "connections," for as soon as the whites could afford a proper wife or went home, the girls would be left with their children in slavery. But no one paid attention to him. It was not uncommon for a woman to have children by five or six different men, and there were some who had children of markedly varied complexions; one fathered by a black, another by a mulatto, and a third one by a white. Sometimes white men purchased slave women and their children, and when the men decided to go home they occasionally would set them free and give them a small house and even two or three slaves of their own, who could be hired out or employed as hucksters.[45] This was enough incentive to other women to follow their example.[46] But usually the woman and the children remained in slavery. Wray noted that even men who had wives in Europe sometimes had children with black women:

> It is astonishing what airs these people give themselves and what influence they get over the white men, often far more than white women over their husbands and they are too often a pest to the estates and the slaves. In these colonies there is not perhaps one white person out of twenty that is not living in fornication and the same is true of the coloured women. Among the slaves such a thing as marriage is not known and very seldom among the free coloured people, except within the last two or three years in Demerary. As far as I can

> learn marriage is far more common in the islands than here. . . . It is difficult to preach against sins of this kind when you know that almost every individual is guilty of them and perhaps some that we are obliged to be with daily and to whom we are under great obligations.

Wray had difficulty understanding the sexual mores in the colony. It seemed strange to him that a "virtuous" young woman would marry a man she knew was keeping another woman, who perhaps had three or four children by him, and that sometimes after she married she took the children into her own house.[47] The many and flagrant violations of "the laws of God" (as he understood them) put Wray in an odd position. How could he explain the fourth commandment, "Remember the Sabbath day, to keep it holy," to a congregation that was systematically forced to work on Sundays? How could he explain the seventh and the tenth commandments, "Thou shalt not commit adultery" and "Thou shalt not covet . . . any thing that is thy neighbour's," to the slaves in the presence of managers, overseers, and clerks who were living "in open and notorious fornication" with the slaves' daughters and wives? How could he tell the congregation that it was a sin to live in fornication, or to have two wives at the same time, or to take the wife of another man, when those sins were being committed all the time by both managers and slaves? "You may conceive of my feelings," he once wrote to the LMS, "when explaining the ten commandments on Mr.————'s estate who often heard me and who has three wives professedly, two of them sisters, both pregnant at that time and since delivered." And he added that on another estate where he had once lived, "the manager had then four wives . . . and perhaps many more had children by him."[48]

Slaves, too, often had more than one wife, something that scandalized Wray, who did not seem to realize how difficult it was for the slaves to abandon the practice of polygamy, so common in parts of Africa.[49] Nor did he lend weight to the fact that (as we have seen) there was a profound imbalance in Demerara between males and females among both whites and blacks. In town, there were more white males than white females, but many more black and mulatto women than black and white men. The total Georgetown white population, including the districts of Robbs Town, New Town, Stabroek, Werk en Rust and Charles Town, included 727 white men and only 377 white women—along with 316 white children. But among the population of color the ratio was inverted: there were 353 males and 2,147 females. On the plantations, the imbalance was reversed.[50] Among the slaves there were many more men than women. In 1817, for every 1,000 African-born female slaves there were 1,724 males. And even if the creole[51] slaves were included there was still an imbalance of 1,000 female slaves to every 1,311 males.[52]

The attachment to traditional practices, the overwhelming power masters had over slaves, whites over blacks, and men over women, the imbalance between males and females, and the example of a few black women who managed to gain status, favors, and sometimes even freedom for themselves and their families because of their relations with white men—all help to explain the difficulties Wray had convincing his congregation to respect the seventh and tenth commandments.

But if there were many problems, there were also pleasures in Wray's mission. Initially, what thrilled him most was to see people walking fifteen, sometimes twenty miles to attend Sunday services. And even on weekdays some walked miles to evening services after working all day. Wray was particularly touched one day when he saw in the chapel a woman who had been carried three miles in a hammock because she could not walk. The chapel soon became a point of gathering where people from different plantations could meet.[53] Wray's evangelical messages stressing the themes of grace and redemption, freedom and love, equality of all before God, and communal solidarity, deliverance, and the ultimate vindication of the people of God had a strong appeal to the slaves. Yet not all of them were driven by religious motives—although Wray would have very much liked to believe they were.[54] Many slaves went to the chapel to meet others, or simply to evade their masters' control. Wray once asked a slave who had walked six miles to hear him preach why he had come so far. "My heart told me to come," the slave answered. Nothing could have pleased Wray more. But when he asked the slave how he knew about the preaching, the slave said that he had seen a great many people coming from town and had followed them. "They must have told him," Wray consoled himself. He would certainly have liked to think that the man had been driven by his piety, and not by his desire to be with others. Wray also took a great pleasure in watching the slaves' "progress," their curiosity about religious matters, and the fervor with which they prayed.[55] To him, such things seemed to be evidence that his efforts were being rewarded.

Wray was intrigued and sometimes embarrassed by questions slaves asked him, and by the things they said. One slave, Quamina, confessed to Wray that he never had heard of Jesus before. He also told Wray how disturbed he was because he recognized he had sinned and believed there was nothing he could do to obtain salvation. Before the missionary had arrived, Quamina—like many other African slaves—had thought that when he died he would go back to his own country, but now he was confused.[56] Another, "a man of colour," came to ask Wray the ancient question: Why, if God knew that man would sin before he made him, did he make him to sin? He also said that he had been told that as soon as Adam and Eve had committed sin, Adam had blamed Eve, and Eve blamed the devil. "Now, Sir," the man asked, "who must Satan blame for his fall?" Others asked if the suffering of Christ had been determined by God. Would the world be destroyed before or after the Judgment? Was the apostle Paul a man on the earth, as they were? What would happen if a person who had changed his heart died before being baptized? Would such a person go to heaven? How was it possible that the body could rise again after it had rotted in the ground? Would those who knew one another on earth and went to heaven know one another there?[57]

Some of the slaves' questions were difficult to answer—just as the same questions had been difficult to answer during the long history of Christianity. But it was even more difficult to make Christian slaves abandon their traditional religious practices and beliefs. Most slaves in Demerara had had some acquaintance with Christianity before Wray arrived, either through contact with priests and missionaries in their place of origin (in the Caribbean as well as Africa) or from hearing other slaves talk about it. But they also had kept African traditions—however transformed by slavery, by slaves borrowing different rituals from others, by the new environment in which they lived, and by their notions of Christianity.[58] To Wray, African rituals seemed as mysterious as his own rituals must have seemed to the slaves. But apparently it was easier for them to accept Wray's beliefs and rituals than for him to accept theirs.[59] Except for those who were Muslims, most Africans—although they did recognize a supreme God—included in their pantheon other gods, spirits, or divinities that stood between man and the ultimate God. Prayers also were familiar to them in the form of requests for health and well-being, and often included "statements of innocence of any evil intention." So they had little difficulty adapting to the missionaries' ideas of prayerful supplications to a supreme divinity. But perhaps the most important characteristic of African religions was that—except among Muslims and in the Dahomey, where there was a established priesthood—there was no orthodoxy, thus no heterodoxy. And although rituals occupied an important place in all African religions, the rituals were relatively flexible, pragmatic, and experimental.[60] This gave African religion its great resilience and capacity to interact with other faiths and practices.

Though profoundly distorted by his ignorance and biases, Wray's careful notes reveal the pervasiveness among slaves of African beliefs and rituals, which were indeed found throughout the Caribbean. He thought the slaves worshiped water and would walk eighteen or twenty miles to do so. He described in detail how, when a man died, the slaves buried him late at night so that his "countrymen" might come to the funeral from different plantations. They poured into the grave the water they had used to wash the body, and believed that if they spilled it on the ground the next day would be a rainy day. According to Wray, just before they nailed a coffin shut they would speak a few friendly words to the dead. If the dead man had been a smoker they would put a pipe and some tobacco in the coffin along with him. They would carry the coffin to the grave site, drumming and dancing as they went. The bearers walked three times around the grave with the corpse, and then put it in. Then the dead man's wife or friends addressed the corpse, promising within a few months to bring enough food to satisfy him and all the good friends he was going to join. They also asked the dead person to give them strength, and to let them know who had been responsible for his death. Finally, they drank rum and returned to their homes. Three months later, they would make a great supper on a Saturday night and invite their friends. They spent this night drumming and dancing, often until the next morning. On the Sabbath morning, very early, they took a tray of provisions, such as fowls, pork, or beef cakes, and placed it on the grave.[61]

Like most whites, Wray saw all slave religious rituals as incomprehensible "magic" or "superstition" that could do only harm. In 1814, he heard that a letter allegedly written by Jesus Christ was circulating in the colony. The word had gone around among slaves and free blacks that whoever had a copy of it would be protected from harm. The letter forbade people to work on the Sabbath and, as a consequence, the servants of one lady had refused to do any work that day. Several blacks had given five shillings apiece for copies of the letter. Some had brought them to Davies to know if they were "good." To show that he did not take these things seriously, Davies had torn their copies to pieces. (Apparently the letters had been imported from Barbados but, inevitably, some whites blamed the missionaries for it.)[62]

Davies—who had steadily expanded his activities and managed to gain the confidence of the governor and to have a chapel of his own in Georgetown—was also preaching to slaves in surrounding areas. He too worried about "strange" practices among the slaves and attributed them to blacks and mulattos "setting themselves up as preachers."[63] He reported to the LMS that these preachers had produced much evil in Demerara. Davies described in detail an episode involving slaves on plantation *Den Haag,* the largest in the colony, where one woman, calling herself the Virgin Mary, had set the slaves into an "enthusiastic frenzy." They had behaved in a rebellious manner towards their manager, and the ringleaders had been imprisoned. Van den Heuvel, the proprietor, perhaps hoping to pacify the slaves, had called Davies to "instruct" them. According to Davies, after he addressed the slaves—a crowd of about 600—he asked them if they were not sorry for the bad things they had done. In one voice they cried aloud: "We sorry Massa. We sorry Massa." However romanticized and self-serving Davies's report may have been, it reveals something of the masters' ambivalence toward the missionaries. Although most masters mistrusted them, they were quite ready to call the missionaries whenever they thought they could be used to subdue the slaves. On such occasions, masters seemed to see Christianity as an antidote to the messianic messages the slaves themselves elaborated from their mixed Christian and African traditions.

Among the slaves' practices that whites feared most was what they called "obeah."[64] While Wray was in Demerara he seldom mentioned "obeah" in his letters or journal (either because these rituals were less common in Demerara than in Berbice, or because when he was in Demerara he was less aware of them). But after he moved to Berbice in 1813 he reported several instances of "obeah." All of them impressed him profoundly, and he went out of his way to learn what he could about it. He even approached Wilberforce, who in a letter to the LMS mentioned that he had received a very interesting letter from Wray on this "delicate" subject. At one time, Wilberforce explained, this "system of superstitions and roguery" was conceived to be of such import that it was even gravely discussed by the agent for Jamaica, and even by the Privy Council in an inquiry of 1788–89. But the colonial assemblies that were questioned about "obeah" practices since "had indignantly replied that no reasonable person in the West Indies now thought anything seriously of obeah, [any] more than the English now did of witchcraft, which was believed in by our forefathers." Wilberforce added that his friend Mr. Stephen, master in chancery, could lend the LMS (confidentially) his unpublished work on the legal state of slaves in the West Indies, which had the best account of "obeah" he had ever seen. He concluded that the subject deserved attention because it was an evidence of "the horrible wickedness and cruelty of the present slave system in the West Indies."[65] This kind of encouragement from so prestigious a source could only convince Wray that it was even more worthwhile than he thought to describe in detail what he saw and heard about "obeah."

Wray was convinced that "light and knowledge alone" would root out "obeah" and other "evil practices." And he was very willing to collaborate with the colonists for that purpose.[66] In Berbice he was called a few times by the authorities to attend slaves who had been sentenced to death for practicing "obeah." In 1819, he reported in great detail the trial of Hans, a slave who belonged to the crown and had been employed at the fort. Hans was accused of having practiced "obeah." Trying to report what he had not seen, and certainly did not understand, Wray—in spite of his obvious efforts to be accurate—gave a confusing though fascinating picture of the rituals performed by Hans and other slaves. Wray's description of the events which led to the man's trial not only brings to light aspects of the secret life of the slaves and their attachment to African beliefs and rituals, but also how difficult it was for whites to understand the slaves' world.

According to Wray, Hans, a Congo,[67] had been called several times by a head driver named January, to a plantation where slaves were constantly sick. January had sent for him in the hope that Hans could find out who was "responsible" for so much sickness. When Hans arrived the slaves were assembled. Hans asked for a tub of water and a white pullet. After killing the pullet by wringing its neck, he pulled out its feathers and stuck them in the children's hair, then washed their faces with water from the tub. Those who were sick he washed to heal and those who were not he washed to shield. He put several things into the water, including a bell, an image, and a piece of root of a wild cane that Hans said he had brought from his "country."[68] Hans made the slaves form a circle, and while he sang a "country song" they danced on one foot, clapping their hands until daylight. (What probably happened is that the slaves danced by stamping the ground with one foot, as was common in some African ritual dances.) Hans asked for "horses" (something Wray erroneously interpreted as meaning men to carry him, but that in fact meant people capable of receiving or being possessed by gods). Four slaves were appointed. Then Hans ordered the slaves to put a pot of water inside the circle. He threw some sour grass in it and stirred it twice. He sprinkled some of the water on one man's face and the man became as though he were "crazy," jumping high off the ground and throwing himself down. Hans repeated the ritual and several people were affected the same way. He then asked the slave Frederick to throw water on some slaves and they also became "crazy." Several fell to the ground. Hans continued to dance, making a variety of motions, and ordered one of the slaves to flog those who were in "trances"[69] with a piece of wild cane. Venus, a female slave who had been particularly affected, continued in her trance. Hans chewed guinea pepper and spat it into her eyes. But this did not bring her to her normal state. She continued to dance, throwing herself upon the ground, and rolling about. She finally struck Frederick on his breast and burst into a "hysterical" laughter, then struck him again. Hans told them that when the "thing overcame," they could walk to the place where the poison was hidden.[70]

Hans then went to Frederick's house to remove "the bad thing that was there," hidden in a sheep's horn, which Hans said he knew was in Frederick's house by the smell. Hans explained that not everyone could detect that smell, but he could, because of a peculiar knowledge he had from God—a knowledge he had acquired in his own country. Hans asked that a child and a "buck pot" with some water in it be brought in. He made every one in the room put a little water in the pot, then took a piece of salempores (a type of cloth) and covered the child's head with it and put the pot of water in her hands. Hans directed one of the slaves to take the child on his shoulder, and he got upon the shoulder of another slave. (Here Wray's report is confusing.

He probably misunderstood what Hans told him. What probably happened is that Hans called on the gods to participate in the ritual by "riding" him and the child, a common metaphor used to define the believer's possession by a god.) Hans ordered a plank to be removed from the floor, dug a hole close to the bedside, and asked the slaves to examine the pot to see if there was anything in it but water. They all acknowledged that there was nothing else there. Hans directed the child to stand over the hole for a while, holding the pot in her hand, with her head still covered with the salempores. He then instructed another slave to take off the salempores. When she did it, they all could see in the pot a sheep's horn with a piece of cloth tied over its open end. Hans told one of the slaves to remove the cloth, and they found that the horn contained blood, hair, nail parings, the head of a snake, and other things of the sort. Hans explained that the stuff in the horn was the "bad thing" that had been destroying the children, but that it would do so no longer.

Wray interpreted all this as a conspiracy against Frederick. Apparently Frederick had been accused before by the slaves of being a poisoner. They had complained to their master, who told them that if they found the poison he would have Frederick punished. "But, they of course, could find no such thing," added Wray, who was convinced that the whole thing was an act of delusion promoted by Hans. Wray wrote that the dance was called Mahiyee or Minggie Mamma dance, and that January had told the slaves that they had to give a gift to Hans. They collected sixty guilders, plus fowls and other things, which they gave to Hans for services rendered.

When Hans was asked during his trial if he was in the "habit" of using his powers, he answered he had visited another plantation at the request of the drivers and the carpenter because of pregnant women miscarrying. There he also had washed the children and the bellies of the pregnant women and this had prevented the children from dying and the women from miscarrying. On another plantation he had discovered two men whom he had identified as poisoners, and through his power he had compelled them to confess that they had poisoned two men who had stolen their fowls. He said that the blacks knew he possessed power to help them, and great numbers had called on him. When he was asked whether he had ever failed in the exercise of his powers, Hans answered that since some people who had applied to him had diseases "above his art," he could not help them. He was then asked whether he had ever been requested to use his powers against whites. "Yes, numerous applications have been made to me but I have always rejected them." These applications, he explained, had been made by slaves with bad masters, who wanted him "to cool their hearts." But that was not part of his art. When he was told that it was contrary to the laws of the colony to exercise such powers, Hans said he was no "obeah man" and did nothing that was bad. "All my art consists in helping the negroes that are sick."[71]

Whites, however, were not persuaded by Hans. They ignored that according to most African religions, human beings would live forever in health and happiness if it were not for the workings of the forces of evil. When disease struck the community, the source of evil had to be rooted out. A diviner would be consulted to discover the device used to bring misfortune or disease. The source of the problem could be a dissatisfied spirit, an ancestor punishing descendants, or an individual venting anger, envy, or selfishness, as in the case of Frederick. So when slaves in Berbice or Demerara called a diviner to put an end to their misfortunes, they were rehearsing old practices in a new environment, using the means available to them to bring some control to their lives.[72]

For whites, "obeah" was a dangerous practice that had to be extirpated. They all tended to condemn Hans's "art." For Wray, such rituals were the result of ignorance—nothing but bad habits that needed to be eradicated. The care with which he took notes and reported Hans's story in minute detail reveals his awareness that the directors of the LMS, and perhaps the readers of the *Evangelical Magazine* where his piece might be published, would be interested in hearing such tales. And by reporting the stories the way he did, Wray was helping to reinforce a view of the world which opposed civilization and barbarism, religion and superstition, reason and emotion, a view that gave to his British readers a sense of superiority that not only hampered their understanding of other worlds, but also made them oblivious to the contradictions of their own. (After all, witchcraft was still a common practice in "civilized" England.) Wray's narrative betrayed his inability to penetrate the magic world of the slaves. His lack of understanding was shown by his indiscriminate use of words such as "obeah," "crazy," and "hysterical" when he described slaves' rituals, by his misreading of the expression "to be a horse," by his conviction that Venus had been faking, and that the whole ceremony was nothing but an elaborate plot by the slaves to incriminate Frederick.

To the fiscal, the ritual Hans had performed seemed an outrage, and an offense against established law. Like Wray, he was unwilling or unable to differentiate between the practices slaves considered evil and the practices they considered good. He lumped all slave rituals under the category "obeah." In his mind "obeah" was a dangerous practice to be severely punished. He requested the death penalty for Hans, on the basis of a colonial law passed in 1801 and reaffirmed in 1810—a law supposedly grounded on the scriptural injunction "Thou shall not suffer a witch to live."[73]

By contrast, the lawyer who defended Hans was a man of the Enlightenment, in tune with the new "philanthropic" trends in the mother country, and thus inclined to see the "criminal" as a victim of a social order in which the contenders were "ignorance" and "knowledge," "barbarism" and "civilization," "paganism" and "Christianity." In his opinion, society, not the criminal, was responsible for the crime. He argued before the court—Wray tells us—that this "poor and ignorant" man had been brought from a pagan country twenty years before and no effort had been made to make him a Christian—which alone could eradicate this evil. And after Wray—whose labors had been beneficial in many instances in doing away with this evil—arrived, Hans had been employed in the fort and deprived of instruction, since there was no chaplain there. All this was meant to demonstrate that it was the colonists' own fault that Hans still believed in "pagan" traditions (traditions the lawyer obviously despised no less than Wray or the prosecutor, and wished to be removed by Christian teachings). The lawyer then argued that the "Jewish [Biblical] law," which had been invoked to condemn Hans, was peculiar to the "Jewish Nation," that its government had been a theocracy long since abolished, and that if the spirit of that law was followed the colonists would have to punish with death all slaves not converted from paganism to Christianity—in which case the country would be substantially depopulated. With respect to the colonial law he argued with imperial confidence that no law of its kind was valid unless confirmed by the mother country, and the statute in question had not been confirmed either by the laws of England or Holland.[74]

Nothing could have pleased Wray more than such praise for his work and the lawyer's stress on the importance of converting the slaves. Wray's position not only led him to sympathize with the lawyer's arguments but also gave him a keen sense of the contradictions of the slave system. In his report to the LMS he commented, with his habitual irony and visible satisfaction, that the lawyer might have added that Sabbath-breaking was also punished with death by "the Jewish law," so that to be consistent, the fiscal should also punish with death those who worked on Sundays or went to the market. "It is remarkable," Wray wrote, "that in 1810 when the law was made to punish the Obeah with death founded on a Jewish law, another law was made that permitted Masters to work their slaves till ten o'clock on Sunday." And he observed that the lawyer might also have said that under the "Jewish law" adultery was punishable with death and if a similar law existed in the colony, most slaves would have to be executed.

Hans escaped death but was sentenced to be whipped under the gallows and branded, to be imprisoned one year, to stand four times in the pillory, and to work in chains for life "for the benefit of the colony." Slaves who had participated in the ritual were whipped, among them January and Venus. Wray said he had heard that Venus was to work for twelve months in chains. "It is said she was stripped naked when she was flogged," he added.[75]

Wray visited Hans several times in prison. He spoke to Hans in the creole he had learned so that he could talk to slaves who could not speak English. He pointed out to Hans the great sin he had committed against God in "pretending to possess these supernatural powers," and tried to convince him that all those involved in that ritual had sinned. Satan had blinded the minds of all. "Multitudes were so blinded as to worship Gods of their own making, Gods of Wood and Stone, and to offer human sacrifices to them." But when the word of the "true God" was known men's minds were enlightened and they turned from evil practices. These and many other pious things Wray told Hans, who seemed to be "a good deal affected," and sometimes "indeed shed tears." After they talked, Hans asked Wray to give him a new heart; he would throw away his old.

Hans told Wray that he had been instructed in Catholicism in Africa, and always prayed to God to bless the means he used for the recovery of people. He insisted that his art consisted in doing only good. When Wray, moved by intense curiosity, asked him to pray in his language, Hans knelt, addressed God by his African country's names for Mary and Jesus Christ, and then made a cross on the ground with his finger. "I was affected to hear him use the name of Christ and not understand the meaning of it," wrote Wray, whose self-confidence was unshakable. Wray also spoke to Venus and January, and both seemed to him very attentive. From all this he concluded that only Christianity could "extirpate this evil" that no persecution could eradicate. "This shows us the necessity of exerting ourselves to spread the glorious Gospel among the poor people and I think such respectable merchants, as Winter, Innes, Englis and Gladstone and others would patronize this work if they were only well informed on the subject."[76]

A few years later Wray reported a story of a slave murdered during a "Minggie Mamma" or "Water Mamma" dance. The details were similar, indicating a familiar pattern, only this time the episodes led to a tragic ending.[77] Once again the slaves sent for someone to discover the person responsible for the deaths of many people on the plantation. They held a dance, with the greatest secrecy in the silent hours of the night, and accused a woman of "obeah." She was tied to a mango tree and severely flogged. The next day, the overseer noticed some blood on her clothes and inquired what was the matter. But, afraid of the "obeah man," she did not tell him the truth. That night she was tied again and beaten until she died. The "obeah man" was arrested and condemned to death. The execution took place on the plantation, with the governor, the fiscal, all the members of the court, the militia, four or five hundred slaves, a great number of white people, and the Rev. Mr. Austin of the Church of England, present. The next day during services, Wray spoke of those events, comparing the "obeah" men to Cain, David Ananias, Sapphira, Simon, and Etyma.[78]

In his struggle against African "superstitions," as he called the slaves' rituals, Wray found allies among the colonists, who were always ready to elicit his support in their efforts to control the slave population. Whenever there was a case of "obeah," colonists seemed to forget momentarily their hostilities toward missionaries and called Wray in the hope that he could exorcise this "evil." Some colonists also called him when the slaves became intractable. They expected Wray to use his prestige to bring the slaves to order. And the missionary often succeeded. Yet in every other way Wray was alone in his struggle to convert the slaves, and his actions aroused more suspicion than sympathy from most colonists.

The colonists were in a quandary. They wanted to use the missionaries to put down slave resistance. But when the missionaries wanted to teach the slaves to read, when they insisted that the Sabbath be respected and that the slaves not be forced to work on Sundays, when they organized night meetings and attracted to their chapels large numbers of slaves from different plantations, then most colonists treated missionaries as enemies and did everything in their power to stop them.

What the colonists seemed not to realize was something Wray knew well. For the missionaries to be successful in their mission they had to gain the trust of the slaves, and for that they had to side with them, particularly in cases in which there was flagrant injustice, when the slaves were clearly victims of unusual violence and oppression. Missionaries had to be willing to hear the slaves' complaints and speak for them when their demands seemed "reasonable." They had to play the role of arbiters when there was conflict between slaves and managers. The slaves had to believe that the missionaries were sincere and honest in their purposes. If the missionaries became mere spokesmen for the colonists, if they always sided with the masters against the slaves, they would have no followers. Wray's chapel (and later Smith's) was always crowded because—among other things—the slaves had come to see Wray as someone who sympathized with their plight, someone fair. Slaves had to see Christianity not as something that added to their oppression, not as something that justified abuse and violence, but as a promise of both physical and spiritual redemption. And it was precisely this that made Wray and Smith so dangerous in the eyes of the colonists.

At a time when people in Britain were condemning the institution of slavery, making speeches about the rights of man and the supremacy of the law, when politicians were discussing ways of improving the slaves' conditions of living and even talking about emancipation, when the cruelty of the masters was being exposed in Parliament and in the British press, and slaves' perception of the social order, of what was fair and unfair, of what was possible or impossible, was changing, no missionary could ignore such trends. Those missionaries who did had no followers among slaves and their chapels were empty. They might please the masters, but they lost the slaves' support. That is exactly what would happen when Wray eventually moved to the neighboring colony of Berbice, in 1813, and Richard Elliot replaced him.

Elliot, conscious of the colonists' opposition to Wray, decided to ingratiate himself with the managers and masters. Soon it must have become clear to the slaves that this new "parson" was not a man to be trusted. Of the five or six hundred that usually attended Wray's services, only six or seven remained.[79] On one occasion, when Wray announced he would come from Berbice to preach at *Le Resouvenir,* he found Elliot practically alone, with only one slave inside the chapel, while a great number waited outside and only went in when Wray arrived.[80]

The lack of trust between Elliot and the slaves became apparent when he found his horse stabbed to death. Elliot immediately suspected Romeo, a slave deacon at his chapel. In a letter to Wray, Romeo complained that Elliot had stolen the fowls of "the negroes," and not only had unfairly accused him of killing his horse but had even searched his house. Wray reported to the directors of the LMS that the slaves' fowls had wandered into Elliot's garden and he had ordered his cook to kill them for dinner. Unsuccessful in their appeal to the manager and attorney, the slaves had decided to take matters into their own hands, killing Elliot's horse in revenge. Such behavior was not uncommon. Slaves often acted that way when they considered managers or overseers cruel and did not manage to have them dismissed. Wray, who had taken a strong dislike to Elliot, complained that Elliot had alienated the blacks because of his good relations with the managers and his indifference toward the slaves.

Wray's antipathy to Elliot was rooted in a clash of personality style and religious commitment, and it became immediately obvious from the moment Elliot arrived in Demerara from Tobago, where he had been sent by the LMS in 1808. Elliot told Wray that although he had been given a lot of support in Tobago, the "negroes" did not care about him and would not come to the chapel or send their children to school. Wray doubted Elliot's sincerity and commitment and thought he had assimilated some of the West Indian whites' prejudices against "coloureds."[81] His relations with Elliot became even more tense after Elliot, hoping that the LMS would nominate him to replace Wray, decided to remain in Demerara instead of returning to Tobago. Meanwhile, the LMS had sent a man named Kempton to occupy that post in Demerara, but when he arrived Elliot simply refused to leave. Wray and Davies were divided on the issue. Davies supported Elliot's pretensions, while Wray, outraged by both Elliot's and Davies's behavior, wrote furious letters to the LMS condemning both of them and supporting Kempton. Apparently forgetting the troubles he had had at *Le Resouvenir,* Wray accused Elliot of wanting to stay because it was a comfortable place: a house, a chapel, and a congregation without any trouble and more than £100 a year. It was an ideal place for a missionary recently arrived from England; and if Elliot wished to stay in Demerara, there was plenty of work for him in other places on the West Coast.

Wray returned to Demerara to try to settle the matter, but instead of calming things down he became involved in futile quarrels with both Davies and Elliot. One day he announced he was going to preach at *Le Resouvenir.* Elliot locked the chapel. When Wray arrived at night with Kempton there were three hundred blacks waiting, but the key of the chapel could not be found.[82] Wray suspected that Elliot (who was staying in Georgetown) had hidden the key.

The whole Kempton-Elliot affair ended with a disgusted Kempton returning to London. The LMS instructed Elliot to leave *Le Resouvenir.* Yet Elliot's presence continued to rankle Wray. In May 1815, he informed the LMS that Elliot remained in Demerara with the approval of whites who "profited by his Ministry." Blacks had given up going to the chapel, except perhaps for about half a dozen.[83] Such bickerings and rivalries among missionaries were not likely to enhance their mission and made them even more vulnerable to the colonists' criticism.

Slaves and masters were engaged in permanent war—a cold war that took place every day in many forms, but from time to time burst into violent confrontation. There was no way the missionaries could avoid being drawn into such conflicts. Sometimes both parties sought their intervention. Once, when he was still in Demerara, Wray received a letter from the attorney of plantation *Success,* about the conduct of "the negroes." On Sunday the slaves had collectively refused to take their allowance of saltfish because it was no larger than a common allowance and it was customary to give more on holidays. They also had refused to feed the cattle in the evening. The managers informed the attorney, who came and talked to the slaves. As a punishment, he prohibited them from going to the chapel in the hope that this would have a better effect than floggings. He then asked Wray not to admit them to the chapel until he gave them leave to attend services.

Wray received the note almost a week after the event, but by that time he had learned of the incident from the blacks themselves. He went to *Success* and told the slaves that they ought to be ashamed of themselves. They should be thankful that the attorney encouraged them to hear the Gospel when so many were being prevented from doing so by their masters. The attorney had spoken to the governor very favorably about preaching, and the governor had told him that he would write a favorable account of it to "King George and his great men at home." By their rebellious conduct, Wray told them, the slaves had made all this sound untrue, and his enemies would say they were not better off for hearing the Gospel. Wray also declared that the slaves had made him ashamed and should not come to the chapel until they had made due submission to their attorney.

The missionary was at a loss to know how to act. Some of the slaves had been baptized and were used to coming to the Lord's Supper. It appeared to him that it would be very improper to exclude them just because they had made a complaint. It was likely that they had spoken "unguarded words" on this occasion, or exaggerated the case to the attorney. "But what will not oppression do?" he asked. "They have made complaints to me before but I have kept them from going to the attorney, but human nature cannot always bear it except oppression has taken away all their feelings." Confused and uncertain, he decided to postpone the Lord's Supper. And he concluded: "I do not wish to encourage their complaints by listening too much to them, and at the same time I am afraid of driving them to despair. I pray that God would give me grace to act with wisdom. It is an unpleasant thing to be engaged as a missionary among slaves."[84]

Wray found himself involved in many similar incidents. Later in 1812, while he was still at *Le Resouvenir,* forty or fifty of the plantation's slaves complained about the manager to the fiscal. (Post had died in 1809 and his widow had married a man named Van der Haas.) The new owner had appointed as manager his brother, who had been in the colony for just two or three years, working during some of that period as Post's bookkeeper. As a consequence of the slaves' complaint, Wray received a letter from the fiscal saying he wanted to speak with him about the problems at *Le Resouvenir.* Wray told the fiscal that although the slaves on the whole seemed attached to Van der Haas they seem not to like his brother. (Sometime before his death, Post had confided in Wray that the man was very cruel to the slaves and for that reason he had dismissed him from the estate. But Post was afraid that when he died the man would be hired again.) Wray also told the fiscal that the slaves had made serious complaints against the manager to him and to his wife. They said the manager flogged them severely for every trifling fault and made no distinction between the weak and the strong in giving them lashes. They also complained that a "woman with child" had to do as much work as any other. The manager had flogged in a "most shameful manner" a slave named Hector; and Quamina, the leading carpenter, had received ninety stripes from the manager, plus more from the drivers, just because he had not been able to complete his day's work. Several months before, Quamina had received a hundred stripes in a "severe manner" and had been put in the stocks to prevent him from going to the fiscal.

Wray added that Sandy, one of the drivers and "a negro of whom Mr. Post spoke well," had been driven out of the "sick-house" before he was well, and had been beaten with bamboo on his already blistered back. Another slave had had his hand broken by the bamboo.[85]

A year later, when he was settled in Berbice, Wray returned to *Le Resouvenir* for a short visit and found that the slaves had more complaints. The stories were very similar. Some told him that on one estate the manager had prevented them from getting together in the evenings and on the Sabbath to preach and catechize. They had complained to the attorney but had gotten no redress. "He would not permit more than two or three to meet and told them that it was against all religion, against the law of England and of the governors." "This is the gentleman," Wray commented, "who once requested me to prevent them from coming to the meeting for a few evenings as a punishment because they and the manager had a dispute. I did not expect better things from him. . . . " The slaves from another plantation complained of the manager for turning them out of the "sick-house" under the pretense that they were lazy when they really were sick. They said that the manager allowed the drivers to flog them "improperly." "He is a young man lately of England, who of course, can know but little of the management of negroes and their work," commented Wray. The slaves also complained the manager condoned the flogging of women with young children. The women were forced to lie down on the ground to receive the lashes and were stripped to their thighs, and in their opinion this was very improper so soon after having children. Another complaint was that the women with infants were obliged to turn out in the morning as quickly as the rest of the people, and if they did not they were flogged. One slave told Wray that his wife would have delivered her child in the stocks if it were not for the compassion of the nurse and midwife, who released her just a few hours before. And when the manager found it out he confined both the nurse and the midwife. The slaves also reported that when they first went to talk to the attorney, he sent them away without hearing what they had to say. When the attorney finally went to the estate to investigate the slaves' complaints, four women who were either pregnant or had new-born children were in the stocks. But he made no effort to see them. At first, when the attorney told the slaves that they should come one at a time to speak, no one came forward. (Wray asked the slaves why they did not come forward when the manager told them to speak out. They answered that those who did were usually picked out as ringleaders and asked questions in such a manner as to confuse them.) A couple of slaves finally spoke. One invited the attorney to go to the "sick-house" to see with his own eyes what was happening, but he refused to go. The matter was "settled" by giving the slaves a good flogging. They appealed to the fiscal, but found no redress. Wray suggested that they appeal to the governor. Commenting on the incident, Wray wrote that the manager denied all those things and was "believed in preference to the negroes for he is a white man and free."

From such incidents slaves had learned that they could play one white against another. They could complain to the manager about the overseer and to the attorney about the manager. And when the plantation owner was around they could appeal to him. They could appeal to the authorities: the burgher officers, the fiscal, and as a last resort even to the governor himself. Most important, now they could appeal also to the missionaries. Although, as we have seen, slaves' complaints were often dismissed, sometimes they found support. Masters and local authorities feared that too much cruelty would drive the slaves to despair. So occasionally they fired an overseer or a manager for "abuse of authority."[86] However rare, such experiences were enough to convince the slaves that somewhere there was a power superior to that of managers and even masters, a power they could appeal to for protection. They hoped that missionaries would speak for them.

Wray's position was indeed far from comfortable for he was caught in the middle of a struggle between slaves and masters that had no end.[87] Long before the Americans and French had risen in the name of freedom, slaves had struggled for their freedom in the colonies of the New World. For them, it was an old struggle, dating back to the moment the first boat carrying slaves had arrived in America. What was new for the slaves was to have any whites on their side, whites who used a language of "human rights" which the slaves could appropriate to achieve their own purposes. That seemed to tip the balance in their favor.

Among those whites—"friends of the negroes," as they were called by their enemies—were missionaries like Wray and Smith. Not surprisingly, the more their congregations grew, the more the missionaries were criticized by masters and managers and persecuted by the authorities, the more the local press accused them of intending to bring about the emancipation of slaves, the more the slaves were likely to see the missionaries as people they could trust. When managers and planters opposed religious instruction, slaves flocked to the chapels. When masters prohibited them from learning how to read, the slaves attached to reading an almost miraculous meaning and struggled with all their means to master this art. When they read the Bible and heard sermons, slaves were particularly touched by those passages that spoke of captivity and deliverance. But the more slaves flocked to the chapels, the more managers and masters became suspicious of missionaries and created obstacles to their work. The war had no end; the missionaries were trapped. Opposed by masters and managers, and attacked by the press, missionaries like Wray, who took their work seriously, became increasingly convinced that the abolitionists were right; while the colonists became increasingly convinced that the missionaries were dangerous. In an attempt to overcome the obstacles colonists put in their way, missionaries did not hesitate to resort to their political connections in London, thus strengthening the ties between the LMS and abolitionist leaders in Parliament. This in turn would make the missionaries even more suspect in the eyes of the colonists.

Typical was the controversy over the slaves' right to attend religious services, a controversy which had started immediately after Wray arrived and from the beginning divided the colony. Most managers, masters, and the local authorities took one side, and missionaries, slaves, and a few managers and masters, the other. The tension between the two groups reached proportions that required the intervention of the home government. Colonists objected particularly to night meetings. They feared that if the slaves started attending services at night there would be no way to control them. In 1811, in response to colonists' requests, Governor Henry Bentinck finally issued a proclamation prohibiting all meetings of slaves after sunset, except those necessary to the functioning of the estate. Once again Wray tried to mobilize the few planters and managers who supported his work. Several signed documents testifying to his good services, including Van der Haas, from *Le Resouvenir,* John Kendall, former manager of *Friendship,* George Manson from *Triumph,* Alexander Fraser, C. Grant, Van Cooten, Semple, James Wilson, and Andrew Black and William Black (both employed by Wilson), plus a few others.[88]

With those documents in hand, Wray sought a meeting with the governor, who refused to see him. But Wray was not a man easily intimidated. Distress seemed only to give him more energy. So he decided to do something drastic. He went to England to plead his cause personally. After listening to Wray and consulting Wilberforce and Stephen, the directors of the LMS contacted Lord Liverpool, requesting in the name of "British values" the intervention of the government to guarantee "the former religious toleration which Mr. Wray has heretofore enjoyed in the Colony of Demerara." Some days later they were informed that the governor of Demerara had been instructed to authorize the slaves "to assemble for the purpose of Instruction and Divine Worship on Sunday between the hours of five in the morning and nine at night, on the other days of the week between half of seven and nine at night." It was a victory for Wray and a defeat for the governor and his supporters. The issue of night meetings continued to be a point of friction between the missionaries and the colonists, but the missionaries had secured the home government's support.

The support the British government gave the missionaries could only enhance its reputation as a protector of slaves. Since the abolition of the slave trade, this notion had spread to every corner of the empire, and everywhere slaves assumed the King was on their side. Wray received vivid confirmation of this on his return voyage to Demerara. His ship encountered another vessel carrying about a hundred slaves. The captain had Spanish papers, but spoke such good English that Wray suspected he was either English or American. The captain said he was going to Cuba. Since he was not steering in that direction Wray concluded that he was probably going to smuggle slaves some place else. When Wray and several other people visited the ship they were greeted by the slaves with cries of "King George! King George!"—so far had spread the King's reputation of being a "friend" of the slaves, a reputation most abolitionists would have questioned. In any case, whatever support the missionaries received from the British government only compromised them even further in the eyes of the watchful colonists.[89]

When Wray arrived in Demerara, much unpleasantness awaited him. He was told to report to the governor. But when Wray met him, the governor asked if had brought any books with him. When Wray answered that many of his books were to arrive soon, the governor demanded to know what kind of books they were. Bibles, school books, and fifteen copies of an *Exposition on the Bible,* for "the most respectable inhabitants," answered Wray cautiously. The governor allowed that he had nothing against Bibles being brought into the colony. Then, abruptly, he told Wray to present himself to his secretary.

Wray's uneasiness only increased when he introduced himself to the secretary. The man asked him more questions: Who was he? What did he do? Where had he come from? Where was he born? Where did he intend to live? During the interrogation a former functionary standing nearby exclaimed with a jeering laugh: "Oh, it is that Methodist Preacher who left the Colony sometime ago." Wray answered that he was not a Methodist. What are you then? the man asked. "I am a Protestant Dissenter, Sir," said Wray. The men, who seemed decided to provoke Wray, insisted on knowing what the difference was. And the secretary continued to ask impertinent questions. He finally told Wray he did not think the governor would allow him to preach to "the negroes," for he often had heard the governor say he did not like to have them instructed. The incident was symptomatic of all the prejudices that surrounded the missionaries. Wray felt abused and harassed. And there was more harassment to come in the days ahead.

After Wray returned from England, a publication entitled *Cushoo: A Dialogue Between a Negro and an English Gentleman on the Horrors of Slavery and the Slave Trade* was found in the hands of one of Davies's students and was taken to the governor.[90] This was one of the many tracts the abolitionists were printing to recruit popular support. In the dialogue the "gentleman" presented all the common arguments in defense of slavery, while Cushoo exposed their absurdity. Cushoo denounced the cruelty of the slave trade, the violence of slavery, and the contradictory behavior of whites who called themselves Christians but commited all sort of sins. He concluded by stressing that the love of liberty, which according to the gentleman, Englishmen had "often spilt their blood to maintain," was a very good thing in England. But "day roast and burn us for dat in the West Indies." "Blacke' man," Cushoo said, "want to be free like white man." In the end, the gentleman, who seemed pained and shamed by Cushoo's words, announced that Parliament had agreed to put an end to these miseries as soon as possible. The pamphlet concluded with a long excerpt from William Cowper's popular poem *The Task,* calling for the emancipation of slaves throughout the empire:

> We have no slaves at home.—Then why abroad?
> And they themselves once ferried over the wave,
> That parts us, are emancipate and loosed.
> Slaves cannot breathe in England. If their lungs
> Receive our air, that moment they are free;
> They touch our country, and their shackles fall.
> That is noble, and bespeaks a nation proud
> And jealous of the blessing. Spread it then,
> And let it circulate through every vein
> Of all your empire, that where Britain's power
> Is felt, mankind may feel her mercy too.

Cushoo could not but enrage the colonists. Not surprisingly, Davies was accused of distributing to the slaves books containing dangerous ideas and summoned before the governor and court. Davies protested that he was totally unaware of the tract. He never had read it and did not know if it had ever been in his house. The governor accused Davies of being unfit to educate children, since he gave them books he knew nothing about, and threatened to revoke his license. Davies insisted that he knew nothing at all of the publication. He had never seen it, much less given it to any child. When the governor asked what sort of books he did have, Davies said he had books on various subjects, but chiefly on theology. The governor then decided to send someone with Davies to check his books. Fortunately, on their way to Davies's house, they met with the child from whom the book had been taken. "Who gave you *Cushoo?*" the man asked. "Mr. Gravesande, Sir," replied the child. The man seemed satisfied with the answer but since he had been ordered to check Davies's books he went on. He inspected Davies's library, took some of the books Davies used in the school, some tracts, and Wray's catechism, and left. Davies followed him back to the court and after they presented the "evidence," everyone seemed to be fully convinced of Davies's innocence. The governor apologized for the inconvenience, and Davies, who never missed a chance to ingratiate himself with the authorities, was quick to say that he was glad his innocence had been proven and hoped that "we shall always show that we are not the promoters of rebellion, but of order."

Governor Bentinck, however, would not let things die so easily. He sent for Gravesande, but since he too appeared to be ignorant of the contents of the book, he was released. Bentinck then decided to send copies of *Cushoo* to every governor in the West Indies to alert them of the danger of such publications. Fearing that this would injure the missionary cause in the islands, Wray reported the whole episode to the LMS and recommended that it be publicized in England, not only in the *Evangelical Magazine* but also in the newspapers which the West Indian governors read. He also suggested that the case be brought to Lord Liverpool's attention. Wray scornfully observed that the person who had first taken *Cushoo* to the governor, although a former member of the Court of Justice, was an ignorant and profane Mason, who once had been a manager or a wood-cutter, but had married a planter's widow and had become "a great man."

The LMS directors created a committee to investigate the matter. After examining the order given by the secretary of the society to send out a considerable number of books to "Gravesande Esq.," the investigators concluded that neither Gravesande's directions nor the secretary's order contained instructions for tracts of that sort. They contacted the person responsible for the shipping, who told them that the tracts had been packed merely to fill the space left in the box—a remark that could only sound suspicious to anyone familiar with the methods of the abolitionists.[91] Informed by the LMS of the incident, Lord Liverpool recommended that the society take care in the future to prevent "such accidents."

The situation became increasingly polarized. While missionaries were rallying their supporters in England, in Demerara several colonists had come forward in support of the governor's repressive policies, and both the Court of Policy and the governor sent to England their justification for the proclamation against religious meetings at night. Wray learned that the governor also had forwarded a copy of *Cushoo* to Lord Liverpool and was accusing Wray of having sent it from England. The animosity against missionaries was so intense that, for fear of the penalties that might be imposed on him, J. Wilson, a planter on whose estate Wray customarily preached on Thursdays, forbade him to preach.

Soon, however, Wray started hearing rumors that Governor Bentinck had been recalled to England. He also saw a copy of a petition that was circulating in the colony among planters, managers, and attorneys in favor of Bentinck. It said: "We are but too well aware that the Government of the West Indies colonies is becoming daily a more delicate and arduous task. The marked prejudices against their inhabitants by zealots ignorant of our local situation, who omit no occasion of misrepresentation are but too well known to the world, but prudential considerations prevent us from enlarging on the subject." The petition went on at length, praising the governor's long experience in the area, his success in putting down an inchoate insurrection in a neighboring colony, and promoting several improvements in the town and countryside. In spite of the colonists' efforts, Governor Bentinck was replaced by Hugh Lyle Carmichael. The new governor assured Wray of his protection and assistance and even promised to communicate to the Prince Regent any suggestion for furthering religion in the colony.[92]

But the atmosphere remained tense. Wray wrote to the LMS that a certain Mr. Cuming, a planter who had signed the petition against the missionaries, had said to his manager that if there were an insurrection among the slaves it would be because of the missionaries. Another had declared that they soon would all be put to death, if slaves were allowed to go to the chapel in the evenings. "People are talking this way every day at their table and at their large parties," commented Wray. "Their servants hear all they say and tell their companions and so the whites put evil design into the heads of those negroes and if an insurrection ever takes place, which I pray God it never may, the white people will have none to blame but themselves."

Recognizing that things were still not settled and night meetings continued to generate much controversy, the new governor issued another proclamation. He tried to be tactful, but tact always fails when people are so polarized. Trying to please both sides, Carmichael pleased no one. He started by saying that since the former proclamation had been "either improperly understood or willfully misapplied," he had thought it proper to issue an explanation. He then went on to say that the sole object and spirit of the British government's order was to permit and encourage the instruction of blacks in religion and morality. But that it was surely the government's intention that the "gentlemen" under whose "authority" the slaves were placed should be left free to chose the time for such instruction. This would probably be most "eligible in the day time," and under the "inspection" of any proprietor, attorney, manager, or other white person that chose to attend the meetings.

Probably trying to dispel the suspicion the colonists had shown of Wray and Davies, the governor added that it should be understood that these missionaries were "not of that sect usually called Methodists." They were "properly qualified and employed by the Missionary Society for the Propagation of Religion." He reassured the public that the directors of the society were of the "first respectability," and that their instructions to missionaries had cautioned them that it "would not be proper but extremely wrong to insinuate any thing that might render the negroes discontented with their state of servitude, or lead them to any measures injurious to the interests of their masters." The principles, catechisms, and character of those "clergymen" had undergone a minute investigation and they had been approved by the Church of England. A fair trial should be given to the "zealous exertions of those preceptors, who had been already useful in many instances to misguided negroes ignorant of the obedience and gratitude they owed to those who feed and maintain them."[93]

The proclamation was on the whole flattering to the LMS missionaries but it also tried to safeguard the masters' authority. Characteristically, Davies welcomed it. Just as characteristically, Wray was disturbed. He soon realized there were loopholes that were likely to create problems for him. And as he predicted, the morning after the proclamation, a planter informed his slaves that the governor had forbidden night meetings. Although he was willing to give his slaves a pass to go to the chapel, he prohibited them from going to Wray's house at any time. Wray was also offended by the governor's remarks about Methodists. He thought that they would displease many people at home and turn them against the LMS.

All this opposition made Wray pessimistic about the future of his mission. To make things worse, Van der Haas died and his widow seemed in poor health. Wray was afraid she would die too. Van der Haas's brother, the manager, was not someone he could trust. Wray was tired of being constantly harassed. He was also tired of living with little money. His family was growing, and (in his opinion) the financial support he received from the LMS and from Van der Haas was not enough to maintain him and the mission, particularly since the other colonists refused to help him. In February 1813 he wrote to the directors that one could not find a single planter on the East Coast of Demerara that would give ten joes a year to keep the Gospel in the country. "Planters have not got the will to support the Gospel. They have a power if they had a will."[94] Then even Mrs. Van der Haas herself began to complain that he was teaching slaves how to read. She also passed on the complaints of neighbors about the late hours of the night meetings.

When he finally received the invitation to move to Berbice in 1813, Wray was ready to go. The offer was very attractive. He was to have a house and servants and £300 a year, with sugar, rum, coffee, plantains, firewood, and the like. He was to preach to the slaves on the crown's own properties (out of the reach of colonists' persecution) and his wife was to maintain a school of "industry" to teach the girls needlework, so that they would learn to make and mend their own clothes.[95] Wray hoped that he would find more support and sympathy in Berbice, and that another missionary would soon come to replace him at *Le Resouvenir*.

CHAPTER FOUR

A True Lover of Man

Whoever replaced John Wray at *Le Resouvenir* would have to face a situation even more tense than the one Wray had found in 1808. In spite of Wray's deliberate efforts to gain the support of the planters and to convert the slaves into an obedient, disciplined, and hard-working people, the actual effect of his work had been to intensify the conflicts that pitted slaves against masters. It had antagonized planters, managers, and local authorities and made them even less willing to accept evangelical missionaries in their midst. And, by seeking and obtaining the British government's support for his mission whenever he met with serious opposition from the colonists, Wray also had deepened the gulf that separated them from the mother country. His disputes with his fellow missionaries left a religious community divided and even more vulnerable to the attacks of its enemies.

Imbued with notions of social control typical of a society of free laborers, Wray often had violated protocols and etiquettes that were essential for the maintenance of the relations between masters and slaves. He had trespassed boundaries of class and race and had challenged the masters' power and authority. At the same time, by stressing the right the slaves had to read the Bible and to attend religious services, by teaching the slaves that Sunday work was sinful, by preaching against adultery, by encouraging them to resort to the authorities in cases of serious conflict with managers—in sum by trying to implement his idea of Christian life—Wray had given slaves new motives for contention, new pretexts for resistance. In the chapel they were reassured of their humanity. They were constantly reminded of the equality of all—blacks and whites, masters and slaves—before God. And they had created new bonds of solidarity. In such circumstances, preaching to slaves had become an even more difficult and risky task.

After Wray moved to Berbice, he continued to insist that the London Missionary Society send a new missionary to *Le Resouvenir.* But he was just as insistent that they be very careful about selecting the right sort of man. In 1815, hearing that the directors were considering someone for the station, he warned them that there was a great difference between easy talk in England about being a missionary and actually working among slaves. "It is a very difficult thing to get the confidence of both master and slave and more especially when a missionary has to live on an Estate. It is also extremely important to the colony; one wrong step among slaves may cost the lives of hundreds of colonists. . . . A great deal of wisdom, prudence, patience, perseverance and coolness under contempt and ridicule are necessary to be among slaves."[1] In January of 1816, Wray wrote again, regretting that the directors found it so difficult to find a suitable person to replace him. Once again he urged caution and with his characteristic sensitivity to class and racial issues he remarked that when choosing a missionary they should remember that he would have to preach to both whites and blacks. It would be impossible to shut the chapel to whites—they would soon say the missionaries were conspiring with the slaves. And if a preacher could not address whites with "some propriety" they would only ridicule him and spread unfavorable reports. In any case, it was "a benefit to have white people hearing the exhortations to the slaves" because the whites might lose their prejudices against missionaries.[2] But time passed, nobody came, and the slaves on the East Coast remained without religious instruction. The directors of the LMS offered the post to several students at the Gosport Seminary, an institution sponsored by the society and devoted to training candidates who aspired to a missionary career. But everyone the directors selected to send to Demerara ended by not going for one reason or another. So it was pure serendipity that drove John Smith to Demerara.

In January 1816, John Smith, a student at Gosport, applied to the LMS, offering himself "for preaching the gospel among the heathen." He was a young man from Coventry, a cabinetmaker by profession, a man of few resources. Under the patronage of the Hampshire Association, he had been admitted to the seminary at Gosport only six months earlier to train as an itinerant preacher. As a reference he gave the name of the Reverend John

Angell James of Birmingham, of whose chapel he had been a member. After they received Smith's application, the directors recommended that the committee of examination inquire whether the candidate was suitable to be sent immediately to *Le Resouvenir.*[3] The committee contacted James (apparently not telling him where they were considering sending the young applicant) and he responded with a very supportive letter. So, at a meeting on January 29, the board of directors resolved that John Smith be sent to Demerara, as soon as equipment could be assembled and his passage procured.

When James heard of this decision, he was shocked. He wrote the society at once, arguing that his "young friend" was not really qualified for such a difficult posting. Whatever talents Smith had, James argued, "are yet in the bud . . . to send him out in the present state of mind will be like plucking the blossoms for the heathen instead of waiting for the ripe fruit." James's objections went deeper, went in fact to the question whether a man of Smith's social class and training, a man "from the humbler ranks of life," could gain the confidence of the Demerara planters. His warnings were reminiscent of Wray's. Although the two men lived thousands of miles apart and may never have met, they shared an acute sense of the protocols of class and race. The LMS might suppose, James argued, that a man like Smith was perfectly qualified to teach "poor negroes." But he could succeed in this only with the consent of the planters, and before such men, Smith would appear "totally unprepared."[4] When the directors persisted, James wrote again, this time even more urgently and with specific reference to the deeply troubled situation Smith was being sent into. He raised the crucial question: Would Smith be able to secure the "patronage" of the "whites" in Demerara?

> Whoever goes out to Demerara should go out as a repairer of the breach—a restorer of paths to dwell in. To do this, it is necessary he should not only be a man of peace, but a man of personal influence. No other is likely to be respected by the missionaries whom you have already there. No other is likely to repair the mischief which imprudence and imbecility have already occasioned in that quarter of the world. The missionary cause stands at this moment on the edge of a precipice at Demerara, and you want a man of great wisdom as well as gentleness to take it off. Mr. Smith, as it respects men and things, is entirely a novice. His timidity, the necessary result of a deficient education and a modest temper, renders him unfit. . . . Are there no other missionary students at the seminary, who have enjoyed more advantages than he has?[5]

James's letter may have convinced the directors; for a while, they stopped talking about sending John Smith to Demerara.[6]

Smith's letter of application had been followed by another in which he told the LMS about himself. Smith's self-portrait was conventional and cliché-ridden.[7] It sounded remarkably similar to many missionaries' autobiographies in the *Evangelical Magazine*. Like many others, he had been born to ungodly parents and had not had "the advantages of a religious education . . . therefore, nothing to counteract the evil influence of a wicked world" until he was eleven, when he did receive a "little religious instruction in a Sunday School." As a youth, "worldly gratifications" had become the object of his pursuits until the providence of God finally rescued him. From then on, his fears had vanished and he had begun to "have some perceptions of the all sufficiency of Christ to save to the very uttermost all them that come to God by him." He had decided to forsake his worldly companions and join the congregation at Tonbridge Chapel.

John Smith was born in the Midlands, probably in 1792.[8] He was from one of the many modest families that migrated to urban centers during the last decades of the eighteenth century. His father died when he was still young (some biographers say in Egypt during the Napoleonic wars) and he was left alone with his mother. In his early days, like many other lower-class children of his time, he did attend Sunday school and at the age of fourteen was apprenticed to a tradesman in London, a Mr. Davies, in St. John's Lane, Clerkenwell, with whom he continued to keep up a correspondence even after he went out to Demerara.

Although his family was not religious, Smith, like so many tradesmen of the day, was attracted to evangelicalism. When he was seventeen, he started going to churches to hear different preachers. The following year, he almost died of smallpox, and in the face of death he underwent a conversion. Then, when he was nineteen, he was admitted as a member of Tonbridge Chapel (Sommer's Town) and became a teacher in the Sunday school. He was already entertaining the idea of becoming a missionary and began reading everything he could about the work.[9] When his apprenticeship ended in 1813, he applied to the LMS.[10] What he did during the two years between this first application and his 1816 approach to the society is not clear. He may have continued to teach in the Tonbridge Chapel Sunday school. But there is some evidence that he became an itinerant preacher, and went to Liverpool, Birmingham, and Gosport.[11] When he applied again and was accepted, the LMS put him and a man named Mercer—who later would join Smith in Demerara—under the tutelage of the Reverend Samuel Newton of Witham, Essex.[12] Under Newton's direction Smith was to be trained as a missionary, while acting as an itinerant preacher in neighboring villages.[13]

Smith learned English grammar, geography, and theology. He was a good student and soon earned Newton's praise for his accomplishments. On July 15, 1816, Smith reported to the LMS committee of examination and expressed interest in going to Africa. But he was told to go back to Essex to pursue his studies. Meanwhile, the society's assistant secretary wrote Newton that Smith was not intended for any part of the world in which the "learned languages" were necessary, so he should confine himself to giving Smith "scriptural and enlarged views of Divine Truth," and to improving his capacity to communicate his ideas to others.

John Smith remained in Essex for a few months more. Then, in October, he asked to be considered for the post in Demerara. The society had still not replaced Wray at *Le Resouvenir,* so it granted Smith's request. His fate was sealed. He was to be a missionary among slaves. As soon as he was informed of the decision, Smith wrote to the directors expressing his wish for "a zealous and industrious female who might be a suitable companion, . . . likely to promote the cause of Christ." He believed he had found such a one. He had known her for about a year and a half, but had not made any "disclosure" of his intentions toward her. Like him, she was a member of the Tonbridge Chapel and had been actively engaged in Sunday school work. Smith was confident that the Tonbridge deacons would be able to answer any "inquiries" about her.[14] Soon, Jane Godden, a young woman of twenty-two, was examined by the directors and found to be an adequate wife for John Smith. Little is known of Jane Godden except that she was from a poor family, taught Sunday school, and was willing to take the risk of making her life in a strange country as the wife of a newly ordained missionary.

John Smith was ordained on December 12, 1816, at Tonbridge Chapel, and a few days later he and his wife were on the *William Nielson* bound for Demerara. Like hundreds of other young missionary couples they carried their bag of dreams. And if they had learned well their Sunday school lessons and had read carefully the *Evangelical Magazine,* they must have had their heads full of resolutions, norms, and warnings. Gravity, countenance, frugality, endurance, temperance, perseverance, self-reliance, moderation—these were qualities to be cultivated. Do not contend unless Providence calls you to it. Do not be angry on account of things not really sinful. Avoid all levity. Never hunt for vainglory and the applause of men.[15] Such norms and cautions were typical of the ethical code preached in every evangelical tract and in every Sunday school. And for the poor, like John and Jane Smith, this code was often the best asset they had to help them make it in the world. Implicit was the promise that if they behaved accordingly they might avoid poverty and degradation, and win a place not only in the Kingdom of Heaven but also in the kingdom of men. So it was not surprising that they should cleave to those notions with the urgency and commitment of people for whom there was otherwise little hope.

Armed with their convictions and formulas, Jane and John Smith were prepared to live exemplary lives, and even to suffer martyrdom if necessary.[16]

The lesson was clear in the pages of the *Evangelical Magazine:* "A missionary is a martyr, the noblest of martyrs, for he courts that martyrdom which others have only suffered; and we doubt not that the moral effect of one such example both on the church and the world, will be greater than that of five hundred sermons."[17] The martyr's reward was God's infinite protection. "No weapon against thee shall prosper and every tongue that riseth against thee in judgment I will condemn." The lesson was clear. The young missionary couple could expect that though difficulties would be many, God would lead them safely through all. He would provide for their bodies and their souls.[18] They might have to live modestly, and suffer for the sake of religion, but in the long run their efforts would not be in vain. This was the promise of the *Evangelical Magazine:* "Wealth we cannot promise as your reward. Like your Master, you will be poor. Worldly honors and dignities we have not to bestow, you may be despised and reproached for your work's sake. Earthly pleasures are not to be expected. You may be called to suffer for the cause of Christ. But we can promise you the affection and esteem of brethren and friends. . . . Every missionary who is faithful unto death shall receive a crown of glory that fadeth not away." Was there anything better to hope for?[19]

To live poorly was something neither Jane nor John Smith would fear. For both, to go to Demerara under the sponsorship of the LMS was only to exchange the uncertainties of life in London for the uncertainties of life in a different part of the world. But as missionaries in Demerara they could hope to be protected by the LMS. Their job as missionaries would at least shelter them from the unpredictability of the labor market in England. Besides, what could be more rewarding for people who had endured poverty and humiliation than to be part of a community of men and women who had as their mission to preach the Gospel to the "heathens," to rescue them from the "degradation" in which they lived, to save their souls from damnation? What could be more meaningful than "a crown of glory that fadeth not away"? To the new world they would take their notions about the dignity of man and equality before the law, their belief in the redeeming effects of education and religion, their conviction that all truth was of God. They carried to a world of masters and slaves their hope that "British liberties" and British notions of justice and fairness might be extended to all people. Most of all they carried a profound conviction that they were in the service of God, and that anyone who opposed them served Satan's purposes.

The instructions the LMS gave Smith were similar to those they issued to every missionary bound for the Caribbean. They were straightforward, but filled with warnings. Smith must have read and reread them with mixed feelings of hope and anxiety:

> You are now going, dear brother, as a minister of Christ, to declare his Gospel to the negroes. Ever remember that *they* are the first and chief object of your ministerial attention; to their conversion and edification must the energies of your mind be directed. You will doubtless have opportunities of preaching the word to the white people also; and we wish you to do this with faithfulness, prudence and affection. . . . Yet remember that as this society is formed for the purpose of spreading the gospel among heathen, and other unenlightened nations, your first, your chief, your constant business is with the poor negroes. You need not be informed that they are deplorably ignorant; you will probably find them mere babes in understanding and knowledge; and that you must teach them as you would teach children. Such discourses as might be well understood in a country congregation in England, would perhaps be unintelligible to them. You must study to exhibit the great things of the gospel in the plainest manner, and with simple easy language. By conversing with them in private you will find out what ideas and words are best understood by them. . . . Similitudes, well chosen, may be very useful. Let them be familiar allusions to what they well understand; but while they are familiar, let them not be so low or vulgar as to degrade the divine truths they are designed to illustrate. . . . The directors have long been of opinion that the negroes are likely to derive far greater advantages from catechizing, accompanied with familiar conversation, than from formal sermons, though they would by no means undervalue them; and doubtless many may hear your sermons to whom you cannot have access in private. Still, however, labour daily and diligently, visiting them from hut to hut, and receiving them at stated seasons, especially in the evenings, when they have done work, at your own habitation (a certain class, perhaps at a time) and repeat, again and again, every important truth of the gospel . . . a few leading truths, both as to doctrine and practice, well learned, in this manner, will be of more real use, than hearing a hundred discourses.[20]

The directors warned Smith that he might meet with difficulties peculiar to colonies where slavery existed: masters were suspicious and unfriendly to the idea that their slaves should be taught, and were afraid that the public peace and safety would be endangered. "You must take the utmost care to prevent the possibility of this evil. Not a word must escape you, in public or in private which might render the slaves displeased with their masters, or dissatisfied with their station." The instructions also had recommendations for Jane. "The Directors hope, that Mrs. Smith will consider herself not merely the wife of a missionary, but a female missionary," to whom the female slaves and the children were entrusted. She was reminded that she should "teach the negro mothers how to bring up their children in the fear of God," and should warn the girls against the "temptations" so prevalent in the colony.

The instructions were clear enough. What was not clear at all was whether

John Smith, or anyone else, could live up to the ideals set before him. Would it be possible to preach to the slaves according to those instructions and still avoid alienating their masters? And if he pleased the masters would he still enjoy the confidence of the slaves? Could he succeed where John Wray had failed?

With the zeal of the newly converted, and the determination and illusions of an artisan who—thanks to his own efforts—had managed to escape the uncertainties of the labor market and carve for himself a modest but relatively sheltered position as missionary, Smith embraced the independent producer ideology that promised rewards to those who cultivated discipline and self-reliance, worked hard, and saved money. Yet his experience of poverty and privation was still too close to him to be simply forgotten. From it he had learned to hate arbitrariness and impositions of any sort, and had developed an acute sensitivity to injustice and unfairness, which in turn he assessed with the yardstick of the producer ideology.

As a man of modest origins who had succeeded in becoming a missionary, Smith could not but feel an enormous pride. After all, this accomplishment had required considerable effort and self-discipline. Not surprisingly, he would demand of others what he had always demanded of himself. This tendency could only be reinforced by his piety. Because his religion had taught him to watch constantly for sin in himself and to suppress it severely, he was also watchful and severe with others. Pride, severity, an exaggerated concern with personal autonomy, an exacerbated sense of fairness, an obsession with sin—those were not characteristics that would help him to accommodate to a society where compulsory labor, personal dependence, and arbitrariness were the norm.

The LMS had warned Smith of the obstacles he would face in the colony. What he did not know was that he had been put in charge of an impossible mission. As time would show, some of the difficulties would come from his values, from his abhorrence of slavery, his commitment to abolition, his deep piety and religious convictions, his strong sense of mission. Others came from personality traits—some of which explained his successful transition from artisan to missionary—his persistence, his intense drive, his almost obsessive commitment to whatever he did, the fervor with which he devoted himself to his work, and his unyielding nature. But the greatest conflicts derived from a fundamental incompatibility between his evangelicalism and his commitment to the independent producer ideology on one hand, and the realities of daily life in a slave society, on the other. In the colony he would rehearse step by step Wray's experience, but in less favorable conditions because people on both sides had become increasingly polarized. Like Wray,

Smith would be trapped in the struggle between the colonists and the mother country; between those who wanted to create a world of free laborers and those who held onto slavery; between men like himself and Wray, who wanted to extend education to all people, and those like Governor Murray, who thought of education as a privilege of the rich and the well-born; between those who believed that all men should be equal before the law, and those who believed that some people had more rights than others. And—again like Wray—Smith would constantly be asked to take sides in the struggle between masters and slaves, a struggle that could only cease with the end of slavery.

Before he arrived in Demerara, Smith was already prepared to see the planters as sinful, godless people, and the slaves as helpless and innocent victims of oppression. That, after all, had been a central theme of the antislavery campaign for several decades. For Smith, as for all those hundreds of working-class men and women who signed antislavery petitions, the abolitionists' rhetoric had a strong appeal.[21] The year before Smith left for Demerara, the abolitionist campaign had reached another peak. The Anglo-French Treaty of Paris in 1814, allowing France to reopen its slave trade for five years, triggered a flood of petitions from all over Britain. The abolitionist networks mounted a vociferous opposition to the ratification of the treaty. Within two months almost eight hundred petitions (containing one and a half million names) denouncing the relevant clauses of the Paris treaty were received by Parliament.[22] And slavery and abolition were widely debated in the press. The mass campaign of 1814–15 culminated in debates around Wilberforce's motion aiming at creating a system of slave registration that would allow the government to monitor the pattern of births and deaths and to ensure that slaves were not smuggled into the British colonies.[23]

Since the abolition of the slave trade, emancipation had carried an aura of respectability. It was supported by members of the nobility and important figures in the Church of England. Antislavery had also come to serve as a pretext used by different politicians to mobilize public support. Most of all it had found a large number of followers among evangelical groups, which used their networks very effectively, first to campaign in favor of the amelioration of slaves' living conditions and then for emancipation. The abolitionists' rhetoric had a particular appeal to British artisans and workers to whom words like exploitation of labor, oppression, and unfairness sounded particularly compelling, at this early stage of industrialization, a time of profound social dislocations and radical mobilization, when the resentment of the poor and dispossessed was fanned by the winds of revolution. Antislavery notions had spread among working men and women who either followed secular reformers who spoke the language of the universal rights of men, or attended dissenting congregations and joined societies for moral improvement and useful knowledge.

As a former artisan-apprentice and as an evangelical missionary, Smith had been particularly susceptible to the antislavery campaign, which had produced the most inspired speeches in Parliament, filled the pages of the newspapers, and generated numerous petitions. Abolitionist discourse provided him with a code with which he assessed the slave experience in Demerara. It legitimized his own feelings of hostility against oppression and arbitrariness. It made him see whim and injustice where others saw necessity, and history where others saw nature. Not surprisingly, he would find confirmation of the abolitionists' depiction of slavery in his day-to-day experience in Demerara. The arbitrariness of slave managers and owners, the abuses of colonial authorities, the violence of the slave system, the immorality of the colonists, and their disregard for Christianity—all would be vividly displayed before his eyes.

On their side, the colonists were also prepared to see him as an enemy. During the year before Smith arrived, all their traditional biases against evangelical missionaries had received a new boost. There were bitter rumblings about a slave revolt in Barbados that had resulted in the burning of almost 20 percent of the island's canes and the deaths of about a thousand slaves killed in battle and executed under martial law. For several months the word of mouth was that "Methodists" had been responsible for the rebellion—until it was proven that there had been no missionaries in Barbados for more than seventeen months. The colonists then blamed Wilberforce, the African Institution, the debates on the "Registry Bill," and the campaign for emancipation, arguing that the controversy over those issues had sown the spirit of rebellion among the slaves.[24] The suspicion with which Smith approached the colonists, and with which they, in turn, awaited him, did not bode well for his mission.

On board the *William Nielson,* Smith could only wait with trepidation for the moment he would arrive in Demerara. A year later he would recall the "distressing fears" that had haunted him and his wife during the voyage, "when all was terror and alarm for our safety." And he was grateful, since "He who hath the winds in the hollow of his hand and at whose voice the waves are obedient," had sustained their hope and their vessel and had brought them safe to Demerara.[25]

The voyage took several weeks. There were days of calm when the breeze died away, the sea subsided for a while, and nothing was heard but a faint rippling against the hull. On those days, when the ocean looked like glass shining in the sun, the Smiths could get out of their small berth and walk on the deck. Sometimes they amused themselves by watching the dolphins playing around the ship. On calm nights, they could watch the falling stars or the moon rising from the ocean. As the passage wore on, they could see Polaris sinking into the depths behind them. But there were days when everything seemed threatening, when the young couple watched with apprehension as the heavy clouds gathered on the horizon, and the ship rocked and creaked in an unusual way, wrestling with the waves, with the captain shouting and sailors yelling. These were days of real discomfort and terror: a dark sky and sea, winds howling, and masts moaning, while the waves crashed against the vessel's frail planking. On such days they wondered if they would survive, and they could only pray and hope that the Lord would be merciful and guide them safely to their station.

Finally, on February 23, after a rough passage, they saw land for the first time in more than eight weeks. Dark blue water gave way to a muddy yellow-brown. At first, all they could see was a low stretch of land covered with thick masses of foliage. Swarms of birds were everywhere. The beach appeared to consist of mud, with a few intervals of sand. A fresh breeze pushed the *William Nielson* into the four-mile-wide estuary, through the channel that crossed the long mud and sand bar that almost blocked the entrance of the Demerara River. Soon they could see large mansions painted white, with red roofs, and clusters of little cottages, all surrounded with cabbage palms (*Area oleracea Jacq.*) and palmettos. Boats of all sizes and types were plying up and down the river, most with crews of half-naked blacks and mulattos singing strange songs.[26] Then, on the eastern bank, bordered by hundreds of masts of merchant ships, schooners, and sloops at anchor, they spotted Georgetown.

After such a long sea trip everything must have seemed welcoming, even the warehouses and wharves, the glaring sun, and the noisy crowds of blacks, they were seeing for the first time. The contrast with the country they had left behind on a cold and gray winter day could not have been greater. But there was no time for astonishment or perplexity. Elliot and Davies, the two other LMS missionaries stationed in Demerara, had come to the harbor to welcome them. The future was full of promise and hope.

Georgetown's main street was crowded with blacks on foot and whites in buggies or on horseback. Most black men wore only a cloth about their loins and many women had nothing on but a short petticoat, tied over the hips and reaching no further than the knees, with breasts exposed. A few men had on cotton jackets and trousers, and a few women were dressed in cotton prints, sometimes white muslin, with their heads enveloped in colorful striped kerchiefs worn turban fashion. The dirt street along which the bewildered missionaries passed seemed interminably long. On each side they saw handsome dwellings, standing in enclosed courtyards planted with shade trees and flowering shrubs. There were oleanders, hibiscus, orange and lemon trees, bananas, and great palms. The houses were built of wood on brick foundations. Venetian blinds covered the windows. In the center of the street was a canal planted on each side with smallish trees, and crossed at regular intervals by bridges. The scene was joyful and lively.

On the waterside stood the warehouses of Robb Stelling, Donald Edmonstone and Co., and many others. Further up the road was a large, old, barn-like building where government business was transacted and where public functionaries had their offices. Near a wooden bridge over a canal was the store of McInroy, Sandbach and Co., one of the richest and most influential commercial houses in Demerara.[27] The town had been originally laid out in lots of 100 by 200 feet, but many had since been divided into half- and quarter-lots, and in those places the buildings were crowded together.

Sunday was market day, when many slaves came to town to sell the produce of their gardens. At the marketplace, plantation slaves sold fruit, vegetables, fowls, and eggs, and town hucksters had stalls offering salt beef, pork, fish, bread, cheese, pipes, tobacco, and many other articles of European manufacture. Most hucksters were free women of color who purchased their commodities from merchants on two or three months' credit and then sold them at retail.

A mile from Georgetown, to which it was linked by an excellent carriage road, was Kingston, a small village with very neat and good houses, also painted white. Situated at the mouth of the river, it had a pretty view and was quite open to the sea breeze. A little farther away were two other small towns, Labourgade and Cumingsburg, the most elegant of all, where most of the rich people lived. Between Georgetown and Labourgade there were Bridge Town, where the free black and mulatto hucksters lived, and New Town, with merchants' stores, retail shops, goldsmiths, watchmakers, hatters, apothecaries, cigarmakers, and other small tradesmen. In New Town there was a large wharf belonging to the merchants called "the American Stelling," where small vessels were loaded and discharged. Everywhere there were canals, wharfs, and warehouses. Georgetown and the neighboring districts gave an impression of intense commercial activity.

The Smiths spent their first night with the Davieses before setting out for *Le Resouvenir.* The next day they called upon Van Cooten, the plantation's attorney (since the owner, Mrs. Van der Haas, was in Holland). Smith delivered his letter of introduction from the LMS, and the attorney welcomed him with an assurance that he would receive the same support which had been given to Wray under Post's will—1,200 guilders (about £100). Smith was pleased. But when he reached *Le Resouvenir,* joy and excitement gave place to grief and disappointment. Both the chapel and the house were in a terrible state. After Elliot had left, the slaves had appropriated the lumber to repair their huts. Smith would have to buy timber and hire a carpenter to do the repairs. The house had no garden, only a small piece of ground where slaves kept their fowls (he thought it would not be prudent to drive them away). There was no furniture, so he had to purchase many things in town. But first he had to learn about the local money. This was no easy task since coins from different countries circulated freely in Demerara. The smallest amount in circulation was 5*d.*, or one bit, and the largest 5*s.*, or three guilders. Sixteen pennings made one stiver (1*d.* sterling), five stivers, one bit (5*d.*). Twenty stivers, or four bits, made one guilder (20*d.*) and twelve guilders made 20*s.* But there were also other coins such as dollars, which represented about 5*s.* each; gold Portuguese coins called ducats, the equivalent of 9*s.* each, and johannes, or joes, which were the equivalent of 36*s.*[28] Smith had to translate all these different coins into pounds, shillings, and pence. But when he did it, one thing became clear: everything was very expensive. He gave a note on the LMS for £100 to be paid in three months. He apologized to the society for spending such a large sum and explained that everything was extravagantly dear in the colony: "Thirty pounds in England would be equal to one hundred in Demerara."[29]

More difficulties and disappointments lay ahead. When Elliot introduced him to the governor, John Smith was met with what seemed to him a cool reception. The governor said he would not allow him to preach without a license—which Smith did not have—and told him never to teach a single black to read. "I have been informed," Smith reported to the LMS, "that the planters will not allow their negroes to be taught to read on pain of banishment from the colony. Here are thousands of people thirsting for the word of God, I can plainly see that past disputes have had a very unhappy effect, upon blacks and whites. May the God of peace restore and preserve peace."[30]

The governor's cool reception was compensated by the warm welcome the slaves gave him. On March 7, Smith told a few blacks he would preach for the first time that night. He asked who had been sexton during Wray's time and the man came and lit the candles and rang the bell. Soon several slaves came to greet him. At half past seven Smith went to the chapel and to his surprise he found four hundred people waiting for him. Two days later, on Sunday, the chapel was again nearly full, even though it was a wet morning. After the service he catechized the people and in the evening he preached again to a large congregation. He was much pleased with the appearance of the people, the women "for the most part dressed in white, with a white or colored handkerchief twisted around their heads like a turban," almost all without shoes, the men in white or blue jackets and trousers.

John Smith

This engraving, adapted from a miniature on ivory (artist unknown), appeared in David Chamberlin, *Smith of Demerara (Martyr-Teacher of the Slaves)*, published in London in 1923—precisely a century after the Demerara rebellion. In all likelihood, the original was a conventionalized miniature of the type the London Missionary Society had painted of all the missionaries it sent abroad.

A few days after Smith began to preach, Wray came for a visit, and before he had been in the house five minutes the room was half filled with slaves who had come to greet him. The next day both missionaries preached, and the chapel was so crowded that many could not get in. Smith felt very happy, but the feeling was not to last. Wray decided to pay a visit to Van der Haas, the manager, and found him not at home. When Wray got back to Smith's house, however, he received a note from the manager threatening to turn him out if he came to his house again. Such blatant hostility toward Wray could easily turn against Smith. He started hearing rumors that some of the managers in the neighborhood were coming to the chapel in disguise to see

Plantation *Le Resouvenir*, 1823

From left to right, the buildings are: the master's house, the mill, Bethel Chapel, and John Smith's house. The unknown artist chose a Sunday, when the slaves were not working, for this somewhat idyllic picture. He also chose to show only slaves, and to have their informal procession from the chapel place a heavy pictorial emphasis on its role in their lives. Reproduced from Chamberlin, *Smith of Demerara*.

whether he attempted "to make the slaves dissatisfied with their condition." He began to worry about his license. He did not have one yet and he knew that many whites in the neighborhood hated religion and all those who preached it, and they would be only too glad to see the governor withdraw his protection. Smith did not want to give them any pretext.[31] Two months after his arrival, he wrote to the LMS that

> Much prudence is requisite in this country. I pray and hope my friends in England pray that I (and all the missionaries here) may be embued with that wisdom and prudence, which is so necessary to direct a missionary in any heathen country, but more especially in such a country as this, where we have *slaves* to instruct, and masters to please, where almost every planter looks upon a missionary as one who aims at nothing less than the entire subversion of the colony. The white people will not come to hear us themselves, but stay at home and say we preach mischief, that we are in league with Mr. Wilberforce and that nothing but the total emancipation of the slaves will satisfy us, and those of the gentlemen who are not so violent against us say that our intentions *may be* good, [but] our teaching will ultimately have a bad effect. So that I have not met with one *single* planter who can be called a real friend to the instruction of the negroes. Mr. Van Cooten, the attorney for the estate, is rather friendly to it, but I am fearful he does not feel a lively interest in it, as the late very excellent and pious Mr. Post did. Mr. Van der Haas, the manager (brother to Mr. Van der Haas the owner of the estate) allows the negroes to come to the chapel but like the rest of the whites never comes himself.[32]

Having heard of the problems Wray had faced while at *Le Resouvenir,* Smith was determined to avoid them. Because he depended on the patronage of the whites, he tried not to alienate them. He carefully omitted from his sermons passages from the scriptures that could lead to misinterpretation by both masters and slaves. Sometimes he worried that he might have made a mistake: "Having passed over the latter part of chapter 13 as containing a promise from the land of Canaan, I was apprehensive the negroes might put such a construction upon it as I would not wish for. . . . It is easier to make a wrong impression in their minds than a right one," he wrote in his diary on August 8, 1817. A few months later, noticing a white in his congregation on a Tuesday night—when whites seldom came—he thought the man might be there to catch "some expression which he might make use of in order to impeach my veracity. . . . But blessed be God, I am animated by a nobler motive than to teach what they call here 'the principles of revolt.' I wish, and pray, and labour to train the souls of the negroes for heaven. These men who look upon the missionaries as dangerous characters are themselves generally influenced by sordid principles, so they measure us by their own rule."[33]

But how could he prevent people from interpreting his words the way they wanted? The metaphorical language of the Bible gave margin to many different interpretations. Slaves heard what they wanted to hear and had definite preferences for certain passages of scripture. Some days they did not seem to remember anything he preached, but at other times they showed an extraordinary memory. One Sunday after preaching, Smith went to catechize the slaves and realized that not one could tell what the text was he had been preaching on just five minutes earlier. He noted with some exasperation in his diary that he had to repeat it twenty times.[34] But on another occasion, when he preached on Isaiah 9:17—which he freely interpreted as meaning that "the wicked shall be burned into hell"[35]—he noticed that the people were particularly attentive. When he questioned them in the evening, they could recall perfectly well the text they had heard at noon.[36]

It was even more difficult to avoid giving the wrong impression to the few whites who occasionally came to the chapel. They were mostly managers and overseers from neighboring plantations, and very few were "friends of religion." Some came only to watch him or to see if slaves had disobeyed their orders and gone to the chapel. And when they spotted any of their slaves there, they were sure to inflict severe punishment. Whites often behaved offensively in the chapel. One day, some whites were laughing while he was preaching. Another day one was shaking his "bunch of seals at the end of his watch chain," making such a noise that Smith felt troubled by it. "The behavior of most of the whites is so unbecoming in a place of worship, that I sometimes wish they would not come at all," he wrote in his diary.[37] He was even more irritated on a Sunday when managers and overseers were playing next to the chapel what Smith thought was "a childish and ludicrous game." They spent the whole day, from nine to five, bowling—making such a noise that he could hardly concentrate on what he was doing.[38] On another occasion, Hamilton, the manager of *Le Resouvenir,* seemed so offended by Smith's preaching on the conduct of the unmerciful servant (Matthew 18: 21–35, which concluded with the warning: "So likewise shall my heavenly Father do also unto you, if ye from your hearts forgive not every one his brother their trespasses") that he left the chapel evidently in a rage, slamming the door after him and remaining outside until the service was over. And, just as Wray had, Smith felt at loss when he had to preach the commandments to managers and overseers who lived so openly in sin.

Preaching under such constant worries and distractions was not easy. But, aside from the feeling of always being watched and of always walking on a tightrope, there were other more mundane things that made his work particularly strenuous. As soon as he arrived in Demerara, he started complaining of "moschettos" bites. "I am much annoyed by the moschettos. They have bitten me so much that my shoes have become too small for me in consequence of my feet being swelled," he wrote in his first letter to the directors.[39] Some time later he wrote in his diary, "Moschettos so troublesome that our service was quite unpleasant. The negroes come to chapel on a week day more than half naked. Their legs and thighs being bare, the moschettos sting them, which caused them to . . . a sudden slap." The "moschettos" were such a plague that sometimes in the wet season, Smith was forced to sit on the bed under a net to prepare his sermons. There were also cockroaches, and beetles which flew about the pulpit when he was preaching, driving against his face and getting down his neck and chest. From the outside came the sound of the crappos and frogs, and sometimes the noise of the cattle mill grinding coffee, the cracking of the whip and the cries of the people, all diverting his attention from his subject. At nights the chapel was gloomy, with only a few candles to light a place that was big enough to seat seven hundred people. At daytime services on Sundays, he was frequently annoyed by fowls and lizards that wandered in. And when it rained the pulpit was often drenched. Quite a different scene from Tonbridge Chapel.[40]

Smith was also driven to exasperation by the gobbling of the turkeys that wandered loose in his yard. They belonged to Romeo, one of his deacons. Smith told him again and again to take them away, and even threatened to kill them, but Romeo's turkeys continued to come. One day when he was preparing his sermon, Smith became so irritated by their constant noise that he killed one—which much displeased the old man.[41] Some months later, someone got into the fowl house and stole more than a dozen of the Smiths' best fowls. "I very much suspect Romeo to be the thief," he wrote. "I believe there are but a few of the negroes who think it much of a sin to rob a white man."[42] Much later he learned that the slaves liked to keep their chickens and turkeys in his yard because they feared the fowls would be taken by other slaves if they were left at the slave quarters while their owners were away.[43] Smith was also troubled by the way holidays were celebrated in the colony. At Easter, Whitsunday, and Christmas, managers gave rum and sometimes tobacco to the slaves, who then spent the whole night drinking, drumming, and dancing. Some got drunk and the next day were severely flogged for not showing up for work on time.

Nothing, however, could discourage the missionary couple. They were driven by a profound commitment to their mission and were determined to devote all their time and energy to the salvation of the slaves' souls. During his first month Smith held meetings almost every day. On Sundays he held three services. At seven in the morning, the sexton would ring the bell and hoist the flag. When the slaves arrived Smith preached, and after the service he catechized them. He preached again at eleven and a third time in the evening. On Monday evenings he catechized the slaves in the schoolroom, while Jane catechized the children in the hall. On Tuesdays he preached to the slaves in the morning, and on Thursdays he taught the people singing. Friday nights he held a prayer meeting and sometimes he also preached. On Saturdays, Jane held meetings for the women while John held a "church meeting." Wednesday was the only day they held no services—although sometimes he catechized on Wednesdays too. On most Wednesdays the Smiths went into town on errands, to pick up mail or to buy groceries and supplies for the chapel. They usually stayed overnight with the Davieses, and John assisted Davies in his chapel. Sometimes they spent the night at the Elliots' instead.

Commitment and disposition to work hard, however, were not enough to guarantee the success of the mission. The missionaries had to show great flexibility and capacity to adjust to the slaves' rhythm of life, to the rules and protocols imposed by masters and royal authorities, and to the practical routines of daily life in an environment for which they had not been prepared.

Initially, when the Smiths wanted to go to town, they would get a ride in the manager's or the attorney's chaise. And before their relationship with their neighbor Michael McTurk went sour, he gave them a lift from time to time. But Smith soon realized he needed a horse to go to town and to visit other plantations. He did not want to baptize slaves without their masters' permission and was eager to gain their approval and support for the mission. Besides, any white man without a horse would earn no respect in the colony. So he bought a horse and a chaise. But because he had no experience with such things, he met with much trouble. Once, his horse backed the chaise into a trench and he was only rescued by the intervention of McTurk and Van Cooten. A few days later his chaise and horse fell again in the trench and it took ten slaves about an hour to get them out. After all the nuisance, he sold the capricious horse and bought another one. Eventually, he did learn to ride and to drive the chaise.

Gradually, the missionaries adapted to their new lives. Smith learned that he could not expect the slaves to attend services more than twice a week. Slaves had too much to do and worked until late on weekdays. He also had to start his Sunday services later because slaves received their allowance of saltfish on Sunday mornings. So he adjusted his schedule to their needs. He decided to hold "church meetings" on Sunday afternoons, after services,[44] instead of Saturday nights, because planters were complaining that he kept slaves until late hours.[45] When he did hold night services he tried to finish no later than nine o'clock, and when he became aware that the slaves at night were too tired to hear long sermons he limited himself to fifteen or twenty minutes. After having been in the colony for a few years he even decided to give up evening services during the three-month-long coffee harvest, since the slaves spent long evenings pulping the coffee berries they had picked during the day.[46]

With the concern and care of a devoted minister and teacher, Smith tried to adapt his teaching to the slaves' interests and understanding.[47] He made frequent comparisons between Biblical events and their own lives, and when he realized that the slaves did not care for religious tracts he stopped reading from them, and chose to read instead from the Old Testament, which the slaves seemed to like better. "I commenced preaching on the historical subjects of scripture," he wrote in 1821. "This I intend to continue occasionally perhaps twice a month. These subjects I have already selected and arranged, about 100 in number. In this way of preaching I trust important truths of revelation will be fixed upon the minds of the people in a more pleasing and impressive manner than if delivered in an abstract form."[48] Sometimes Smith read from *Pilgrim's Progress,* sometimes from the *Evangelical Magazine* or other journals, always selecting texts he thought would make his preaching more attractive to the slaves.

He also told them stories about the progress of the Gospel in Africa, something he thought they were curious about. His instinct seemed to be correct, for he could write in his diary that "Having given notice on Tuesday of my intention to read some religious intelligence respecting the advancement of the Redeemer's cause among the Heathen nations, a great many people came to the chapel."[49] Once, when he preached from Isaiah 49:23—"and Kings shall be thy nursing fathers"—he read to the congregation letters from the King and Queen of one of the Sandwich Islands to the King of Madagascar.[50] Another time he read the congregation a letter someone had brought in. It was written by W. F. Corner, a missionary in Africa, to his mother, Nancy Corner, who lived in Demerara. The story went that he had been born a slave on plantation *Thomas* in Demerara. His father had been offered freedom but had asked the master to free his son instead. The master had agreed and Corner had been sent to England, where he had joined a missionary society and later had gone to Africa to preach the Gospel.

There were many interesting stories to tell, and Smith recorded them carefully. One of the most exciting was that of Dora, whose son had been kidnapped in Africa. After a number of years, she herself had been kidnapped and brought to Demerara. Years later, her master bought several slaves, including her son. Mother and son had been finally reunited. (Dora, a "Dutch slave" who understood little English, was still living in Demerara in 1821.)[51] Such stories could not but appeal to the slaves, and Smith's willingness to use them was one of the reasons for his success as a preacher.

Following the LMS's instructions to the letter, Smith always tried to illustrate his lessons with examples taken from the slaves' lives. He selected texts which seemed most appropriate to their circumstances[52] and spoke in simple ways. He always quizzed them to find out what they had learned and encouraged them to ask questions of their own.[53] Preaching to the children was one of his favorite activities. On Easter Sunday, 1818, Smith addressed five hundred children and was pleased to see that although some "fell to fighting, others sat with their mouths wide open and scarcely took their eyes" off him.[54]

As an Independent and a good congregationalist, Smith gave great autonomy and initiative to the members. The month after his arrival at *Le Resouvenir* the congregation chose five deacons—Romeo from *Le Resouvenir,* Barson and London from *Beterhope,* Bulken and Quamina from *Success.* Deacons played an important role. They served as mediators between Smith and the slaves. It was their responsibility to bring to Smith those who wished to be baptized. The missionary then assigned a slave who could read to teach them the catechism. Once they had learned it, they were examined by two deacons, and then by Smith, who, if he found the candidates ready for baptism, set a date for the ceremony. Within six to twelve months after they were baptized, slaves could apply for full membership in the congregation. Once again they were examined by the deacons and by Smith and, if approved, were invited to a special church meeting in which the missionary and several chapel members, including the deacons, spoke on their behalf. Finally, at Smith's request, the whole congregation voted by raising hands whether to accept them.

Deacons were also in charge of keeping the missionary informed on the state of the congregation. If a deacon reported that a member of the chapel had misbehaved, Smith would not allow the person to participate in communion. At the sacrament, the deacons were to hand bread and wine to their brothers and sisters. They were also responsible for collecting money for the chapel. The collection was made once a month, when the Lord's Supper was administered and only the communicants were expected to contribute. It was also the deacons who appointed the catechism teachers on each plantation, although in principle their appointment had to be confirmed by Smith. All such functions gave deacons a privileged position.[55]

Quamina, a slave carpenter from *Success,* soon became Smith's favorite deacon. He reported to Smith on the slaves' behavior and on the problems they faced from day to day. He would tell why some slaves had not come to the chapel, who was in the stocks, who had been flogged, who had committed adultery, or what managers had prohibited the slaves from attending services. Smith always consulted Quamina when he wanted to know something about a member of the congregation. In time, both of the Smiths came to trust Quamina's wisdom and admire his piety. No week would pass without Quamina bringing them some useful information. He was the most loyal, well-behaved, trustworthy and pious deacon. Quamina seemed to be capable of the kind of devotion Smith always admired, and sometimes even envied.

Following a common practice in Protestant churches, Smith encouraged male slaves to pray aloud, and he was often moved by their fervor.[56] As an evangelical he saw in the slaves' emotion an evidence of the strength of their faith. At times, Quamina was so overcome with emotion that he could hardly finish his prayer. One day, after a prayer meeting during which the two deacons, Quamina and Romeo, prayed, Smith wrote in his diary that he was much pleased with the affection and simplicity of their prayer.[57] Another time, after he administered the Lord's Supper, he could not but contrast his own coolness with the profound emotion experienced by the blacks. He wrote in his diary, "Oh Lord, take away this heart of stone and give me a heart of flesh." He talked about their "countenance," so expressive of grief and sorrow for their sins and the suffering of Christ, and confessed that this sight excited profound remorse in him. "Oh, how slow are my affections, how earthly my mind, how insensible my heart! These people who can have only an imperfect idea of the nature and extent of the sufferings of Jesus are dissolved into tears, while I, who read and meditate and preach upon the sorrow and tears of the man of grief, can scarcely feel one tender emotion."[58] A month later, he wrote that during baptism several adults were so affected that they could not hold up their faces, so he had to pour the water on their bowed heads.[59] A few months later, he wrote again:

> This morning one of the negroes engaged in prayer and while thanking God for presenting his Gospel here, so precisely described the circumstances of my leaving my friends, the tender anxiety of my mother, my own feelings together with the uncertainty of future events in a strange land that this descriptive thanksgiving affected my mind more deeply than anything had done since I have been in this country.[60]

Smith had gone to Demerara to convert the slaves to the cause of Christ but he was being converted to the slaves' cause. He was always profoundly touched when the slaves described in their prayers the persecutions they suffered for the sake of religion. Once, when a slave prayed that God would put an end to the planters' opposition to religion he gave so many details about the various "arts" employed by managers and masters to keep slaves from the house of God, and to punish them "for their firmness in religion," that Smith could not help thinking on the teaching of Exodus 3:7–8, that the time was not far distant when the Lord would make it manifest by some signal judgment that "he hath heard the cry of the oppressed."[61] Such shared emotions brought him even closer to his congregation, so close that sometimes he seemed to forget he was a white man, and referred to "the whites" just the way any black in Demerara might have done. Once, after meeting a man known for his hostility to religious instruction, Smith wrote:

> I conversed with him on the subject and he appeared friendly to the object, rather than opposed to it, but there is no knowing what his real sentiments are upon this for the white men sure will speak fair to one's face and when in company with others like themselves they use thin tongues against us like poisonous arrows.[62]

However cautious he may consciously have been of not wanting to alienate the whites, Smith's heart was with his flock. In spite of his patronizing attitudes, his obvious sense of superiority, and occasional bouts of racism, which once even led him to compare some half-naked slave women who were washing clothes in the river to orangoutangs,[63] Smith sympathized with the slaves' predicament and condemned their oppressors.[64] And there was no shortage of passages from the Bible that could be used to condemn the inequity, cruelty, and immorality of managers and masters, and to bring solace to slaves. If Smith did not use them, his congregation would.

For the most part, slaves seemed to reciprocate Smith's feelings of sympathy. They came to the chapel in increasing numbers and responded generously to his appeals for donations. From the money they earned selling produce from their gardens or performing small services on Sundays, they sometimes managed to gather more than £100 for the LMS.[65] They collected money to repair the chapel, and brought him gifts from their gardens—yams, potatoes, fowls—as tokens of their appreciation. One day several slaves brought him a pig. They said they had jointly purchased it for a few dollars. He suspected they had stolen the pig, but hadn't the heart to question them further.[66] Like Wray (perhaps even more than Wray), Smith seemed to have gained the slaves' trust—although, as events would prove, they never trusted him entirely. And (again like Wray) he was inevitably drawn into their many conflicts and afflictions.

"A missionary must in many circumstances act the part of a civil magistrate," Smith wrote in his journal just a few weeks after he had arrived at *Le Resouvenir*. On that day, he had been asked to settle several quarrels. Many husbands and wives had complaints to make. Some were jealous, some complained of being abused. Gingo and his wife had come to the Smiths' house in the evening to settle a dispute. The Smiths and Wray (who was staying overnight) were to act as mediators. Gingo and his wife belonged to two different plantations, and although she was allowed to visit him, he was not allowed to visit her. This created a lot of stress. Gingo's wife said that he wanted another wife. Gingo responded he had discovered his wife with another man. After hearing both sides, the missionaries concluded that they were both at fault, and after an hour of talking with both of them, the couple finally agreed to stay together—an outcome that seemed to please Smith.[67]

But this was only the beginning of his work as a "magistrate." What he did not realize was that, as with Wray, the slaves had cast him in a role common in many African societies, that of mediator.[68] In the years to come he would be asked to play this role time and again. Every month, sometimes every week, there were similar cases. A few days after Gingo's episode, Joe of *Success* came to complain that his wife had left him. Apparently he had beaten her and she had run away, so the manager had punished both of them by giving her fifty lashes and him thirty. Joe wanted another wife.[69] Four days later, Smith found himself embroiled in another case. Hector brought his wife Juliet to speak to Smith. Hector had two wives. He had taken up with Juliet when she was very young. After having five children by her—all dead by then—he had taken another wife, and had brought her home to live with Juliet. The two women were continually quarreling, which made Hector unhappy. Hector said he loved Juliet but could not send the other away because she was with child. Smith handled this case with remarkable flexibility. "This is a difficult case," he wrote in his diary, "I told them they must make peace and live without quarreling, which they promised to do."[70] Some time later he complained that he found it very difficult to have to settle disputes between wives and husbands. A man who had once been a member of the chapel came to him with his head cut "in a most dreadful manner." He told Smith that his wife had attacked him with a shoe brush. In spite of Smith's advice, he went home saying he would "bring blood out of his wife's head." Both came back later, and it took three hours for Smith to calm their dispute.[71]

The missionary was also asked to intervene in quarrels among male slaves. Sometimes, it seemed to him that the slaves were always fighting with each other. One day, Emmanuel and Bristol from *Chateau Margo* complained to him of Coffee from *Success*. "I declined hearing the tale out until I can see Coffee," Smith wrote in his diary. "When shall Christians love each other?"[72]

One of the most disturbing cases Smith had to solve in the first months after he arrived had to do with a slave he had hired as his own servant. The man was incorrigible. He already had three wives, but was always chasing other women and always getting into trouble. Welcome, a slave on a neighboring plantation, came one day to complain that he had caught the man three times with his wife Minkie. A few days later, on a Sunday just before a holiday, Jane Smith found the man and the maid "in a secret place," mending their clothes for the holiday.[73] So many were the problems the servant created that one day Smith lost his patience and had him put in the stocks. After that he decided to send the man back to his master. To Smith, the whole incident must have been very annoying—perhaps even a little frightening. There was nothing he resented more than the violence he saw around him, and there he was, doing the same thing he had always condemned in others.[74]

But the greatest scandal Smith found himself involved with arose when

Susanna was "seduced" by Hamilton, the manager of *Le Resouvenir.* Susanna's affair agitated the whole congregation. It was hardly the first time that a slave woman was sexually involved with a white man. There had been others before, like Genney, a fifteen-year-old who had gone to work in town and had been "seduced" by a Mr. Jemmet.[75] And there would be others after. But the case of Susanna was the first to happen to a woman in Smith's congregation. He first heard of it from Quamina, in April of 1821. Several members of the congregation had met to discuss the best way of telling Smith about the affair. They finally had decided that Susanna herself should tell him. But when Quamina told him the story, Smith was so furious that for a moment he thought that if Susanna did show up he would turn her out of the house. He also worried about how her husband, Jack Gladstone, would take the news. Disappointed with Susanna, Smith thought he could have confidence in no one.[76]

The days passed and Susanna did not come to see Smith. Having calmed his hostile feelings, he sent for her. When she arrived, he told her to quit her relationship with Hamilton and threatened to exclude her from the congregation if she persisted. But he saw no sign of repentance in her. She said that Hamilton had gotten her into his bedroom, shut the door, and by "absolute force abused her." Smith thought how ironic it all was. He remembered that on Christmas day the previous year, Hamilton had told him he had given a slave 150 lashes for forcing a "negro girl to uncleanness." Still, Smith blamed Susanna for not complaining of Hamilton and for staying in the house afterwards.[77] Since the conversation with Susanna had no result, a month later she was excluded by the unanimous voice of the congregation after a tense meeting. One or two of the female members "spoke with a view of extenuating her fault." One of the deacons who had been in charge of persuading Susanna to leave Hamilton made two or three attempts to speak during the meeting, but his tears prevented him. Smith was particularly saddened because he had always thought that Susanna was in every respect one of the most promising members of the congregation.[78]

One year later a similar case happened, this time with an outcome more pleasing to Smith. Before the assembled congregation, Bill accused his wife Betsy of adultery. She responded that for a full year her manager had been trying to entice her to "cohabit" with him. She had always refused, and partly because of the harassment she had asked a friend to request her master to hire her out. Hearing that she was about to leave the estate, the manager called her into his house in the evening. When he had got her in, he shut the door and "by absolute force ravished her." She was too ashamed to tell anyone. She was ready now to beg her husband's pardon and to promise that nothing of the kind would ever happen again. She dropped to her knees and, weeping, acknowledged "her error" and "her fault" and promised "all that

Bill could wish," a behavior that not only Smith but the whole congregation seemed to find both appropriate and moving. Smith wrote that no one could hold back tears.[79]

"Adultery" and "fornication" among the slaves were even more common than such incidents involving managers and slaves. So, in an attempt to discipline his congregation, Smith decided after a while to establish rules for exclusion from and admittance to the chapel of people who "lived in sin." He stipulated that no one living with a man or a woman without being married should be admitted to the chapel unless they consented to be married. Any man or woman who put away his wife or her husband without having the consent of the congregation would be excluded, and any single person who took a companion without being married would also be excluded.[80] But such rules only seemed to add more problems for him to solve. Now every instance of "adultery" had to be brought before the whole congregation, and some of them were difficult to resolve. The case of Felida and Hay was one of them.

Hay had been sent to work in town. He was there for six weeks and during that time Felida "cohabited" with another man. Hay was told of Felida's affair and pressed Felida, who admitted it. Hay wanted to know whether he was obliged to "put her away." They had lived together many years and she had had twelve children by him. Hay did not wish to part from her, but was ready to do so if the scriptures obliged him. After about an hour of conversation it was decided that there were no laws in scripture that obliged Hay to send Felida away. Smith argued that "the notoriety of the offence seemed to require it, yet, as she gave satisfactory evidence of repentance," there would be no sin if Hay took her again. It was finally agreed that Hay could keep his wife and they went home together.[81]

Amarillis's case was more difficult to settle. She wanted to be baptized, so Smith decided to investigate her life. When he asked Amarillis's mother who Amarillis's husband was, she replied: "Me no know. Me no been watchman for her." Two days later, Amarillis herself came to see Smith, but he was not much taken either by her appearance or by the conversation he had with her. Amarillis's husband lived in town. She saw him perhaps once every three months. Smith was convinced that Amarillis would not be faithful to her husband while he lived at such a distance, and that the man was not likely to return, so he advised her to make an amiable separation, get another man, and be married.[82]

Smith's efforts to institutionalize the slave family according to his own religious standards were constantly defeated by the power structures, constraints, and stresses generated by slavery, and the lack of control slaves exercised over their lives, as well as by their own notions about sex and family. Not that the slaves did not have families, or that their families were unstable. Many had kept the same partner for twenty years or more. But, following African traditions, some slaves had more than one wife. In Africa, however, those relations were based on clearly prescribed mutual obligations between wives and husbands and among the wives themselves. In a situation of slavery, it was practically impossible to respect such obligations, so that the advantages that might have resulted from such family arrangements diminished considerably while conflict increased. Besides, slave couples often lived apart, on different plantations, and those who lived together were always under the threat of being separated from each other and from their children. The missionaries' demand that the slaves live according to the strict standards of Christian morality added new tensions to an already problematic situation.

For the missionaries, such tensions and conflicts only confirmed their notions that there was a fundamental incompatibility between Christianity and slavery. For people to live as Christians, they believed, slavery had to be abolished. This belief could only be reinforced by affairs involving white males and black females. As Wray had been, Smith was scandalized by the way white men treated black women and the children they had with them. They usually kept the women as mistresses, refusing to marry them. Some were "cynical" enough to justify their behavior by saying that they treated the women better than the black men did. But when they had children with black women, whites often kept them as slaves. Such behavior outraged Smith.

The missionary was also shocked by the abusive way managers treated slaves. It seemed to him that they forced slaves to do too much work and punished them cruelly. Slaves were kept until late at night, particularly during harvest. When they did not finish their tasks during the week they had to finish them on Sundays. Their houses were built of frail materials, thatched with leaves, and typically enclosed with wattles plastered with mud. Occasionally they were built of better materials, with shingled roofs and boarded walls, but the houses were very low and seldom had windows. Masters only provided the slaves with an iron pot for each family and a blanket for each individual. Everything else—stools, tables, and the like—had to be supplied by the slaves themselves. Their diet was also poor and their clothes allowances insufficient.[83] Slaves were in rags during the week and children were left naked until they were eight or nine years old, unless the parents bought clothing for them.[84] From the age of twelve, children worked as many hours as the adults. And when slaves were sick they were put in what was called on every plantation the "sick-house," which to Smith seemed more like a charnel-house, with the sick lying on blankets spread on a "sloping kind of platform, elevated two feet above the floor." Medication was precarious at best. When patients were in danger they were given barley or fowl soup and wine with sugar.[85] But for Smith the most appalling features of slavery were the punishments the slaves were constantly being subjected to, and the "persecution" the Christian slaves suffered from managers.

Time and again he complained in his letters to the LMS that slaves were mistreated, that masters, managers, and local authorities had no respect for the Sabbath—that they were always making their slaves work on Sundays—that they did not support the work of God and persecuted blacks who came to the chapel. He hated all the arbitrariness, all the unfairness, all the violence that seemed to be inevitably associated with slavery. One day, after seeing some slaves working in irons, one of them with his back flayed by the whip, Smith lamented:

> O Slavery, thou offspring of the Devil, when will thou cease to exist? Never I think was my sense of mission more disgusted with the degradation of the human species or my feelings more keenly touched. I hail the day when slave-masters shall be imbibed with the feelings of Christian men, and the slaves enjoy their birth right; they are treated worse than brutes. Thanks to a kind providence there are a few (and alas a very few), masters who do treat their slaves as though they had common feeling.[86]

Everything Smith cared for—"Justice," "Christian feelings," "Human Dignity"—was "degraded" in slave society.

Smith was profoundly affected by the tales of cruelty he heard from the slaves. No month—often no week—would pass without some one recounting some story of persecution. Only a few weeks after his arrival at *Le Resouvenir* he was already hearing reports that the managers were "opposed to the Gospel." A slave told him that Mr. Pollard had promised to give a hundred lashes to any slave who went to the chapel, unless the parson sent him a notice. Smith did not know whether the story was true, but was inclined to believe it since he had heard the same stories from Wray.[87] A month later, the overseer of *Goed Verwagting* complained to Smith that the master was very severe with his slaves. They had to work from five in the morning until seven or eight at night and were allowed only three pounds of saltfish per person each fortnight. He said he had seen the master order the slaves to suffer fifty or a hundred lashes without any apparent provocation. When either master or manager was in a bad humor he would vent his spite on the slaves.[88] The overseer of *Success*—who had been in the colony for just a few weeks and probably saw things a bit the way the Smiths did—told them similar stories. One night when he dined with the Smiths, he complained of the way the manager, Mr. Stewart, treated the slaves, often giving them a hundred lashes and routinely overworking them.[89]

Reports like this quickly found their way into Smith's diary. He had been in Demerara for less than a year when he wrote: "This evening a negro beonging to ——— came to me complaining of the manager's cruelty. . . . I believe the laws of justice which relate to the negroes are only known by name here, for while I am writing this the driver is flogging the people and neither manager nor overseer is near."[90] The next day he wrote again: "The first thing as usual which I heard was the whip. From ½ past 6 until ½ past 9 my ears were pained by the whip. Surely these things will awaken the vengeance of a merciful God."[91] Two weeks later he was writing again: "This afternoon my heart was very pained on hearing of the cruelties practiced by a manager on this coast." It appeared that a slave complained to his manager that he was unwell. The manager did not believe him and ordered the man to be put into the stocks, and beat him with a stick. The man died in the stocks.[92] One week later, Smith confided to his diary that a slave had complained his manager would not allow slaves to come to the chapel, and had put in the stocks those he had found there, and had continued to do this every Saturday evening and Sunday morning, keeping them confined all day.[93]

The complaints had no end and (once more rehearsing Wray's experience) Smith sometimes wished the slaves would not tell him of their troubles with the managers—"as it is not my business to interfere in such concerns and only obliges me to treat such conduct with apparent indifference and behave with coolness to those who relate it." But coolness and indifference were exactly what he lacked. He was always wondering how the slaves could tolerate such oppression and sometimes he thought that sooner or later they would run away or rebel. "Mr. G. . . . called this evening," Smith once wrote. "He was going to town to search for negroes who had run away. No wonder . . . Mr. G. told me today that Mr. V. B. treats his slaves with great severity and cruelty, that for every little offense he withholds their fish and plantain. . . . Milder treatment would be more politick."[94] When he heard that an expedition to the bush was being organized to "fetch" runaways, Smith could not avoid thinking how ironic it was that the Indians—the "Bucks" as they were called—and even a few blacks had joined the expedition. He asked one of the Indians whether he would sell his son, "a fine boy of eight or nine." The man, oblivious of the taunt, replied, with apparent surprise, and probably not without pride: "No. I am a Buck."[95]

So it was in Demerara. How long would such oppression and stupidity last? Smith wondered. How long would the slaves endure mistreatment and persecution? "I observed in the slaves a spirit of murmuring and dissatisfaction," he wrote in March of 1819, "nor should I wonder if it were to break into open rebellion. However, I hope it may not."[96]

Worse than hearing the slaves' complaints and speculating about possible rebellions was to hear the sound of the whip, and to imagine the wounds it left on people's bodies. It drove him crazy. He would sit in his house obsessively counting the lashes, hating the arbitrariness and violence of the system, wishing that slavery were abolished and people were free from degradation.[97] Jane Smith shared her husband's outrage. One morning, just before sunrise, she woke him up to listen a dreadful flogging one of the slaves was receiving. When he was sufficiently awake to listen with attention, he noticed that two drivers were flogging the slave. He could tell this just by the sound of the whips. He counted eighty-two lashes, and was sure from the regularity of the strokes that they all had been given to a single person. "This is a slight specimen of the force of the law in this country. The planters laugh at the law," was his bitter comment.[98] The law authorized only thirty-nine strokes, and required that either the master or the manager be present at the time the flogging was inflicted on the slave. But no one seemed to care.[99]

Once when he saw an old man who took care of the manager's cow being whipped on his thighs, Smith again commented with bitterness, "The way here is not to tell a negro of his error, but punish him first and then tell him what to do."[100] A month later he woke up again to the sound of the whip. When the flogging was over, his wife asked him from the adjoining room, "Did you count those lashes?" "Yes," answered Smith. "How many did you reckon?" she asked. "One hundred and forty-one." Had she been counting too? he asked. "Yes, I reckoned one hundred forty." Both were outraged. And for Smith the record and repository of his rage was his diary: "Ah, the men who spend the Sabbath evening over the bottle and glass, divert themselves with cards and backgammon, are haunted with hideous dreams and fearful forebodings during the hours of their slumber. Then they rise to vent their arbitrary malice and authority . . . , it may be upon the innocent."[101]

With time, the Smiths would grow increasingly bold, and instead of passively sitting counting the lashes, or regretting that slaves had been put in the stocks, they tried to intercede on their behalf. When Asia, an old and sick woman who had been the driver of a gang, was put in the stocks because she refused to work in the fields, Jane Smith wrote a note to the manager warning him that if he continued such severe treatment she would feel it her duty to inform Henry Van Cooten, the attorney. The manager wrote a terse response saying he would keep Asia in the stocks until she agreed to work in the fields. Jane Smith then went to Van Cooten begging him to consider Asia's case. Van Cooten seemed displeased. He said that the manager had always kept him informed of what he did. He was sure Asia was not in the stocks. But he promised to go to the estate to check. Commenting on Van Cooten's behavior, Smith said that the old man seldom visited the estate and when he did he only talked to the manager. Only rarely did he go to the "sick-house" or speak with the slaves. "And of course, all is right and just as it ought to be." Yet Van Cooten surprised Smith. After a few days, he did go to the estate, liberated Asia from the stocks, and ordered that she should do some "light work" like picking cotton—which to Smith did not seem much of an improvement.[102] Such interventions by the missionaries could only increase the managers' hostility toward them. If an attorney became convinced a manager was systematically unfair and abusive, the manager could lose his job. But the Smiths were growing more and more upset with what they saw around them, and more and more inclined to challenge what seemed to them stupid rules and restrictions.

Like Wray, from the outset Smith deplored the way masters and managers opposed his teaching slaves to read. The arguments they used were very familiar to him. They were the same "futile" arguments that had been used in England against Sunday schools. The colonists feared that teaching the slaves to read would subvert the social order, but to Smith this seemed to disguise the real issue: "Under these plausible pretences of zeal for the good of the colony, lie couched the old and hateful principles of enmity to God. . . . [F]rom this source originates the various oppositions which are made against the religious instruction of the negroes."[103] After being in the colony for some time Smith, instead of merely lamenting the planters' and managers' opposition, resolved to teach the slaves anyway.[104] In a letter to the LMS he mentioned that he was teaching the "negro children" of *Le Resouvenir* to read, and since this was forbidden by the governor he had do it by "stealth."[105]

Night meetings and Sunday work were two other issues that pitted Smith against planters and managers.[106] Since Wray's time, night meetings had caused serious confrontations. And as soon as Smith started preaching at night, managers and planters started complaining.[107] But the really controversial issue was the Sabbath. Wray already had had a lot of trouble trying to convince both managers and slaves to keep the Sabbath. Smith faced the same difficulty, and for the same reasons. His diary filled with complaints about slaves being forced to work on Sundays, finishing tasks they had not finished during the week, loading a boat which was leaving, working in the sugar mills, or doing other tasks. Every manager he talked to, even those he considered his friends—like a certain Kelly, a man who often dined with the Smiths and invited them to his house in return—seemed convinced that Sunday work was inevitable. The managers argued that on sugar plantations when they ground cane late on Saturday, the liquor had to be boiled on Sunday or it would sour. The boilers also had to be cleaned on Sundays, and this would take several slaves four or five hours. Sugar that was boiled on Saturdays had to be packed into hogsheads on Sundays. This occupied carpenters and coopers for hours. Slaves also had to work on Sundays when there was sugar and rum to be shipped and a vessel was leaving. So there was always something to do on Sundays, especially on a sugar estate. But on cotton plantations the situation was not fundamentally different. Although there was less work, the slaves were also asked to perform many tasks on Sundays, such as drying, cleaning, and ginning cotton or baling and loading it for shipment. On coffee estates, they frequently had to dry and pulp coffee on Sundays. When boats brought cargo to any plantation on a Saturday night, managers felt obliged to unload them on Sunday. And often new buildings were put up on Sundays.

It appeared to Smith that managers were inventing work for slaves to do on Sundays. On one plantation near *Le Resouvenir* they had removed the slaves' houses about a mile and a half farther from the sea. The new building was big enough to accommodate ten families, all under the same roof. But since partitions to separate one family from another had not been built, the slaves were forced to do it on Sundays. On another estate, a temporary lack of plantain forced the slaves to walk seven or eight miles on Sundays to fetch it. It was also on Sundays that the slaves' weekly food allowances were distributed. As a consequence of such practices, slaves often missed Sunday services. And although the chapel was always full, individual slaves' attendance was very irregular.[108]

Even when masters and managers did not assign special tasks, the slaves themselves found things to do on Sundays. They cultivated their gardens, went to the market, or hired themselves out to get money. One day the slave Azor hired someone to work for him on a Sunday. When Smith objected, Azor explained that his wife was about to have a child and her house was not finished. Since she was a free woman his master would not give her a house. So Azor felt obliged to help her. But he could not find anyone to do the job except on Sundays. Such situations were impossible to solve. How could Smith tell Azor that it was not right for him to hire someone to work on Sundays? "I scarcely know what to say to it" was his perplexed and dispirited comment.[109]

Sometimes Smith pressured the slaves to respect the Sabbath. But the result was often disappointing, as when he tried to convince Jack Ward the butcher, one of the members of the congregation, not to sell meat on Sundays. Ward was a slave, and he had a stall in the market. His frequent absences from the chapel on the Sabbath led Smith to investigate. "I found that the market was his chapel and the stall his altar, and his chief prayer, that customers might be many." He warned Jack Ward of the consequences of such "wicked conduct," and told him if he continued acting that way he would exclude him from the congregation. The man's answer was not satisfactory. Smith finally suspended him from communion and urged him to reflect and pray. Jack Ward finally returned with his answer, which Smith concluded amounted more or less to this: "If he could serve God and Mammon he would gladly do it, but if not, he preferred the latter alone."[110]

Managers and planters must have felt like so many Jack Wards, and if pressured by Smith would probably have given the same answer. Smith's ethical code, his insistence on legitimizing slave marriages, his strict sabbatarianism, his attempts to teach the slaves how to read, his night meetings, and his interference with managerial decisions would in the long run expose the incompatibility between Smith's commitment to God and masters' commitment to Mammon. And the repeated conflicts between the missionary and the colonists would confirm both sides' worst expectations. As a result, the colonists became more and more hostile to the missionary, while Smith drew closer and closer to the slaves.

For the slaves those conflicts indicated that they had found an ally, and this gave them more strength and hope. They assumed—as they had with Wray—that Smith had powerful friends in England, friends who were also on their side. The day he preached a funeral sermon on the deaths of George III and the Duke of Kent, many slaves showed up in black as a sign of loyalty. "They suppose every missionary must be personally acquainted with the King. On Sunday just before service one of the negroes came to ask me who was the next great man to Mr. Wilberforce. He seemed to have thought that the next change would bring Mr. Wilberforce to the throne," Smith told the LMS.[111] However confused these ideas may have been, they indicated that the slaves had a feeling that sooner or later power would come to the hands of men like Wilberforce and the slaves would be free. The chapel had created a new social and moral space, forged new bonds of community, and given them new pretexts to challenge their managers and masters. And the more the masters and managers created obstacles, the more the slaves battled for their right to go to chapel. In this process, masters and managers, missionaries and slaves, became increasingly radicalized. And it is not surprising that in spite of the many pains, obstacles, and disappointments Smith's mission prospered. In a few years he could boast that there had been a "Great Awakening" on the East Coast of Demerara. But it was precisely the progress of his mission that would bring him trouble.

On March 5, 1818, Smith reflected on his first year's experience in Demerara: "Tis twelve months today since we came to this house. Oh my Lord what shall I say to express my obligation to thee? Thou hast smiled upon me, and her whom thou has given, as a help meet in thy cause. All thy faithful ministers in this colony have met with opposition, insult and reproach, but I have (at least to my knowledge) escaped all these things."[112] In October, after he received a letter from his colleague Mercer describing the opposition he had met from the government in Trinidad, he concluded that the situation in Demerara was better: "Thanked be God, we are not opposed by our colonial Government." But in the years that followed, he would meet with increasing success among the slaves and growing opposition from masters, managers, and local authorities.

Encouraged by his success, Smith soon started thinking of expanding his mission. He first tried to build a bigger chapel closer to the road. With his habitual energy, diligence, and entrepreneurship, he set himself to convince Van Cooten, the attorney of *Le Resouvenir,* to support his project. The attorney contacted the proprietor, Mrs. Van der Haas, but she told him she would not spend a single penny on it. Smith would have to do it on his own or with the help of the LMS. Smith went as far as to dig out Post's will from notary records, and even consulted a lawyer to find out what rights the LMS had over the chapel; but to his distress he realized that the building belonged to the estate. And since the proprietor was not willing to help him, he had to content himself with making improvements in the old building. His efforts, however, were finally rewarded. The support he had not found among whites he found among slaves, who generously donated almost £200 for repairing the old chapel. The LMS provided the amount lacking.[113]

Smith then decided to extend his mission to Mahaica, where he hoped to get a donation of a piece of land for the building of another chapel. In time the LMS could send another missionary to fill that post. This seemed to him particularly urgent because the Methodists were thinking about opening a "station" there and Smith, spurred by sectarian rivalry, did not want the LMS to miss this chance to expand its own missions. "Since we have been engaged in preaching at Mahaica," he wrote in May of 1818, "Mr. Mortier (the Methodist preacher) has taken the liberty to solicit those gentlemen who had promised us their support, to subscribe towards the erection of a place of worship for the use of Methodists, but our friends were not to be brought over."[114] Mortier was preaching in a private house only about two hundred yards from the place Smith and Davies had been preaching. Smith was proud to notice that only the free "coloured people" and whites attended Mortier's services. The slaves came to hear him and Davies. A few weeks later he was complaining again that Mortier was "using all his endeavours to win our friends over to his party."[115]

Ironically, Smith's jealousy did not extend to the Reverend Archibald Brown, the minister of the Scottish Presbyterian Church, who arrived in the colony in September of 1818. Brown was to preach in town, not on the estates as Mortier and Smith did. As soon as Brown arrived, Smith paid him a visit. He found Brown to be a very communicative young man, wished him well, and prayed that the Lord would bless his labors. Little did he know that some years later, both Mortier and Brown would side against him. But it would be the "very communicative" Archibald Brown, not Mortier, who would conduct in the local press a most demoralizing campaign against the LMS and its missionaries.

Occasionally, during the first two years, in spite of the success of his mission, Smith felt discouraged and on such days he wrote bitter letters to the directors. He stressed the daily problems he faced in his mission and lamented that he was not receiving the support Wray had when Post was still alive.[116] Van der Haas, the manager, was an "enemy to religion," and if it were not for Van Cooten, Smith might not have been able to stay on the plantation. He also complained of the opposition from the planters and the attacks of the press against missionaries. These feelings were not new. In October of 1817, only a few months after his arrival, Smith had written: "Satan is rallying his numerous and malicious forces against us, and what fair play cannot accomplish he is endeavouring to effect by falsehood and fraud. Our character as a body of missionaries is represented in the newspapers here in the blackest colours." He then added sarcastically that he knew how to please those who so much opposed the missionaries. He just had to preach only to whites, neglect the slaves, and banish from his sermons any reference to the third, fourth, seventh, and ninth commandments.[117]

The greater his success among slaves, the more resentful of the planters' opposition Smith became. He wished that Wilberforce and other "great men" in England would make clear their support for his mission. For a while he nourished the idea of writing an anonymous letter to the *Philanthropic Gazette* describing the situation in the colony, in the hope that it would arouse public opinion at home.[118] And indeed, it was a letter about the slaves' living conditions he sent home in 1818, parts of which were published in the *Evangelical Magazine,* that brought about his first confrontation with local authorities.

In the letter, Smith complained that some planters forced their slaves to work on Sundays. Following the publication, he was summoned by the fiscal, who demanded to know the planters' names. Smith refused to identify them. To him it seemed absurd that he be required to prove something that was obvious to everyone in the colony. He defiantly told the fiscal he did not feel obliged to identify anyone, but hoped his letter would have its proper effect on the "magistrates" and that they would prevent such "shameless abuses of the Sabbath." The fiscal replied that the colonial authorities were vexed that such a fact should be made public. The affair ended, but it left its mark. Some months later Davies was writing to the LMS warning that much caution was necessary in selecting extracts from the missionaries' letters for the *Evangelical Magazine,* particularly since some of the most determined opponents of the missionaries had just been elected to the College of Kiezers.[119]

In spite of their attempts to please the whites, Davies and Elliot had been involved in a series of confrontations with them. Davies, who preached in Georgetown, was harassed by a group of youths, who sang lewd songs and made other noises near the chapel while he was preaching. He initiated a lawsuit against them, and in retaliation they filed a suit against him too. Elliot was also in trouble, accused of corruption. The three missionaries often met to discuss the best strategy to follow and to give moral support to each other. They decided to create an Auxiliary Society and to hold annual meetings. Wray, who was in Berbice, joined. But soon the missionaries started fighting among themselves. Smith liked Wray, and Wray seemed to appreciate Smith's devotion to his work, but both men disliked Davies and Elliot. And although Smith often made efforts to placate animosities and tried to bring them together—or at least he thought he did—he often found himself at odds with them, and returned home from their meetings with a heavy heart. Smith disapproved of Elliot's and Davies's conduct. He thought they were too concerned with money and status. Most of all he disliked Elliot's tendency to gossip and slander.[120] Elliot's wife's manners and independent temperament also scandalized Smith. One day she dared coming from town alone with Captain Ferguson to spend the day at *Le Resouvenir* with the Smiths.[121] Smith did not think this an appropriate behavior for any woman, much less for a missionary's wife. Jane shared her husband's feelings, and sometimes the two women quarreled.

Before the end of his first year in Demerara, Smith was already completely disenchanted with missionaries who did not share his commitment to preaching to slaves and who seemed more concerned with worldly matters. "I view missionaries now in a very different light than what I did before I came to Demerary. I thought they must be true lovers of Man, but as far as my acquaintance extends, I cannot find that they speak well, even of those who support them." In time, his relationship with them became increasingly difficult, and eventually they dissolved the little organization they had created.[123] Still, in spite of conflicts and disappointments, the missionaries had to stick together. So they continued to visit each other and to pray together from time to time.

Although his relations with Davies and Elliot were always difficult, Smith always enjoyed Wray's visits. "It is indeed pleasant to converse with a real Christian friend. It is like oil to one's bones," he wrote after one visit. But the two men lived miles apart, and they saw each other only once in a while. Smith felt lonely. He complained that he had no true friends. "I feel the want of a Christian friend and counsellor," he wrote in November of 1822. "We have missionaries from the same society, but fortunately for the colony, though unfortunately for the cause of religion and just rights, the Governor and the court have bought them, the one for 100 joes and the other for 200 per annum."[124]

Smith felt increasingly depressed. He had no friends and his work was difficult and full of challenges. The directors of the LMS did not seem to realize his need for support and acknowledgment, and they seldom replied to his letters. Jane had been sick since she arrived in the colony, and spent half of the time in bed. The couple had few contacts with other whites, with whom they maintained mostly distant and sometimes hostile relations. Except for a few managers and overseers who occasionally came for supper, and a few boat captains like Ferguson, who showed up from time to time, their social life was minimal. Only Van Cooten, the attorney of the plantation, was always kind and supportive. With the others Smith felt he had to be constantly on guard. One could never predict what they might do. From time to time he received a letter from his mother or from some friend in England. But that only made him feel more lonely and homesick. He waited with anxiety for the London newspapers, which arrived only after weeks of delay. They were his only contact with the world outside.

In truth, aside from Jane, the only friends he had were the slaves. Although a whole world of experiences and meanings separated him from his congregation, the slaves gave him a feeling of accomplishment. "We cannot expect to have the enjoyment of society here. I should be sorry not have a relish for the company of our white neighbours. To the black ones we cannot have access, except when they come for instruction and on these occasions I experience a degree of happiness, I never thought of enjoying among them," he wrote in December of 1821.[125]

So he devoted himself almost obsessively to his congregation:

> Yesterday I was so engaged all day as scarcely to have time to eat anything. At morning service I examined 14 or 15 candidates for baptism which took me till half past ten. I then went to my breakfast. Before I had had time to eat my breakfast, eight or nine negroes came round the door desiring to converse with me on the concerns of their souls. I spent about an hour with them, then went to the chapel to preach. After service I wrote down the names of about fifty people who wished to serve God, as they say. Having dismissed the people by about half past three I went to *Lusignan* to see Mr. Brown who was very ill and sent for me. Mr. B. is or at least was a professor of religion and I believe a member of the Wesleyans, but has degenerated into the spirit and conduct of the planters, so as no longer to refuse working on the Sabbath, etc. Conscious of his backsliding state he feels very uneasy in his mind. After conversation and prayer with him, he said he felt more comfortable within. . . . I reached home by 9 o'clock and felt myself very unwell and so sick as to vomit abundantly. I suppose this was owing to much exertion and little eating.[126]

The disease that undermined Smith's body was making visible progress. He was always coughing, he had a pain in his chest, difficulty breathing, and sometimes could hardly preach. But his illness rather than discouraging him seems to have given him a renewed and even stronger sense of mission.

Smith's efforts were amply rewarded. A surprising number of slaves joined the congregation.[127] His work became so overwhelming that he decided to divide the slaves into groups, according to their plantations, and to appoint some of the more knowledgeable slaves to serve as teachers. In September of 1821 there were about thirty such teachers actively instructing the adults, and six or eight others instructing the children under Jane Smith's supervision.[128]

Slaves also took initiatives of their own and started meeting at night to read the Bible and rehearse the catechism. In Demerara as elsewhere they created a black church of their own, which would "witness to Christ in their own way."[129] Slaves' gatherings inevitably aroused the suspicions of plantation owners and managers and they tightened control over the slaves. Some managers went so as far as to forbid the slaves to hold religious meetings. The restrictions led to growing confrontation. Pollard, the manager of several plantations (*Bachelor's Adventure, Enterprise,* and *Non Pareil*), punished severely the slaves he caught catechizing each other.[130] Two months later, Cuming, the proprietor of *Chateau Margo,* gave several slaves fifty lashes for refusing to work on Sundays. The slaves from *Bachelor's Adventure* were also constantly complaining of persecution and unfair punishment. In March of 1821 the conflict between the manager and the slaves reached such proportions that first the burgher officers McTurk and Spencer and then the fiscal were called to intervene. The slaves complained to the fiscal that the manager had nothing against them but their religion, and they asked him whether it was not their right to go to the chapel and to hold meetings for prayer and catechizing. The fiscal told them they were at liberty to go to any place of worship and have their meetings on the estate[131]—a statement that must have displeased the manager but certainly pleased the slaves. Analogous conflicts took place on other plantations. A month later, a slave told Smith that two men from *Hope* had come to the chapel to spy on the slaves who were attending services. They were supposedly writing down their names for punishment. And several months later Smith was told again that three slaves from *Hope* had been flogged and put in the stocks just for going to the chapel. In September 1822, one of the Rogers brothers called Smith because some slaves had defiantly told the manager of *Clonbrook,* when he was flogging them, that he might kill their bodies but he could not kill their souls. When Smith questioned the slaves, they explained that they had been flogged for attending services, and while they were being whipped the manager was taunting them to go to the parson, since he was such a "friend." Smith concluded that Hugh Rogers, as well as his brothers, was "an implacable enemy to the instruction of the negroes."[132] A few months later a slave from *Le Resouvenir* told Smith that the manager had ordered the driver to flog the people who had been in the chapel. Smith investigated and discovered that many of the slaves from *Le Resouvenir* had slept at *Success,* and many others in the chapel, to avoid the manager's reprisals. Apparently the manager had wanted the slaves to dance and "to stir them up he gave them a pail of rum, besides their allowances." But the Christian slaves had refused to dance, triggering the manager's rage. Such challenges to their authority irritated masters and managers and aggravated their ill-will toward the missionary. They started complaining to the governor. Sooner or later a confrontation between Smith and the authorities was bound to come.

In October of 1819, Smith was involved in an unpleasant incident with his neighbor, the planter and burgher officer Michael McTurk. Some slaves at *Le Resouvenir* fell sick with smallpox. Van der Haas, the manager at *Le Resouvenir,* immediately warned the neighboring planters.[133] Fearing the spread of the disease, managers forbade the slaves to go to the chapel.[134] At first, the slaves did not take the order very seriously. Quamina reported to Smith that he had told the manager of *Success* that there was no danger in going to the chapel because the people had already either had smallpox or had been inoculated. But the manager insisted that his orders be obeyed. So the following Sunday, very few slaves attended services.

To Smith, the managers' ban appeared to be a plot against religion. They were always happy to find any excuse to prevent slaves from going to the chapel. Only three slaves were sick and they had been removed to a house more than three miles to leeward of the chapel. In his opinion there was no more danger if the slaves went to the chapel at *Le Resouvenir* than there would have been for people to attend services at Tonbridge Chapel if three people had the smallpox in a solitary house on Hampstead Hill. Besides, there were cases of smallpox in town and yet most of the planters willingly gave passes to their slaves to go to the market on Sundays, and neither the governor nor the fiscal seemed to be troubled by it.

A few days later Van der Haas was discharged from *Le Resouvenir* for his cruelty and replaced by John Hamilton. But the restriction continued.[135] Smith received an order from the fiscal to shut the chapel to all slaves not belonging to *Le Resouvenir,* until the smallpox passed. He complied, reluctantly, and turned the slaves away. He was convinced that it was his neighbor McTurk who had persuaded the fiscal that such a measure was necessary. He believed McTurk was using the smallpox as a pretext to keep the slaves away from the chapel. To verify McTurk's claims, Smith decided to go to the "sick-house." He found five men: two were entirely recovered, one was almost well, another had a bad case of "locked jaw," and only one was still sick with smallpox. The missionary was appalled by the conditions he found the men in. They were in a wretched hovel, whose entrance was no larger than the door of a dog kennel. There was no room inside to stand up. The place was dark and the only light came through the door. The roof was leaking and the sick men had nothing but a litter of leaves to lie on. The doctor had not come to see them at all.[136]

A few days after Smith's visit, McTurk, who was acting as doctor, went to see the slaves. Finding them recovered, he sent them home and ordered the clothes and the hut to be burned. McTurk then wrote a letter to Hamilton at *Le Resouvenir,* asking whether any slaves from other estates had been attending services. Hamilton showed the letter to Smith, who advised him to tell the truth. Hamilton informed McTurk that Smith had followed the fiscal's instructions. But McTurk wrote back asking for more details. Meanwhile, the ban on the chapel continued. The whole incident enraged Smith: "It is surprising what a malignant enmity that Man has to the cause of religion. . . . If such a burgher officer be not a disgrace and a curse to the polity with which he is connected I am much mistaken."[137] On Christmas eve, several slaves told Smith that their managers had given them orders not to come to the chapel any more. They said the order was from the fiscal and was carried from one plantation to another by "a man in a red jacket." Smith promised to investigate the story. He finally discovered that McTurk had requested the fiscal to issue the order, arguing that the smallpox might still be "latent" at *Le Resouvenir.*

That Christmas of 1819 was the worst the Smiths had passed since they left England. No slave from outside *Le Resouvenir* attended services. Smith was terribly upset. He complained to Van Cooten about McTurk's intolerable behavior. The attorney promised to speak to McTurk and two days later the missionary received a letter from Van Cooten saying that McTurk had promised to ask the fiscal to lift the ban on the chapel. Almost two weeks passed before Smith heard again from the attorney. Van Cooten said he had received a letter from the fiscal informing him that the governor was ready to lift the ban, but only after the slaves had been examined twice by "Doctor" McTurk. After much hassle, during which Smith and McTurk exchanged violent words, the ban was finally lifted. But the relationship between the two men never improved, and the enmity between them would, in the end, prove fatal for Smith.

McTurk's plantation *Felicity* was next to *Le Resouvenir.* Smith was always watching McTurk's slaves working on Sundays, and every time he had any chance to talk to one of them he insisted that they come to the chapel instead. McTurk's slaves told him that their master did not like religion and that he was always threatening to punish them if they went to the chapel. But Smith was relentless. One day he met a slave from *Felicity* on the road. He asked whether the slave had ever been to the chapel and the man answered that the "doctor" did not allow him to go. Smith insisted. "I suppose yourself no want for come, you no want for serve God your maker?" "Hugh, Massa! We want for serve God, but doctor tell we, if we go to chapel, he will cut we a——," answered the slave. "I believe what the negro said was true," Smith wrote in his diary, "for though McTurk's estate joins *Le Resouvenir,* I never heard that one of his negroes attended the chapel, and I have made many enquiries. They are frequently if not always, at work, while others are at chapel."[138]

A few months later, Smith gave a lift in his chaise to one of McTurk's slaves. Smith asked the slave if he attended religious services (knowing perfectly well what the answer would be). Predictably, the slave told Smith that his master would not let him go to the chapel. "On Sundays, you can go to other estates where there are Christian people and beg them to teach you," Smith suggested. "Sunday and working day all alike to we people," was the man's answer. "What do you mean? Slaves must not keep Sunday?" asked Smith. "No me no mean so. But the doctor make we all work on Sundays," the man said. "Are you a house boy?" asked Smith. "Yes, now the other one is sick the doctor take me in the house," returned the slave. "Well then if you are a house boy, you often go to town. And when you walk on the road with some sensible people you should ask them to teach you the catechism, that you may learn to know who God is," Smith insisted. But the slave replied in a serious tone, "Massa, the Doctor don't like us to know God. One time he heard me say God knows and he said 'O, you know God, do you?' And he made me eat the soap he was washing with and gave another boy a horse spur and made him spur me because I knew God." That was exactly the sort of answer Smith expected to hear.[139] But if the conversation between Smith and the slave had reached McTurk's ears, it would have infuriated him. He would have become convinced, if he were not already, that the missionary was deliberately undermining his authority and instigating his slaves to disobey him.

McTurk was not a big planter. *Felicity* was a small to medium plantation. Many plantations in Demerara had two or three times as many slaves. McTurk must have bought *Felicity* around 1815, when it had been advertised for £15,000 sterling, with an initial payment of £8,000 and the rest in four annual installments. The plantation had 900 rods of cotton, 800 of plantain, 100 of pasturage, 90 slaves, a large dwelling house with a brick water cistern, a kitchen, a cotton "logie," a gin "logie," a stable, a sick-house and store, two slave houses 96 feet long, and three gardens. It was offered without the slaves for £8,500, and with or without two milk cows, ten sheep, and sixty hogs.[140] McTurk probably had a mortgage on *Felicity,* and like any other planter he was very eager to extract the maximum labor from his slaves so that he could pay his debts as quickly as possible and start making a profit.

As a resident planter, McTurk had managed to be elected to several public offices and accumulated several functions. In 1821 he became a member of the Court of Policy.[141] He was also a burgher officer for the East Coast, and a "way warden" whose job was to oversee the roads and bridges within his district (which included *Le Resouvenir*). He also claimed to be a doctor. As a member of the Saint Andrew's Society,[142] a religious fraternal association that gathered many of the Scots in Demerara, McTurk belonged to a powerful network of planters and merchants, which included men like Lachlan Cuming, the proprietor of *Chateau Margo,* John Fullarton and Evan Fraser, members of the College of Kiezers, and other men who carried titles like "Esquire" and "Honorable."[143] In sum, he was a man no one in Demerara would like to have as an enemy.[144] He was also a harsh and obstinate man, not likely to forget the smallpox incident or to forgive any missionary's insolent behavior. From his position as a member of the Court of Policy, McTurk could do Smith great harm.

McTurk's opposition to the missionary became blatantly clear once again when Smith applied to the governor for a piece of public land for a new chapel in Mahaica. This happened when Smith's old companion Mercer arrived in Demerara from Trinidad. Smith was very pleased when the LMS decided to transfer Mercer from Trinidad—where he had met great opposition from local authorities. Smith immediately thought that Mercer should take charge of a new mission in Mahaica—an idea Smith had been entertaining for at least a year and a half. With that purpose in mind, he went with Mercer to *Clonbrook,* about sixteen miles east of *Le Resouvenir,* to ask John Rogers if he would consent to the erection of a chapel on the "company's path," adjoining his plantation.

The company's path was government property. Between every two plantations there was a path, about 60 feet wide and leading several miles from the sea into the interior. Smith had thought that a plot in the path could be used for a chapel. Rogers said he would make no objection, but told Smith and Mercer that they should talk to Charles Grant and to Hugh Rogers from *Bachelor's Adventure,* to be sure they were agreeable to the plan. After they talked to Grant and Rogers, Mercer applied to the governor. But he was told that the governor had received "serious complaints" against Smith, and wanted to investigate the case further before making any decision. After a month passed without Mercer's hearing anything, Smith decided to talk to the governor.

The governor met Smith and Mercer, but was evasive, saying vaguely that the complaints had something to do with night meetings. He said the missionaries would have to bring him the written approval of the "gentlemen" in the neighborhood where they wanted to erect the chapel. And when the missionaries expressed their fear that some of the proprietors might not be favorable, the governor replied rudely that he surely could not "cram" a chapel on them. Although they suspected it would be fruitless, Smith and Mercer tried again. They first approached Van Cooten, who had always been very supportive, and asked him for a letter of recommendation. With the letter in hand they called on two planters who subscribed to the LMS. But to the missionaries' surprise and distress both refused to support them, saying it was because of the night meetings. The missionaries did not give up. Having heard that a planter who had recently arrived from England was favorable to the instruction of the slaves, they called on him. But he too refused to support them, for fear of displeasing the other planters.[145] Mercer finally had to abandon his plan. After a year he obtained permission from a landowner in Leguan to occupy temporarily an old house on his property until he could find a more permanent residence in Essequibo. But the man insisted that no services be held on week days, forbade him to teach the slaves to read, and also barred slaves from other plantations from attending religious services on his property. And once, when Mercer held a night service, the man threatened to expel the missionary from the estate.

But Smith would not abandon the idea of finding a position for Mercer in Demerara. He continued to approach different people until finally, in September of 1822, the owner of *Dochfour,* John Reed, told him he would give a piece of land on one of his estates about twelve of thirteen miles east of *Le Resouvenir* for the erection of a chapel—but only if the governor gave his permission.[146] "This will give me much trouble, and will be an expense and perhaps prevent my having the land for I have reasons to believe he is no friend of missionary exertions. Policy may induce him to make a show of friendship in some cases and to some individuals," Smith fretted. Two days later he went to see Governor Murray, but he was told that the governor had already left his office for the day. This was just the first of many fruitless trips to Georgetown. With characteristic obstinacy, Smith pursued Murray week after week without success. Sometimes he was told that the governor had already left, other times that he had not arrived and that no one was sure whether he would come. Finally, Smith was informed that the governor had given his petition to McTurk for review. "This leaves no doubt what the result will be," Smith thought. "McTurk is one of the greatest enemies to the instruction of the negroes on the whole coast." As he predicted, his petition was rejected. But Smith stubbornly came back with a new one,[147] and the frustrating visits to the governor's office began again. For months, Smith repeatedly returned to town but never managed to see Murray. Always told to come back the next day, his frustration turned into hatred:

> Just returned from another fruitless journey; have been for the answer of my petition but was again told by the governor's secretary that his Excellency had not given any order upon it, but that I might expect it tomorrow. I imagine the governor knows not how to refuse, with any colour of reason, but is determined to give me as much trouble as possible in the hope that I shall weary

> of applying, and so let it drop, but his puny opposition shall not succeed in that way, nor in any other ultimately if I can help it. Oh that this colony should be governed by a man who sets his face against the moral and religious improvements of the negro slaves! But he himself is a party concerned and no doubt solicitous to perpetuate the present cruel system and to that end probably adopts the common, though not false notion that the slaves must be kept in brutal ignorance. Were the slaves generally enlightened they must and would be better treated.[148]

The fifth time Smith went to the King's House to see whether there was any answer to his petition, the secretary told him that the governor had given him a report drawn up by McTurk, containing a series of heavy charges against him. Smith demanded to see the report. The secretary promised to talk to the governor about it. But, although Smith tried a few more times, he never heard about the petition again. Finally, after many months of effort and exasperation, he had to give it up. It was obvious that Governor Murray would not allow the creation of another chapel.

Smith's stubbornness could only irritate the governor, who was in a precarious position, constantly pressured by missionaries on one side and colonists on the other. Murray knew that the missionaries had powerful connections at home, and that any attempt to curtail their work would be condemned by the British government. He had been admonished by the home government nine times for one reason or another, and had to be particularly careful not to alienate its support. On the other hand, the colonists had grown increasingly hostile to the missionaries. And in spite of the opposition, missionaries continued to arrive—first Wray, then Davies, then Elliot, then Smith, finally Mercer (not to mention the Methodists). Nothing seemed to discourage them. The colonists' irritation was mounting. As a planter himself, Murray could easily sympathize with their feelings.

The colonists were watching the movement toward emancipation with growing apprehension. They also followed with anxiety debates in Parliament over the colonial trade. Although they still had power enough to protect their interests, the colonists felt increasingly threatened by the new tide that menaced both their slave property and their profits. Wherever they turned, the colonists found that their interests were contradicted by some other group of interests in the mother country. East India traders, London merchants, Manchester manufacturers, British ship owners, British distilleries, British consumers, abolitionists, and dissenters—all seemed to be in one way or the other conspiring against the planters. Even the British government seemed to be turning against them. Feeling threatened and helpless, they vented their irritation on the evangelical missionaries. When the slaves rose up in 1823, the colonists would find their opportunity for revenge.

CHAPTER FIVE 

Voices in the Air

Got one mind for white folks to see
'Nother for what I know it is me
He don't know, he don't know my mind[1]

In 1831, in response to those who accused the friends of emancipation of instigating slave revolts, the American abolitionist William Lloyd Garrison wrote:

> The slaves need no incentives at our hands. They will find them in their stripes, in their emaciated bodies, in their ceaseless toil, in their ignorant minds, in every field, in every valley, on every hill top and mountain, wherever you and your father have fought for liberty—in your speeches, your conversations, your celebrations, your pamphlets, your newspapers—voices in the air, sounds from across the ocean, invitations to resistance, above, below, around them![2]

Garrison scorned those who saw an outside instigator behind every revolt. But one had to be an abolitionist like Garrison to understand the point so well. Slaveowners, managers, and royal authorities would have found it too threatening to recognize that the seeds of rebellion really were everywhere, grounded in the slave experience itself. It would have been even more threatening to admit that the slave experience transcended the boundaries of the plantations to encompass a larger world of symbols and meanings, market fluctuations and imperial policies, struggles for power and revolutionary ideologies—a world they could not hope to control.

Like oppressors of all times and places, slaveowners, managers, and royal officials preferred to think the oppressed would rise only if instigated. By converting a historical process as complex as resistance and rebellion into a conspiracy promoted by a few men, they sought to preserve the illusion that they could control what was in fact uncontrollable. Only if rebellion could be equated to the possession of a healthy body by an evil spirit, could it be exorcised. The first task then was to identify the culprits behind the rebellion. The rituals of exorcism would follow: trial, flagellation, and death.

So it was in Demerara in 1823 when the slaves rose up. Planters, managers, and local authorities spent many days trying to find the instigators of the rebellion. They blamed British abolitionists, evangelical missionaries, and the "reformist party" of Wilberforce for the destruction of life and property. They called hundreds of witnesses—slaves, managers, masters, officers of the regiment, missionaries, anyone who might bring evidence which would serve their purpose. The investigation generated an extraordinary collection of documents, which offers a vivid picture of the slave rebellion. But the trial records need to be interpreted with care. No doubt many facts were distorted. Most of the evidence was given by whites thirsty for vengeance, or by slaves who were either terrified at the prospect of incriminating themselves or trying to ingratiate themselves with the authorities in the hope that they would be spared. There are also many contradictions and omissions. Witnesses answered only the questions they were asked. They did not speak of what they may have thought relevant, but of what those who conducted the inquiries thought counted. Some slaves were actually told that they would be seen with more sympathy by the court if they gave evidence against John Smith—a fact that came to light later. And they did. Some tried to excuse themselves by blaming others. Most probably lied at one time or another. Since some spoke no English, their testimony was heard through interpreters who, although under oath, may have bent the truth. And when they did speak English, the voices of the slaves could hardly be recognized after being transformed into "readable" English by zealous bureaucrats.[3]

The record prompts many questions. Did the slaves always refer to each other as "the negroes," as the transcript and other papers produced by whites suggest? When they talked about their wives did they say they "kept a girl"? Did they talk about new laws "coming out" from England? Or were those the expressions used by whites as the verbal tokens of slave testimony? To what extent were slaves deprived of their own speech in this process? Is it not in the nature of domination to claim to speak for another? How much of their original speech had the slaves already lost, even before they came to testify? How much of the whites' ways of talking had the slaves already absorbed?

There are still other problems. The testimony is sometimes vague about questions of time, and it is difficult to tell with precision when something happened. One week, or three weeks before the rebellion? One year or two? As one perplexed slave, whom Smith once asked how old his child was, put it: slaves were "not brought up to dat."[4] Indeed, why should they care about "time" as the whites defined it? Only masters and managers cared about it: they wanted to be sure that the slaves began work on time, that they did not take any extra time at lunch, and that they did not quit work early at night. Managers and masters had to worry about the times ships were coming or going, the time they had to report on the estate's production, the time to pay taxes and bills, the time they received their money. Slaves did not, so why should they care? For them there was only work time, which belonged to their masters, and "free time"—which belonged to themselves.

But, in spite of all the many possible distortions, imprecisions, and gaps to be expected from evidence collected in trials and inquiries of this type, the record is so voluminous and detailed that it is possible (with the help of cross references to other documents) to get a vivid and probably pretty accurate picture of what happened—as accurate as a historian's picture of any event can be. In fact, it is astonishing that under so much pressure the slaves managed to be as precise as they were, that they could recall so many details, and that when their testimonies are compared they coincide in so many ways. Having been brought up for the most part in pre-literary cultures in which oral tradition played a crucial role, slaves could remember with great accuracy sermons they had heard, conversations they had had, places where they had stood, a particular pew in the chapel, a certain step on a porch, doors left opened or closed, people passing by. And when they learned how to read and write they still could repeat almost word for word letters they had written or received. In the end—from the hundreds of pages of witnesses' accounts, from the trial records, missionaries' journals and letters to the London Missionary Society, the fiscals' records, minutes of the Court of Justice, memoirs and narratives of militiamen, and from the colony newspapers—the picture that emerges confirms the truth of Garrison's words. There were indeed voices in the air.

Despite the voluminous testimony and widespread discussions of the events that took place in Demerara no one could tell when the idea of an uprising formed in the slaves' minds. Nor could any one tell exactly what their original purpose was. Did they intend from the beginning to rebel and take over the colony—as some people later insinuated? Or did they only intend to strike—"lay down their tools," as they put it—to force the governor to implement the "new laws" that had come from England favoring the slaves, "laws" the planters and managers seemed reluctant to implement? It was never clear, either, whether the slaves thought the "new laws" made them free, or whether they just expected to have two or three days a week for themselves, so that they could cultivate their gardens, go to the market with their produce, and attend religious services. All the different versions of their goals appear in the documents, and sometimes the same witness gives first one version and then another. This seems to indicate that not only had the rebels disagreed from the beginning about the goals to be achieved, but in the course of events many changed their strategies and purposes.

Some slaves testified that the plot had been in the making for more than a year.[5] In fact, in December of 1820, word had spread that some slaves on the East Coast were involved in a conspiracy. A few had been arrested. One slave, Bill, was taken to prison "for conspiring against the white people." This had surprised Smith, who had always thought Bill "quiet and well disposed." But Bill's house had been searched and a gun found—or so it was claimed. Three other guns were found somewhere else in the neighborhood.

Bill belonged to the large Rogers family. The owners of *Clonbrook* and co-owners of *Bachelor's Adventure*. Polidore, another of their slaves, had run away, and when he returned he informed his master that he had been concealed at *Success,* where he had learned that the slaves there were plotting an insurrection and that the slaves of the Rogers family intended to join it. They were saving money to buy arms, and Bill and some other slaves had guns in their houses. Polidore's master sent the information to the fiscal, who ordered a search. Nothing was found at *Success,* but Bill was imprisoned.

When Smith heard that Polidore had been hiding in the house of Jack Gladstone, a slave who occasionally attended services at Bethel Chapel, Smith reprimanded Jack for harboring a runaway slave. Jack protested: he had not even known that Polidore had run away. Perhaps because Jack was Quamina's son, Smith believed him and concluded that Polidore had fabricated the whole story to please his master and avoid punishment.

At the time of Bill's arrest, the colonial authorities were convinced that there was imminent danger of an uprising, but after the first shock they seem to have dismissed Polidore's tale. Such rumors were routine in a slave society and whites had learned to live with them. After being flogged and spending some days in jail, Bill was released and everybody forgot the incident.[6] But during the 1823 trials, the slaves Sandy of *Non Pareil* and Bristol of *Chateau Margo* connected this episode to the rebellion. Bristol also mentioned that from time to time his brother-in-law Jack Gladstone and two or three others said that everybody should fight the whites, and "if they could do no better, they would go to the bush."[7]

Several alarming episodes occurred in 1822. First several buildings burned in Georgetown on consecutive days. On March 28 the governor offered emancipation to any slave (other than an accomplice) who would give the fiscal information leading to the detection and conviction of whoever had committed the arson. A month later he raised the reward, offering in addition 1,000 guilders. But no culprit was found and the case remained a mystery.[8] Then, in August, seven slave houses "accidentally" burned on plantation *La Bonne Intention.* A week later, the residents on the East Coast were alarmed by the ringing of bells and blowing of horns and shells. The boiling house at plantation *Mon Repos* was on fire.[9] It was never determined whether the slaves were responsible for the fires, but it is possible that these isolated acts of sabotage were expressions of the slaves' growing discontent.

Not much attention was roused when several slaves from *Success,* including Richard, one of the drivers, took off in 1823. Slaves were constantly running away and there was no reason to believe that this flight meant anything unusual. No clear connection was ever established between their flight and the rebellion, although Richard did come back the day the rebellion broke out, as the leader of the most aggressive group of blacks, those who assaulted John Stewart, the manager of *Success,* and put him in the stocks. No one seems to have attributed much significance, either, to the fact that at about the time Richard had run away, many slaves had started turning out late for their work.

No one spoke of rebellion any more until a few weeks before the slaves decided to rise (or if they spoke they left no records). This scarcity of evidence, however, does not necessarily mean that the conspiracy was not deep-laid and long in the making. It may only mean that those who collected the evidence at the time of the trial were not interested in searching deep into the past. All they wanted was to find the culprits of the rebellion, and they preferred to believe that the slaves had not planned long and entirely among themselves.

It is possible, however, that Bristol and Sandy were right, that the succession of incidents, and many others never recorded, were not just typical fleeting gestures of slave resistance, but, in fact, were connected. When Jack said that the slaves should fight, his words may have been more than the occasional outburst of a slave dreaming of freedom. The idea of rebellion must have been always latent in Demerara, as it was in slave societies everywhere—not as a clear and well-shaped notion, but as mere possibility: an aspiration to be free that circumstances could crystallize into a concrete plot. And the plot may have been taking shape for a while. This interpretation seems even more plausible in light of the fact that the downward trend in commodity prices and the transition to sugar on some plantations of the East Coast had caused masters to intensify labor exploitation and to encroach on slaves' customary rights precisely at a time when the slaves' notions of rights and commitment to freedom had been enhanced and expanded by British abolitionists' rhetoric and the preaching of evangelical missionaries. If so, it would be only a question of time and opportunity for the slaves to rise. In Demerara the opportunity came when two contradictory things happened, one that threw many slaves on the East Coast into despair and resentment, the other that triggered excitement and hope. The first came in May 1823, when Governor Murray reissued Bentinck's proclamation forbidding the slaves to attend the chapel without passes. The second occurred a few weeks later, when rumors of emancipation began to spread in the colony.[10]

On May 25, 1823, Murray sent to the proprietors "an extract of a despatch containing the instructions of his Majesty's Government relative to the religious worship of the negroes on estates." It actually was an old order directed to Bentinck by the home government many years before and reenacted by Governor Carmichael. Like many of the instructions concerning slaves, it had been disregarded. Murray now felt the need to resurrect it.

The preamble explained that Murray had decided to reissue the instructions

> in consequence of his having become acquainted with the existence of a misconception, of a very serious nature, among the negroes in some districts, and more particularly on the estates on the East Coast; leading them to consider the permission of their master unnecessary to authorize their quitting the estate on Sundays for the purpose of attending Divine worship—a misconception of so injurious a tendency, as to render the most active measures necessary, effectually to eradicate it.

But the instructions also stressed that "considering the beneficial consequences which could not fail to result from the general and judicious extension of religious sentiments among the slaves," nothing less than a very urgent necessity "should induce the planters to refuse passes to the slaves who wished to attend Divine worship on Sundays." To avoid possible "abuses arising out of these indulgences," the governor recommended that the slaves be accompanied to the place of meeting by an overseer, or some other "white person." This would have the advantage of enabling the planter to judge of the doctrines that were being held forth to his slaves.

After this introduction came the instructions that had been sent to Governor Bentinck in 1811. When it was first issued, this document had greatly upset Wray, then at *Le Resouvenir*. It was now likely to upset Smith even more. The document began by stressing the importance of providing the slaves with religious instruction, but then went on to define a series of rules aimed at giving planters and managers the power to control the way that instruction was to be given. It was a masterpiece of colonial legislation, designed not to hurt either the missionaries' work or the planters' interests. And like many other masterpieces of colonial legislation that tried to satisfy antagonistic interests, and to conciliate what could not be conciliated, it failed. Commenting on these instructions, Smith wrote in his diary, "The circular appears to me designed to throw an impediment in the way of the slaves receiving Instruction, under colour of a desire to meet the wishes or rather to comply with the commands of His Majesty's Government."

The instructions read:

> It must in the first place be understood, that no limitation or restraint can be enforced upon the right of instruction and of preaching on particular estates; providing the meetings for this purpose take place upon the estate, and with the consent and approbation of the proprietor or overseer of such estate. Secondly, as it has been represented that on Sundays, inconvenience might arise from confining the hours of meeting in chapels or places of general resort, to the period between sun-rise and sun-set; it may be proper that on Sundays, the power of assembling should be extended to certain hours of the day, viz. from five in the morning till nine at night—and on the other days of the week, the slaves should be allowed to assemble for the purpose of instruction or Divine worship, between the hours of seven and nine at night, on any neighbouring estate to that to which they belong, provided such assembly takes place with the permission of the owner, attorney, or manager of the slaves, and of the owner, attorney, or manager of the estate on which such assembly takes place. Thirdly, to prevent any possible abuse, it may be advisable that all chapels and places destined for Divine worship of public resort, should be required to be registered. The names of the persons officiating in them, should be made known to the governor, and the doors of the places should be opened during the time of public service or instruction.[11]

The reissue of these 1811 instructions could only cause trouble. When Governor Murray took office, everyone seemed to have forgotten the regulations, until he—pressured by the complaints of some of his burghers and observing that the fiscal was constantly having to intervene in conflicts between managers and slaves about religious services—decided in 1823 to dig the instructions from their bureaucratic oblivion. What triggered the governor's decision was a conflict on one plantation under John Pollard's supervision. Pollard was the manager of several plantations located on the East Coast. He had always been known to the missionaries and to the slaves as a man who constantly persecuted slaves who went to the chapel or held religious meetings in their houses. A group of twenty male and female slaves (among them Sandy and Telemachus, who would later play an important role in the rebellion) finally set off for Georgetown to register a complaint with the fiscal. On their way the slaves stopped to talk to the Reverend Wiltshire S. Austin of the Church of England.[12] Perhaps they hoped he would intercede for them or simply give them some good advice. But they spoke with such a force that they would sooner die than give up their religion, that Austin took it on himself to suggest to the governor the convenience of issuing some sort of clarifying regulation. This Austin certainly did with the intention of avoiding conflicts, but the result instead was even more conflict.[13]

The governor's proclamation generated first confusion and then discontent among the slaves who attended services at Bethel Chapel. On May 25, the burgher officers Michael McTurk and James Spencer ordered the managers in their respective districts, which included Smith's chapel, to "wait upon them" with four of the "principal negroes" from each estate, to hear the circular containing the governor's instructions. The slaves were told that no longer should any one go to the chapel without a pass, nor hold any religious meetings on the estates, without permission from their managers. Spencer was very aggressive, particularly with Sandy of *Non Pareil,* a slave who held meetings in his house for teaching the catechism. Spencer threatened him with punishment if he held any more meetings against the manager's will.

The proclamation prompted a series of confrontations on several estates. There was harassment by managers and planters, resistance on the part of slaves, indignation on the part of Smith, to whom the slaves went to complain. One manager refused to give a pass to a slave merely because he did not like him, or so the slave said. Another told an old woman that slaves who could not work could not go to the chapel. The manager of *Clonbrook* vowed to punish severely any slave who did go to Bethel Chapel. On some plantations the slaves had to wait for hours for their passes, missing the Sunday morning service. On others, only a few slaves got passes. As John Smith wrote in his diary, if all planters acted on the governor's recommendation, there would be no congregation on Sabbath morning.[14] But, while some planters and managers used the governor's instructions either as a pretext to persecute slaves they particularly disliked, or to prohibit all from going to the chapel, others, aware of the trouble that might flow from such restrictions, ignored the instructions altogether. After a few weeks of confusion and harassment everything returned to normal. A month later, Smith could write in his diary that the service on Thursday evening had been well attended by as many as three hundred people. And there were even more people on the following Sunday.[15]

Yet the seeds of discord had been sown. The same day he celebrated the large attendance, Smith wrote in his diary that Isaac, a slave from *Triumph*, had come to ask if the governor's "new laws" forbade slaves to meet together on their estates in the evenings to learn the catechism. The manager had threatened to punish them if they did. Smith told him the law gave the manager no such power, but he advised Isaac to give it up rather than "give offense and be punished." Such incidents, of course, could only irritate the slaves. Smith's intervention exposing the manager's arbitrariness and unfairness could only intensify slaves' discontent and aggravate managers' hostility toward the missionary.

It was in this setting that rumors of freedom began to circulate. As the talk had it, some papers that made the slaves free had come from England, but masters were hiding the truth.[16] The rumors were grounded on events taking place in England. In March of 1823, Thomas Fowell Buxton, one of the leaders of the antislavery campaign and close associate of Wilberforce, had presented a motion to the House of Commons declaring, among other things, "that the state of slavery is repugnant to the principles of the British Constitution, and of the Christian religion; and that it ought to be abolished gradually throughout the British Colonies, with as much expedition as may be found consistent with a due regard to the well-being of the parties concerned."[17] His purpose was to extend to slaves the protection of British law, to curtail abuses of the masters' power, and to free all children born from a slave mother after a certain date. Buxton made clear that although his aim was the extinction of slavery throughout the British dominions, he wanted it to happen through a series of preparatory measures aiming at "qualifying the slave for the enjoyment of freedom." All restrictions to manumission were to be removed and slaves authorized to purchase their freedom a day at a time. Slave marriages were to be enforced and sanctioned. The Sabbath was to be reserved for rest and religious instruction, with another day set for the cultivation of provision grounds. Slaves' testimony would be accepted in court, and measures would be taken to restrict abusive forms of punishment and replace the driving system with another more humane.

In spite of its conciliatory spirit, Buxton's motion had met with great opposition from the West India lobby. Trying to moderate the tone of the recommendation, George Canning, the Foreign Office Secretary, had skillfully rephrased it, putting more emphasis on preparing the slaves for freedom than on abolition, and making abstract references to civil rights and privileges, private property and the safety of the colonies. Canning stressed the need to adopt "effectual and decisive measures for ameliorating the condition of the slave population . . . such as may prepare them for a participation in those civil rights and privileges, which are enjoyed by other classes of His Majesty's subjects." And he added that the House was anxious to accomplish that purpose at the earliest period "compatible with the well-being of the slaves themselves, with the safety of the colonies, and with a fair and equitable consideration of the interests of private property." With these softenings and cautions, the motion was carried. After government's consultation with a subcommittee of the Society of West-India Planters and Merchants,[18] several resolutions designed to ameliorate the conditions of slaves were approved by the King-in-Council and forwarded to the colonies, though only as recommendations.[19] It was clear, however, that if the colonies did not follow the instructions the British government would intervene to enforce its implementation.[20] In a dispatch of May 28, 1823, addressed to Governor Murray, Earl Bathurst, the Colonial Secretary, said that for the moment he did not intend to deal with all the proposed reforms and would confine himself to two which he was sure the Court of Policy would adopt: the prohibition of the flogging of female slaves and of the use of the whip in the fields. This dispatch was followed six weeks later by another that included the promotion of religious instruction, the banning of Sunday markets, the encouragement of marriage and family, a prohibition against separating husbands, wives, and children under fourteen, and improvement of the conditions of manumission.[21]

The first of these dispatches arrived in Demerara in the first week of July, 1823, followed by a letter signed by John Gladstone and several other plantation owners who resided in Great Britain, recommending the implementation of the measures without delay, but also warning the colonists to be prepared in case of slave disturbances.[22] The government's recommendations caused great irritation among the colonists, who saw them as just another unwise and undue interference in their lives. The Court of Policy met to discuss what to do, but for several weeks reached no decision. It finally agreed in the first week of August to the proposed reforms, but failed to make the decision public, probably because several planters continued to oppose it. Meanwhile the slaves heard about the British government guidelines, which they referred to as "the new laws"—which some understood as a grant of immediate emancipation. (Canning's casuistry had been clever enough to bemuse the members of Parliament, but not the slaves, to whom emancipation was the only thing that really mattered.)

Inevitably, the first to hear about these "new laws" were slaves who worked as household servants. Already in 1813, John Wray had noticed that the slaves were well acquainted with everything respecting slavery and its abolition from gentlemen's servants and others who had been in England and from their masters' everyday conversations at their tables. Whites spoke freely on these subjects before their servants, who then passed on to others what they heard. That was precisely what happened in 1823. Ironically, one of the first to spread word of the new laws was Joe Simpson (or Packwood), the "boy" of Alexander Simpson, a slave who later betrayed the conspiracy to his master. Apparently Joe overheard a conversation among his master and some friends about the government's measures. Confused and excited, he went looking for Quamina to tell him the news. Unable to locate him, Simpson called on Cato, a free black man who lived in the neighborhood of *Success*. Later, Cato told Quamina about Joe's visit. Intrigued, Quamina sent two slave boys to Joe, with a letter asking for details. Joe Simpson answered that he would search his master's papers and send word as soon as he knew more.

Meanwhile, Susanna, the slave who was living with John Hamilton, the manager of *Le Resouvenir,* told Jack Gladstone that the slaves were going to be free. Susanna, who had been expelled from Smith's congregation two years earlier because of her affair with the manager, had remained Hamilton's mistress.[23] Now, Hamilton had been fired by the plantation's attorney and was preparing to leave. Confronted with his departure, Susanna asked Hamilton to buy her freedom and the freedom of her children. But he refused. That would be like "throwing money in the trench," he said, since they all soon would be free. A few days later Susanna gave this "news" to Jack Gladstone.

The rumor was quickly spreading among slaves both in the town and in the countryside. Any careless remark by a master, a manager, or an overseer that could be interpreted as a reference to the "new laws" was quickly transformed by the slaves into evidence that they were free and their masters were withholding the good news from them. A few days after Susanna spoke to Jack, an overseer flogging a slave at *Le Resouvenir* shouted in a fit of rage: "Because you are to be free, you will not do any work, nor wait till your freedom is given you, but wish to take it yourself." Such words fell on eager ears and were soon reported to Jack.

Roused by the idea of imminent freedom, Jack went to town looking for Daniel, the governor's servant, who was in a better position to know whether the rumors were true. He could have overheard some conversation or might have read about it in the governor's papers. Daniel (who later in the trial admitted he was in the habit of reading the governor's papers) promised Jack that he would investigate. During the conversation Jack must have hinted at a rebellion, because Daniel cautioned him that he had seen a paper about the "Barbadoes war"—a reference to the 1816 rebellion there—where there had been the same talk about freedom and the slaves had made a "foolish" war and a great many lost their lives. Then, perhaps looking for guidance in the Bible, Jack read to Daniel the fifth chapter of Romans. And Daniel in turn read the third chapter of the second book of Timothy, where he found a verse very appropriate to the occasion: "Ever learning and never able to come to the knowledge of the truth." Jack told Daniel that their reading at the chapel on the previous Sunday was: "All things working together for Good for those who love God." Such words must have sounded prophetic.[24]

Jack Gladstone was about thirty years old at the time. He stood a full six feet two inches and made a striking impression on whites, perhaps because of his rather European features. John Wray thought he had a "lively yet thoughtful countenance which gave him a noble expression." And the notice of a thousand-guilder reward for Jack's capture after the rebellion described him as "handsome" and "well made," with "a European nose."[25] Jack had a reputation of being a "wild fellow," and during his trial it became obvious that he was a clever and daring man indeed. Although he went regularly to Bethel Chapel to meet his friends and knew John Smith well, Jack had little to do with him. Jack had been baptized and sometimes played the role of a "teacher," but he did not belong to the congregation. He was too restless to accommodate to its rules. Jack had lived with Susanna until she became Hamilton's mistress. After Susanna left Jack, he remarried (this time he was married by John Smith) and his wife was a slave at plantation *Chateau Margo,* where he visited her regularly. But he was still involved with other women—a habit which sometimes caused him and others trouble and pain.

Both Smith and the manager of *Success* had tried unsuccessfully to restrain Jack. On February 5, 1822, the manager, John Stewart, wrote Smith a revealing note, saying that he had received an unpleasant letter from Lachlan Cuming, the owner of *Chateau Margo,* concerning Jack and "his wives." Stewart added that aside from the women Cuming mentioned, George Manson, Cuming's manager, had told him he knew two or three other women at *Chateau Margo* with whom Jack occasionally "cohabited":

> As for my woman Gracy, he has ruined her; she hardly does anything for me and I shall now be under the necessity of sending her to work in the task gang, although I know she has not been accustomed to such a work, but I cannot help it; I have told Jack and her repeatedly of the impropriety of their connection and once punished them by confining them both in the stocks for some time, when they declared they should never be guilty of the same crime after that period, and I believe I have been told, you once or twice gave them a severe lecture on this subject. His father, Quamina, seems to be very much hurt at his son's shameful conduct, although he acted rather hastily to me yesterday morning, for which I found it necessary to confine him for a short period, and if I had punished him, I would have been justified, and certainly his son was the original cause of all this. Under all these circumstances, I feel it incumbent on me to request you to chastise Jack and Gracy, as far as becomes you as a minister, the idea of a married man turning out his wife, and bringing in another woman in her presence, and to her own bed, is to me horrid; with hopes you will excuse me for thus troubling you, and that you will consider it as meant for both Jack and Gracy's future good.[26]

For Jack, this must have meant just another intrusion into his life, another attempt to restrain whatever limited freedom he enjoyed. Most of all he must have resented being put in the stocks. But he well knew what it was to be a slave: to be closely watched; to be punished for things managers and masters were free to do with impunity; to be put in the stocks; to be sent to work in the task-gang; to depend on the whims of managers and masters; and to have his wife and children taken away.[27]

Jack lived with his father Quamina at *Success,* only about a mile and a half from *Le Resouvenir.* Quamina was the plantation's head carpenter, and "first deacon" at Bethel Chapel. He was a man much respected by both slaves and free blacks, and his reputation went far beyond the boundaries of *Success.* The missionaries considered him a wise man and had only praise for his devotion and piety. Yet his initiation into the mysteries of Christianity dated from the time Wray arrived at *Le Resouvenir* in 1808. Until then, Quamina was not a Christian. But he soon became very interested in religion and an assiduous member of the chapel. He told Wray that when he was young he had been a house "boy" and had "fetched" girls for the pleasure of the managers. At that time, he had seen no wrong in doing such things. But Christianity had changed him. Quamina soon gained Wray's confidence and became a deacon. When John Smith replaced Wray in 1817, Quamina had been quick to gain his affection and appreciation. Smith was profoundly impressed with Quamina's devotion and commitment and deeply moved by the emotion with which Quamina prayed. As a deacon, Quamina actively participated in the activities of the chapel and became a sort of broker between Smith and the other members of the congretation. He reported to Smith on the behavior of slaves belonging to the congregation, assisted him in religious services, and helped him to settle disputes.[28]

Quamina was a proud man, and a hard worker. He took a special interest in his people and was always ready to stand up and speak for them. Like many slaves in Demerara, he was an African and must have been sold as a slave when he was still a child, together with his mother, who died a slave in Demerara in 1817.[29] Like many African slaves, Quamina had had several wives. But he had lived with Peggy (identified in one document from the 1820s as a free woman) for twenty years, until her death in 1822. Like most slaves, Quamina had been humiliated and severely punished—once so harshly that he had been confined in the "sick-house" for six weeks. At one time or another, to his great distress, Quamina had been forced to miss religious services because he had been sent by the manager to do some work. On the day his wife died, he had been sent to work at a considerable distance. When he returned he found Peggy dead. Such experiences were a constant reminder of his status as a slave, and strengthened his feelings of solidarity with others in the same condition.

Quamina and Jack Gladstone were very close. Working on the same plantation, father and son met frequently. When they heard about the new laws, they told each other the stories they heard and together they pondered what to do. There was, however, a great difference between the two. Quamina was a man of reason, Jack a man of passion.

For a slave, Jack had a relatively privileged life. He was a skillful artisan—a cooper—and did not have to work under the direct supervision of a driver and the constant threat of the whip. He enjoyed a relative degree of freedom to move about and often went to town or to other plantations to visit friends and relatives. But, like everyone else, his life was affected by the rhythms of the market and the acceleration of the pace of labor, and he was subjected to humiliating constraints, interferences in his personal affairs, and abusive punishments. Jack was tired of being a slave, and the idea that some papers had come from England making slaves free and that masters were keeping the news from them was intolerable. He would not rest until he knew what was really going on. So, after he talked to the governor's servant Daniel, he visited Tully, an African who lived in Georgetown and belonged to a carpenter named Hyndman.

Jack wished to know whether Tully (also called Taddy at the trial) had heard anything about freedom "having come out." Tully asked Jack who had told him that. "Some of my friends," answered Jack, discreetly. Then he added more boldly that he would get freedom by force if he could not get it otherwise. (Later, in his trial, Tully reported that he asked Jack whether he was able to do such a thing and that Jack had answered yes; but Tully had seen "no fight on his eyes.") Tully had known Jack since he was a little boy. He had a girlfriend at *Success,* where Jack lived, and he used to go there to visit her. He trusted Jack, so he told him that indeed he had heard some rumors in town. And he may have agreed to talk to his friends about Jack's project. (In his testimony, Jack said that Tully had promised that if Jack and his friends decided to rise up, he would send all his friends from town—something Tully denied.)

The idea of rebellion was growing in Jack's head. He repeatedly inquired about "the papers that had come out from England" and kept hearing new stories that seemed to confirm his suspicions. The stories circulated rapidly from one slave to another through a network of friends and relatives. York reported to Jack that he had heard from Damon and Providence, of *New Orange Nassau,* that their overseer had told them the slaves were free, that "all the great men at home had agreed to it except their masters," and that "he dared say that they would rather give the slaves three days [a week] than freedom." Then Gilles added further evidence. Arriving at *Success* from the West Coast, where he lived, he told Jack that his master had read to the slaves in the field a paper saying that no women were to be flogged, and that if men did anything wrong the master himself should "lick them" and lock them up. Since then, no driver on his plantation had carried a whip and no woman was flogged.[30] Yet another slave informed Jack that he had overheard Michael McTurk say, at dinner with Lachlan Cuming, he did not know what made the King so "foolish" and "partial" as to give the slaves freedom; it would have been better to give them three days a week.

Jack continued to gather information and to talk with his friends about what he heard; and the more they talked, the more confident they became; and the more confident they became the more they thought rebellion might succeed. The best place to meet people was at the chapel on Sundays. In recent years, an increasing number of slaves had been going to the chapel. In spite of the governor's proclamation and the masters' attempts to restrict attendance, the chapel was always so full that many had to stay outside, hanging around in the shade of the palm trees. Some days there were more than six hundred. Slaves would come from distant places, as far as plantation *Orange Nassau,* sixteen miles away. After service they would meet at the *Success* middle path and chat.

At the beginning of August, Susanna came with a new story. "Mr. Hamilton, says we are to be free," she said, "but he does not think they will give it to us unless all the sensible people went by force about it, and would not give it up without a positive promise from the governor."[31] Susanna also said that Hamilton had told her that the governor had estates himself and would not give the slaves freedom, and that the planters had made a large subscription "to keep it back." Slaves would only be free if the King bought all the "gentlemen's estates."[32] Jack had no means to know whether all this was Susanna's fabrication, or if Hamilton had really said such things. But he was intrigued. "What do you mean by force?" he asked. "Are we to fight or what?" No, the slaves were not to fight, said Susanna. "They were to take the arms from the whites and drive them to town." She added that Hamilton wished the slaves would give him some time till the coffee harvest was over, so that he could sell his things, but "if one or two sensible men wanted to talk to him he would explain how they should do it."[33]

Jack could hardly wait to speak to Hamilton. But although he tried several times to see the manager, he had no success. Every time he went looking for Hamilton, the manager was either busy or out. Susanna continued to repeat her story, adding details and instigating the slaves to act. She said that when the governor came to see the reason for it all the slaves should come forward and speak out. If they were not "a parcel of cowards" they would have had freedom already, "for it had been ordered for some time." This was neither the first nor the last time Jack would hear the word "coward." One day, when Jack and Cato were talking about the papers that had come from England, Cato also called Jack a coward: "You see things plain before you and won't search for it." "Coward" was a charged word among slaves, a word that spurred them to action. In a culture that made stoicism a necessity, no one wanted to be called a coward.

As time passed, more and more slaves started talking about the new laws and in the course of their conversations the logistics of the rebellion began taking shape. Jack was busy enlisting people to participate in the uprising. He had gone around the fort and had seen too many guns there, but Paris, a slave who had worked in the powder magazine before, promised Jack that "if he could get three daring fellows to assist him," he would have the powder magazine secured. Jack had some doubts. He thought that if Paris interfered with the fort, the troops would murder him and anyone who helped him. But Paris insisted that was the only way to proceed.[34]

On Friday, July 25, 1823, Quamina and Bristol decided they had better ask Smith whether he had heard the report that "the King had sent orders to the governor to free the slaves." Smith told them he had not heard of it and if such report was in circulation, it should not be believed because it was false. Quamina insisted. He was sure there was something happening and he wished to know what it was. Smith asked him from whom he had "imbibed" this fancy and Quamina answered that his son Jack had heard it on the previous Sunday from Daniel, the governor's servant, who had heard his master talk with some gentleman about it. Besides, several "negroes" had heard the same thing in town. Smith replied that it was likely that some orders had been sent out to the governor, since the government in England did intend to make some regulations for the benefit of the slaves, though not to make them free. Smith's answer did not satisfy Quamina.[35] Half-measures were not what he hoped for. The days passed without the slaves hearing from the governor about the new regulations. Some started doubting they could trust Smith. After all he was a white man, why should he "deny his own colour for the sake of black people"?[36]

And, indeed, a few days later Smith mentioned to John Stewart, the manager of *Success,* that there were rumors among the slaves about some instructions the government had received. They were saying that "freedom had come out," and a number of them had asked him about it, Quamina in particular. Before a week had passed, Richard Elliot came from the West Coast to visit Smith, who shared with him his concerns. Both missionaries called on Stewart, and Smith spoke again about the slaves' agitation.[37] Stewart reported the conversation he had with the missionary to Frederick Cort, the plantation's attorney, and Cort decided to talk to Smith. So, on August 8, Stewart and Cort paid a visit to the missionary. Cort asked him if the slaves really believed that they had been made free. Yes, answered Smith, several had asked him about it. How could they have found out such a thing? Cort asked. They might hear it in various ways, Smith replied. They could hear it from sailors when they came down from the estates with produce. After all, sailors were constantly teasing them about what fools they were to be slaves. They could also hear it from hucksters in town. Cort, suspicious, pressured him further. "Did he know of any other person who might have told the negroes?" (He was most certainly thinking of Smith.) Smith said he had no idea. He admitted that he had considered telling them from the pulpit that the rumor was unfounded, explaining to them what he believed to be the truth. Cort warned Smith not to assume such responsibility. Whatever he said "might be exaggerated to his own prejudice."[38] Sooner or later the Court of Policy would reach a decision. The only reason for the delay was that one member of the court had been sick and another was out of the country. Perhaps convinced by Cort's cautions, Smith kept quiet.

Meanwhile Jack and Quamina had asked another slave, Dumfries, to go up the river to *Rome* (one of the largest plantations on the West Coast, with about 600 slaves) to recruit support. Dumfries was to tell certain slaves there that the King had "sent out their freedom but the white people did not want to give it to them." The slaves on the East Coast were "going to make war with the white men." If the people from *Rome* did not help them, they should at least not help the whites. Dumfries, who was also a cooper and worked with Jack in the same shop, was a Coromantee.[39] So were Smart, the slave who went with him to Rome, and Quamina, Jack's father. At *Rome,* Dumfries and Smart talked to two other Coromantee slaves, Quashy and Quamine[a], who told them there were two "Papa negroes," Fuar and Namitta, belonging to *Peter Hall*—a neighboring plantation—who knew where the maroons were.[40] If the people from the East Coast waited a few months, they could send for the "Bush negroes" (the name given to the maroons) to help. The next day, Smart met Jack and told him the people at *Rome* would send for the "Bush negroes."[41]

It is possible that, at one time or another, some of the slaves had been in contact with the maroons. A number of slaves in Demerara were originally from the Corentin area in Berbice, where there were several maroon communities, and maroons sometimes came down the river to trade with the colonists. Some maroons had even been spotted in the Sunday market in Georgetown.[42] Besides, the Corentin was not far from maroons who lived in the interior of what is today Suriname. The maroons descended from slaves who had managed in the eighteenth century to sign a treaty with the Dutch guaranteeing their freedom. Since then they had lived more or less in isolation on their own land, away from the whites. Their history was wrapped in mystery and legend. And the mystery and legend could only stimulate the slaves' dreams of freedom.[43] To complicate things, the word "bush" was used to define the wooded areas surrounding the plantations, and slaves who ran away were described as having gone to the bush. In Demerara there were many small bush camps. This may have led to some confusion. Later, during the inquiries, Namitta swore that although he had been in the bush with one friend, he did not know anything whatever of the "Bush negroes." Whether he was telling the truth is something we will never know, but the fact is that, with good reason or without it, the word that the "Bush negroes" were coming to help them spread swiftly, and this rumor only increased the enthusiasm for the rebellion.

Among the many people Jack talked to was his half-brother Goodluck.[44] Jack's father, Quamina, had lived with Goodluck's mother, and Jack and Goodluck had been friends for a long time. Goodluck belonged to Peter McClure, a free black who lived in Georgetown. But Goodluck had a wife up the coast at *New Orange Nassau,* and he used to spend nights there from time to time. Four years earlier, in September 1819, Jack's brother-in-law Bristol, a deacon at Bethel Chapel, had introduced Goodluck to Smith as a man addicted to evil things, and a "disturber of the Christians." Since then Goodluck had rarely missed a service and in January of 1820 he applied to become a member. That day Smith wrote in his journal that Goodluck had confessed he had done all sorts of mean things to his fellow men. To gratify his master, he had told many lies, and had caused many slaves to be unjustly punished. He was the one to execute the undeserved punishment, and did it so brutally that he "seldom failed to stain the ground on which he stood with the blood drawn from the innocent sufferers by his merciless lashes." But now he wanted to change. From then on he came to the chapel regularly.[45]

One Sunday, when he was returning from the chapel, Goodluck stopped at the house of Mary Chisholm, a free black woman who lived in front of *Success* and made her living baking bread. Her house had become a gathering point for slaves who lived in the neighborhood. There Goodluck met Quamina, Seaton, Bristol, and Manuel—all members of the Bethel congregation. Jack had just returned from town convinced that the slaves were to have at least three days for themselves. Daniel, the governor's servant, had told him that he had seen it on a paper that was on the governor's table, but he could not show it to Jack because the governor was in the house. Jack said he had told Daniel that the "negroes were going to try to get it" (meaning "their rights"). But Daniel advised them to wait, for if it was "a thing ordained by the Almighty" they would get it. After he had heard Jack's story, Quamina asked Goodluck to make some inquiries in town. While they were talking, a white man appeared on horseback, and they cautiously dispersed.

Goodluck returned to *New Orange Nassau,* and went to town the next morning to make his inquiries. The first person he spoke to was Alfred, a slave belonging to Johanna Hopkinson, a mulatto woman and the mother of

John Hopkinson, the proprietor of plantations *John* and *Cove*. Alfred told him he had also heard the same rumors. He had had a conversation with a Mr. Garret, who apparently was so irritated with the news that he had said to him, "Damn it, I wish that what has come out for you, you can get it and go and eat one another." Alfred told Goodluck that he had answered, "Why, after [all] the white people don't eat one another, how do they expect we should eat one another?" Goodluck and Alfred laughed at the joke and rejoiced at the good news. From then on, Goodluck did more telling than inquiring. Whenever he had any opportunity, he spoke about the "good news" and talked to the slaves he trusted about the plot that was developing on the East Coast. When he was rebuked he just let the subject die. When he told John Langevin—a slave belonging to a free colored man—that the people on the coast were going to put down their tools, Langevin replied he would not join them. Goodluck dropped the subject and never talked to Langevin about it again.

When Mandingo George and Congo George, two slaves from *Endeavour* in Leguan—an island on the coast—came to Georgetown in a boat, Goodluck asked them how things were going in Leguan. Mandingo George told him that "they had taken the whip from the driver."[46] Goodluck saw this as another evidence that the whites knew that something "good had come out" for the slaves. Then he told the two men from Leguan that "the Negroes up the coast" were going "to lay down their tools." George, who had come from Grenada,[47] must still have had in mind the bloody rebellion of 1795–96, in which hundreds of slaves had been killed.[48] He seemed scared: "What are you going to do? Like the brigands in Grenada?" Without waiting for an answer, he jumped into the boat and took off. Goodluck did not get discouraged. He saw the punt from plantation *Kitty* lying at Fort Stelling (Kingston) with two of *Kitty*'s slaves. They too had heard some rumors. "Bass, is it true? I hear something has come out very good for us?" one of them asked Goodluck. Goodluck did not miss his chance: "Yes, I have heard it myself." The slave then told him that the "old man" (probably the manager or the attorney of *Kitty*) had said that if they behaved well they would get three days. Goodluck once more passed the information that the "negroes up the coast" were all going "to lay down their tools and to see the Governor about it." A week later he met the same slaves from *Kitty* again. He had gone to buy fish at Stelling when he saw them and went on board their boat. After asking what was the news and being told that "the same news was going on still," Toney, one of the slaves on the boat, confirmed that indeed "everything was true . . . for their engineer had read it to them out of the paper."

Such information reinforced Goodluck's conviction that the masters were hiding something from the slaves. The next time he went back to *New Or-*

ange Nassau to visit his wife, he spread the word. On Sunday, August 3, he went with his wife to *Thomas,* where she was to pay a visit to a friend. There he met a brother from the chapel and started talking about the latest news. The man had also heard about the slaves having three free days a week. But he said they were going to wait and see what the Lord would do for them. "The Lord says we must help ourselves and he will help us," replied Goodluck, always ready to find reassurance in the Bible. And then he told the man what the slaves "up the coast" were going to do to help themselves.

One week later, Goodluck and Jack met at the chapel. Jack asked if Goodluck was ready, and he said he was. "Do you think we are to live all the days of our lives in this way, and have people cutting, cutting up our skins in this way, and know that there is something good for us, and not take it?" asked Jack. "No," Goodluck answered, "before I will live so, to have my skin torn up, I will sooner die."

Thus, from one slave to another, the word went around. The same stories were told time and again. Some slaves understood that the new laws entitled them to have three days free every week, others believed they had been made free. Although they all felt that their "rights" had been violated in one way or another by their masters and local authorities, not everyone welcomed the idea of an uprising. Some slaves thought the plan was risky and refused to support it. They preferred to wait. Others offered to help. Gradually, a leadership emerged from among the slaves who were linked to Jack and Quamina, and tasks were agreed upon. The slaves from *Rome* and their friends from *Peter Hall,* were to bring the "Bush negroes." Tully was to gather his friends in town, and Goodluck was to establish the necessary contacts in Leguan and Essequibo. Gilles was to take care of plantations on the West Coast, and Paris to control the powder magazine.

Jack was busy trying to organize friends on several East Coast plantations. His plan was to have at least one man on each plantation in charge of informing the others when the time came for them to rise. His choice naturally fell on slaves who were catechism "teachers." When he met Jacky Reed, a slave "teacher" from *Dochfour,* at the chapel, he tried to get his support. He told Jacky everything he had found out. Jacky, who lived about sixteen miles from Georgetown, had already heard something, and he commented that on his plantation the slaves who worked in the mill had slowed down. The coopers were "turning one puncheon" instead of two a day, just to see what the manager would do.[49] On his way back to *Dochfour,* Jacky Reed met Bristol, Jack's brother-in-law, who confirmed Jack's words. Present at this conversation were Manuel from *Chateau Margo,* and Benny and Harry from *Dochfour.* They all knew about the plot. Jacky Reed confided to them he did not feel at ease with the idea of rebellion. Neither did Bristol, who like Quamina was a deacon at Bethel Chapel. Bristol had told Jack several times to drop the scheme and wait. But Jack would not hear him. Every time Bristol tried to dissuade him, Jack called him a coward.

Later, when Jack, Goodluck, and a group of slaves met in Bristol's house at *Chateau Margo,* Bristol insisted again that they drop the affair. This time it was Goodluck who protested. He turned to Jack and said, "You hear that coward Bristol advising to stop the business: what am I to do with all the people I have spoken to in Essequibo?"

Such talk may have been bravado, but it was true that things had gone pretty far. Jack and Quamina had sent to several of their friends on different plantations messages asking them to come to talk. They had met either at *Success,* where both lived, or at *Chateau Margo,* where Jack's wife was, or on Sundays at the chapel. So an increasing number of slaves had become acquainted with the conspiracy—even a few from the West Coast. One of them was Gilles.

Gilles had many friends and acquaintances, partly because he had been sold many times, had worked in a task-gang, and had lived in many different places. He had been head driver at *Endraght,* and was sold off that estate for "knocking down the overseer." He had been bought by a man named McKinean, for whom he worked for three years in a task-gang, going here and there. He was then sold to the West Coast to a James Allan, who kept him for about five years, but sent him off the estate, again for insubordination. Gilles had returned in June to his master's house under a promise of good behavior. While he had been away from the estate he had been hired by Cato, the free black man who lived near *Success,* and had become friends with Jack and Quamina.

When Gilles received a message from Quamina saying that he wanted to talk to him, he came immediately to the chapel. Gilles had four children at plantation *Endraght,* and his wife, who lived at *Cuming's Lodge,* was pregnant. On the pretext that he wished to see his children at *Endraght,* but really intending to see Quamina, he obtained a pass from his manager. On the road he met Hay, a slave from *Success,* who also had a pass to visit his son, who worked for a mason. Gilles told Hay that his son had been sent to Wakenhaam and was out of reach. So Hay decided to return to *Success.* On the way, Gilles explained that he had received "a message" from a black butcher in town (probably Bob Murray) that Quamina wanted to see him. When the two men reached *Cuming's Lodge,* Gilles stopped while Hay went on to *Success.* The next day they all met at the chapel. It was settled that Gilles would do what he could to get the estates on the West Coast to join. The next morning, when Hay met Jack in the cooper's shop, he told Jack the whole story.

Thus, across a dense and complex warp of loyalties based on friendship, family, work companionship, affiliation with the Bethel Chapel (though not necessarily membership), and ethnic or linguistic identities, the threads of the plot were woven. Jack Gladstone was Quamina's son, Goodluck his half-brother. Bristol was Jack's brother-in-law and had known Goodluck's mother, who, in Bristol's own words, had been very good to him. Susanna had been Jack's wife.[50] Susanna's son, Edward, a boy of nine or ten years old, worked with Jack as an apprentice. It was Edward who wrote one of the letters Jack sent to Jacky of *Dochfour,* since Jack knew how to read but could not write. It was also Edward who took his mother's messages to Jack. Henry, a boy of twelve who was under Quamina "by way of employment," also wrote letters for Jack and Quamina. Isaac, who also helped with the writing, had run away from his master and had been hiding at *Success* for a year and a half, without being detected.

Attila, who was later accused of having had an outstanding role in the rebellion, although he was never mentioned as one of those who plotted it, told Colin, his half-brother, about the rebellion three weeks before it broke out. Attila was a slave at *Plaisance,* and Colin lived as a slave at *La Retraite,* but was spending a few days at *Plaisance.* Their whole family was involved in the rebellion, and two of the brothers, Louis, who belonged to *Friendship,* and Attila, were later hanged as ring-leaders. Colin was acquainted with Cato, the free black man who lived in the neighborhood of plantation *Success.* Both Colin and his brother Paul had been baptized by Smith and both attended services at Bethel Chapel.

Gilles had worked for Cato, and Cato was a good friend of Quamina and Jack Gladstone. Tully (or Taddy), the African who lived in town, had known Jack since he was a little boy, because "he kept a girl" at *Success.* Manuel, Bristol, and Primo lived at *Chateau Margo,* where Jack had a wife, and they often met him there. Sandy from *Non Pareil* and Telemachus from *Bachelor's Adventure* often went there too. Hay, Seaton, Active, Dumfries, and Smart all belonged to *Success.* Hay, Dumfries, and Jack worked in the same shop.[51]

Most of the people involved in the plot had been at one time or another in the chapel, although not all of them belonged to the congregation. Many had worked together. Some had lived on the same plantation before being sold away. Many lived on one plantation but had a wife or a relative on another, and visited them regularly. Jacky, from *Chateau Margo,* had a wife at *Northbrook.* Toney and his father and brother lived at *Elizabeth Hall* but Toney often slept in his wife's house at *New Orange Nassau.* Bob Murray, the butcher, lived in town but "kept a girl" at *Plaisance.* And he sometimes slept at Cato's house. Gilles's children were at *Endraght* and his "girl" at *Cuming's Lodge.* Bristol's "girl" was at *Chateau Margo* and his father Cambridge and his brother Dick at *Kitty.*

Many of the plotters were artisans. And it was not uncommon for artisans to be hired out from time to time when there was little to do on their own plantations. Others, like Gilles, had worked in the task-gangs that went from one plantation to another doing jobs the proprietors did not want or could not afford to entrust to their own slaves. Some, like Goodluck, had been sold many times, moving from one plantation to another, extending their network of relations. A few, like Cato, had once been slaves and now were free, but kept strong links with the slave community. There were also slaves like the boat men who were always meeting new people because of the nature of their job. And the house servants, like Daniel and Joe Simpson, who had access to their masters' papers and overheard their conversations, had friends among artisans and field laborers.[52]

Most of those involved in the conspiracy had known each other for years. Sam, Toney, Cato, Tully had all known Jack since he was a little boy. Theirs were long-standing relationships, involving years of shared pleasures and pains. Daniel, the governor's servant, knew Jack well. He also knew Bristol "from living with Governor Bentinck." Since the missionaries had arrived, slaves had added new bonds to those they already had. Every week the bonds of solidarity were ritualized and celebrated in the chapel. Conversion to Christianity, however, could have contradictory effects. It could create new bonds, but it could also destroy old ones. Converted slaves may have stopped playing their drums and practicing their ritual mysteries in nearby woods—although it is almost certain many did not. They may have abandoned some of their old ways of celebrating the dead,[53] but they acquired new ones. And they continued to come together for funerals, or on holidays such as Easter, Whitsunday, or Christmas—holidays that the white man had forced slaves to incorporate into their culture. More important, Wray and Smith had given them new, legitimate reasons (from the point of view of the British government) to assemble.

Smith's system of designating one slave on each plantation to be the teacher and to assist the others in learning the catechism gave the slaves a new pretext for meeting. Soon many self-styled teachers appeared everywhere. Although it would be a mistake to think that all those who plotted the rebellion were teachers and deacons—as was insinuated during the trials—it is still true that several were implicated in the plot. Apparently this was the case with Seaton, who was a teacher at *Success,* and William at *Chateau Margo,* David at *La Bonne Intention,* Jacky Reed at *Dochfour,* Luke at *Friendship,* Joseph at *Bachelor's Adventure,* and Sandy at *Non Pareil.* It is also true that Jack Gladstone used this network to communicate with the slaves on various plantations. These new arrangements only multiplied the many opportunities slaves already had to know each other, trust each other, communicate with each other, and finally to plot with each other.

As became obvious during the trials, the slaves who plotted the rebellion moved around with a great deal of freedom. At nights they sneaked off their plantations and walked miles to visit their friends. They also invented many excuses to leave their plantations during the day. In the dry season many slaves went to fetch water at *Le Resouvenir,* where there was a permanent spring. And since wives and husbands, parents and children were often separated, sold away, and lived on different plantations, slaves found pretexts of all sorts when they needed a pass for a day or two. They also received permission to go to Sunday markets to sell the produce of their own gardens and the fowls and pigs they sometimes raised. Or they were sent to town on errands. And despite persecution and harassment from masters, large numbers of slaves had met at the chapel.

Although women had always played an active part in day-to-day resistance, no woman—except for Susanna—appears to have participated in the planning and plotting of the rebellion. Some historians have argued that women were in an even better position than men to promote rebellion, not only because of the mobility they enjoyed as "hucksters" but also because European notions about women denied them any capacity for leadership, placing them beyond suspicion.[54] The lack of evidence of women's participation in the conspiracy in Demerara may indeed have resulted from a blind spot in the eyes of the whites who assembled documents about the rebellion. But it is also possible that—either because of some African tradition or because they feared some women who had close relations with whites might betray the conspiracy—male slaves deliberately excluded women.[55] After the rebellion broke out, however, women enthusiastically joined in. During the trials some were accused of verbally abusing overseers and masters, of going so far as whipping them with bamboos and slapping their faces.[56] Others were seen waving their handkerchiefs, cheering the men when they passed carrying their weapons: "Niger make Buckra ran to day."[57] And at least one woman, Amba, belonging to *Enterprise,* was spotted carrying a musket on her shoulder and was said to have urged slaves who were trying without success to take a fowling piece from a white man to kill him: "You allow one Buckra man to knock down so many of you? Take for me gun and shoot him." (Yet she herself would not shoot him.) Isaura, from *Bonne Intention,* was said to have set on fire the heap of grass behind the slave quarters. And several women were listed as rebels. But whatever their role may have been after the rebellion started, and whatever knowledge they may have had of the plot—Susanna certainly knew of it and may have even been the main agent of Jack's restlessness—there is no evidence that women played an active part in the conspiracy.[58] The rebellion was planned by men.

Aside from the more obvious links among slaves, there were others more difficult to trace, either because they escaped the perception of the whites and so left few documentary traces, or because they were not crucial after all.[59] But it is intriguing to notice that among the slaves who plotted the rebellion, and later among the rebels, many were Coromantee.[60] It is difficult to say how many. As we have seen, Dumfries, the man Jack sent to plantation *Rome,* said in his trial that he was a Coromantee.[61] So was Quashie, the man he was sent to talk to at *Rome.* So were Smart, who went with Dumfries to *Rome,* and Quamine (also known as Quabica), the other slave they talked to. Was this a coincidence? Was it a coincidence that Richard, a slave from *Success* who had been in hiding for several weeks and reappeared as the leader of a small gang when the rebellion started, was a "Gangee" himself, but "spoke Coromantee," and that Jack Gladstone's father Quamina was described as a Coromantee? What about Amba, the aggressive female? What about all the Quashies, Cudjoes, Quabinos, Quacows, Quaws, Cuffees, and Quaminas who were in one way or the other accused of being involved in the plot or the rebellion? There were Quacco from *Chateau Margo,* Quacco from *Success,* Quamina from *Noot en Zuyl,* Quabino from *Chateau Margo,* Cudjoe from *Porter's Hope,* Cudjoe from *Lusignan.* All these names that appeared on the list of men arrested, punished, or executed for participating in the rebellion were typical Coromantee or Akan week-day names, and were common proper names in many communities on the West Coast of Africa.[62] And indeed, immediately before the abolition of the slave trade, many slaves sold in Demerara were advertised as coming from the Gold Coast—not to mention others who had being transported from Jamaica, and the Corentine in Berbice, where there had been a great concentration of Coromantee.[63]

The Coromantee were known for their "rebelliousness" and had been responsible for a great number of uprisings in the New World.[64] One of the most dramatic was the slave rebellion of 1763 in Berbice. And after that rebellion was suppressed, there had been continuing sporadic conspiracies and small uprisings, not only in Berbice but in Demerara and Essequibo as well. The presence of so many Coromantee among those who plotted and joined the 1823 rebellion in Demerara makes more plausible the hypothesis that, aside from all the other forms of loyalties born out of a shared experience of oppression in the new world, traditional ethnic loyalties may have played a significant role. Historians and anthropologists have emphasized the importance of descent groups and lineage in Africa, and have stressed how difficult it would have been for Africans to keep any semblance of traditional loyalties based on kinship relations under slavery. But the documents suggest that in Demerara loyalty toward kin, however redefined, was important in the making of the rebellion.[65] Equally important seem to have been bonds created by language.[66] It is possible that even those who were from ethnic groups constantly at war in Africa but who spoke the same language would become allies in the New World.[67] This hypothesis is all the more plausible because African languages would have counted for much in an area where half of the slaves were African-born, an area of great linguistic confusion where some slaves spoke only Dutch, others English, and others a patois that included Spanish, English, Dutch, and sometimes even Portuguese and French words.[68]

All this makes us wonder whether behind the seeming transparency of the documentation assembled by whites, from whom the reality of slave experience was always hidden, there was a deeper and elusive "African" reality difficult to grasp, a reality that went much beyond the realm of religion, art, and folklore (to which most of our knowledge is confined). Does the predominance of carpenters and other skilled slaves among the leaders indicate the survival in Demerara of artisanal secret societies which were common in many parts of Africa?[69] Or should it be explained by loyalties formed in the work-place? When Cato referred to Quamina as "Daddy Quamina," did this mean that Quamina had a special role in the community, something like a "fetishman," or a "conjurer"?[70] Or did this expression only signal the respect for elders characteristic of many African groups? Could it be that when Quamina was chosen by the Bethel congregation in 1817 to be a deacon, the congregation was only redefining and confirming a traditional role? That might explain the extraordinary respect Quamina seems to have had from his peers, his preeminence, and his role as a mediator whenever there was a conflict—the role which made John Wray describe him as "a peace maker." Or did Quamina's prestige derive entirely from being a deacon? If so why did Romeo or Bristol or Seaton, who were also deacons, not enjoy the same authority? When Jack Gladstone called for a council to discuss the strategies to be adopted during the rebellion was he re-enacting some African tradition?[71] But when he called his father an "old fool," was he not violating traditional rules of respect and obedience due to older men?[72] Can we infer from Jack's behavior, and from the fact that many of the ring-leaders were not only young men in their twenties and early thirties, but also creoles, that the second generation of slaves was already moving away from African traditions (that required the young to address the old with respect), challenging the older men's authority and creating a new culture in which elements of both the English and the African traditions were combined?[73] And were not even the older people being forced to move in the same direction and to accept the new ways, contributing to the creation of a creole culture with its own rules?[74]

Perhaps one day someone will be able to solve such mysteries by walking backward through the precarious and uncertain paths of oral tradition in search of a lost past, a past that lies hidden behind layers of documents produced by men who, like masters everywhere, refused to understand the souls of their slaves.[75] When we know more about African societies in the nineteenth century and about the ways slavery forced Africans to redefine their cultural inheritance, we may answer such questions with more confidence.[76] But even the little we know now is enough to make us wonder.

There is no doubt that language and kinship were important forms of bonding, among the many others that brought some slaves together and sometimes may also have separated them from others.[77] Religion, also, could bring them together or set them against each other. Scattered bits of evidence suggest that there were tensions between Muslims and Christians. The driver from *Brothers,* Bob, was a Muslim, and was known as the "Mahometan." He clearly opposed the rebellion, and was quite ready to supply the authorities with a list of names of the rebels who had come to his plantation. This could indicate a rift between Muslims and Christians or some traditional rivalry brought over from Africa. There were also tensions between Christians belonging to different chapels. On at least one plantation, slaves who were plotting the rebellion deliberately excluded some others from the conspiracy because they belonged to another chapel.[78] Occasionally, there were conflicts between Africans and creoles or between blacks and mulattos. When Manuel from *Chateau Margo* and Jack Gladstone had an argument about what to do, Jack dismissed Manuel saying that Manuel was a man from Africa whereas he himself was a creole.[79] Moreover, some slaves, mostly drivers and domestic servants, out of fear or calculation, or some sort of devotion toward their masters or hostility toward their peers, opposed the rebellion from the beginning and later sided with their masters. There were also unwilling participants who were dragged into the rebellion at the last minute. These incidents that reveal divisions among slaves are indicative of the difficulties slaves had to face in planning a large-scale rebellion. Yet, as the rebellion showed, divisions could for the most part be overcome.

Among those who rose up, there were "head people" and field workers, men and women, Christians and non-Christians, free blacks and slaves, Africans and creoles, blacks and mulattos, young and old, Coromantees, Kongos, Popos, Mandingos, and probably others whose identities are lost to us. If a common experience united them, it was slavery.[80] And slaves in Demerara had had a long history of individual and collective day-to-day resistance, which had strengthened their commitment to their "rights" and had helped to consolidate links of solidarity and to create leadership.

Particularly important in the making of the rebellion were the networks that brought the plotters together.[81] They visited each other, had breakfast together, and gathered at night to rehearse the catechism and to learn to read. They met along the roads, in shops, in the fields, on the rivers, at the harbors, in the chapel, and at the Sunday market. They talked and talked, and dreamed and dreamed, reassuring each other, repeating the same tales and the same news over and over again, sounding each time more convincing. They plotted. They feared and dared and cautioned each other, reciting the little they knew about slaves in other places—slaves who had struggled for freedom and had met with repression: Grenada, Barbados—the same sad story everywhere. But the dream was too powerful to dispel, and the evidence too compelling to dismiss. So little is needed for so much hope.[82]

It was this dream and this hope that finally brought a number of slaves together on Sunday afternoon, August 17, to make final preparations for the rebellion. Around three or four o'clock, after services, they met at the *Success* Middle Walk to decide the strategy they should follow. According to participants' later accounts, Jack Gladstone played an important role. He ordered the slaves to stand together, facing each other, those from the same estate standing by each other, "to hold a council about taking hold of the white people."[83] He then read a letter from Joe Simpson saying that from what he had seen in his master's papers, the slaves were to be free, that Wilberforce was doing his best for the slaves, and that a new governor was expected very soon, so the slaves should wait. Several people spoke. Some said that they should rise, others that they should wait. Later, the slaves who were arrested gave slightly different versions of what happened in the meeting, supplying details that reveal some fear, hesitation, and disagreement among them. Quamina of *Nabaclis* brought up the example of Barbados, where the people had risen in the same manner, and many had been killed. Sandy suggested that the slaves lay down their tools and then go to town to ask for another free day besides Sunday. Such a strategy was familiar to slaves who sometimes collectively refused to work as a sign of protest and often went to the fiscal to complain. But Sandy's suggestion was not accepted, perhaps because most people wanted freedom and not just another free day a week. They agreed with Paris when he argued that if they simply laid down their tools, the slaves, men and women, would be shot "like fools."[84] Joseph, of *Bachelor's Adventure,* and Bristol proposed to break up the bridges, but it was decided that they only should break the bridges if they saw the troops coming. After much debate the slaves had finally agreed to begin the uprising Monday evening and to confine managers and overseers in the stocks, taking away their arms and ammunition.[85] The rebels intended to force the governor to give them freedom, or at least some days a week.[86] The firing of guns on the coast was to be the signal. Billy and Jacky Reed were to command the Mahaica region. Joseph, Telemachus, and Sandy were to start at *Bachelor's Adventure* and work their way westward, down the coast, till they met Paris, and eastward, on the Mahaica side until they met Jacky Reed. Mars and Azor were to lead as far as they could toward town. Joe was to take charge at Simpson's plantation. Slaves from *Thomas* were to come up the coast in the direction of *Success* and not attempt to go to town, where the military power of the colony was concentrated.

Quamina tried to stop Jack but he dismissed his father roughly: "You are an old fool; the thing people have got, you won't have them have." Before the meeting, Quamina, Seaton, Shute, and Peter had gone to Smith's house, as they usually did on Sundays after service, to tell him goodbye. Smith overheard Quamina and Seaton talking in "a low tone of voice" about the "new laws," and asked them what they were talking about. Quamina answered evasively. "Nothing in particular, sir, we were only saying it would be good to send our managers to town to fetch up the new law." Smith tried to dissuade them, arguing that such behavior would irritate the government and have the opposite effect. Quamina promised they would do nothing they could be sorry for. With this, the slaves left.

Smith's words may have affected Quamina, but it was too late to stop the uprising. After the meeting at *Success,* the slaves returned to their plantations and told others that they were to rise up on Monday night.[87] They had started a process they could not control. On Monday morning Quamina insisted that Jack postpone the plan until Smith could get hold of a letter which had been sent to the governor. (Quamina was probably referring to the instructions sent to Murray by the British government.) Jack replied that "all negroes of the estates had gone to work," and there was no way to reach them. Quamina did not give up. He asked Peter Hood, a carpenter who was working at *Le Resouvenir,* to send word "to the estates down the Coast." Hood sent the message through the boy Cupido. That same morning, Azor, a field slave belonging to Van Cooten, sent his son to ask Quamina what they were to do. Quamina told the boy that they must stop.[88] Quamina's attempt to postpone if not prevent the rebellion may explain why slaves did not rise in any of the estates toward town, except at *Plaisance* and *Brothers.* But he was unable to stop it elsewhere.

The rebellion started at *Success* and quickly spread to neighboring plantations. Beginning around six in the evening, to the sound of shell-horns and drums, and continuing through the night, nine to twelve thousand slaves from about sixty East Coast plantations surrounded main houses, put overseers and managers in the stocks, and seized their arms and ammunition. When they met resistance they used force. Years of frustration and repression were suddenly released.[89] For a short time slaves turned the world upside down. Slaves became masters and masters became slaves. Just as masters had uprooted them from their traditional environment and culture, appropriated their labor, given them new names, forced them to learn a new language, and imposed on them new roles, slaves appropriated their masters' language and their symbols of power and property. Slaves spoke of laws coming out from England. They spoke of "rights." They spoke of the King, of Wilberforce, and of "the powerful men in England." They used their masters' whips and put their masters in the stocks. They broke doors and windows, destroyed furniture, set buildings afire. They whipped managers and masters, stole their clothes and money, drank their wine. And when whites fired at them, they shot back. By the middle of the night, the old African shells and drums were silent. Only the sound of European guns was heard.

Telemachus, Sandy, Jack Gladstone, and a few others were seen on different plantations restraining the rebels, trying to stop looting, to prevent acts of violence against the whites—all in an attempt to maintain order and to lead people who by then had already taken justice into their own hands. The leaders had been recruited mostly among skilled slaves—artisans and drivers, but also boatmen and engineers.[90] Among those later arrested as ring-leaders, Mars and Azor from *Vryheid's Lust,* Quamina from *Nabaclis,* Peter from *Le Resouvenir,* Prince from *Ann's Grove,* William from *La Bonne Intention,* Attila from *Plaisance,* Active and Quamina from *Success,* were all carpenters. Paris of *Good Hope,* Quaco from *Chateau Margo,* and January from *Clonbrook* were boat captains. Jack Gladstone and Dumfries were coopers. Seaton was a boiler and Dick an engineer at *Success.* Ralph was a jobber on the same plantation. Several slaves later implicated in the rebellion—Quamina from *Success,* Bristol from *Chateau Margo,* Paul from *Friendship,* Jacky Reed from *Dochfour,* Joseph and Telemachus from *Bachelor's Adventure,* and many others—were deacons or "teachers" in Smith's chapel. Their participation in the slave rebellion led the authorities to suspect Smith of involvement. His role, however, seems to have been more that of pacifier than instigator. It was probably due to his influence and to the active role several members of his chapel played in the organization of the uprising that the level of slave violence was so low when compared with previous rebellions. Serious confrontations resulting in the deaths of whites took place on only a few plantations, in striking contrast to the bloodbath of 1763 in Berbice. Only two or three white men were killed, and except in one instance, no injury was done to any white women.

For the most part, the rebels showed considerable restraint and discipline, although occasionally the leadership had to use threats and even violence to enforce it. The events at *Foulis,* as they were reported by Hubert Whitlock, the manager, and Biddy Cells, a woman who lived in his house, were typical. On the night of August 18, when she was already in bed, Biddy Cells heard a noise. She got up and opened the window and saw a great number of blacks—from two to three hundred—surrounding the house. She recognized only one, a slave named Caleb from *Paradise.* The slaves demanded that she open the door. She refused, and the slaves broke the door open. The manager tried to fire his musket, but was stopped by Biddy Cells who begged him not to. The slaves snatched the musket from Whitlock's hand and searched the place for arms. After they left, a second gang arrived. Among them was Telemachus. This time the slaves dragged the manager down the stairs and locked him in the stocks, but told him they would release him in the morning.

At *Friendship* something similar happened. But there, Smith, a slave from the plantation, took the opportunity for a personal revenge. He entered the plantation house with several others, struck the manager in the face, and helped to carry him to the stocks. "You had me in the stocks yesterday," Smith said, "I have you now." Apparently, some time before the rebellion, Smith had been caught distributing allowances of stolen fish to the other slaves and the manager had put him in the stocks. On Monday, August 18, just before the rebellion began, the manager had threatened to put Smith in the stocks again. To his surprise Smith answered defiantly: "Come down yourself and put me in the stocks." The manager ordered Ned, the driver, to lock Smith in the stocks. But Ned and Smith simply walked away together toward the slaves' quarters. Later, after the manager was put in the stocks, the rebels brought in another white man. "I am sorry for you, old gentleman," Smith said apologetically, "but you know about the war; as for you [turning to the manager] you ought to be put in hands and feet, in the stock, and if it was left to me I would take your head off." When the manager asked for some water, the only answer he got from Smith was that he would have no water for three days and three nights.

Smith kept guard through the night, and must have been in a state of panic because when he heard a noise in the room, he threatened to shoot the prisoners through their heads if they tried to escape. When the manager pleaded to be released for just a short while, Smith responded that he had orders to keep him there. "Who gave you orders?" asked the manager. "Quaco, and he has five thousand 'Bush negroes' with him," Smith replied. He told the manager that if did not obey Quaco's order he might lose his own head. Yet in the morning he allowed the manager to take one foot out of the stocks and then spoke "a great deal" about a plan the slaves had had for three months to raise the whole country, though the slaves from Essequibo had not joined them as expected. Smith explained that all managers and masters were in the stocks; and the slaves intended to keep them for three days. After they took Mahaica Post they would go to the governor and demand to have three days free.

One of the most dramatic confrontations occurred at *Nabaclis*. Mary Walrand, the wife of the plantation's part-owner Francis Alexander Walrand, testified later that around four in the morning of the 19th, she heard gunfire and the noise of people breaking into the house. While her husband ran downstairs to try to defend the house, she opened the window upstairs and begged the slaves to stop. From the crowd someone shouted: "Look at the lady at the window, fire at her." She was shot in the arm. She drew back, but when she returned to the window to talk to them, she was shot again, this time in her hand. Desperate, she ran to the stairs, where she met her terrified "boy" Billy, who insisted that she remain upstairs. "They have killed Mr. Facker, wounded Mr. Forlice severely, and my master, I believe, is killed, I saw him dragged on the ground," he said. Billy took her back to the bedroom and locked the door. Soon the slaves who had invaded the house rushed upstairs and broke into the room. Some carried guns, but they promised not to hurt her if she showed them where the powder and shot were stored. They searched the room, opening trunks and boxes and taking everything valuable. One of the slaves, who identified himself as Sandy, the head carpenter of *Non Pareil,* said that he knew Walrand was an excellent master and she was a very good lady. "I know that you go to the sick-house, give the people physic, and attend to them." When she asked him what they had done to her husband, Sandy told her that Walrand had been put in the stocks. "I must go there too," she said anxiously. "Oh, no, you must be guarded in the house," he answered. While she was talking to Sandy, Joseph, the driver of the plantation, appeared. She begged again to be taken to her husband. At that point, a tall black man went to the window and shouted at the slaves downstairs who had broken into the "logie" and were drinking the wine. "Make haste away to the Post [Mahaica]; you are losing time." He then rushed to join them, leaving a guard to watch her.

After much insistence she managed to convince her guard to take her to her husband. On her way she saw Tucker's body lying on the ground. He was dead. When she got to the "sick-house," where her husband and Forbes the overseer were imprisoned, she found the overseer badly wounded. She offered to dress his wounds, but he replied that he would rather die. "They have taken all my clothes, and all the little money that I had been toiling for, and this is now no country for a poor man to get his living. . . . If this act passes unpunished, what have we to expect? I lie here murdered by the hands of those wretches. Our Prince gave me a blow in my head. . . . I wish Wilberforce was here in this room, just to look on me, for we may thank him for all that has happened, that the same might be dealt to him by some hand!" Forbes died that night.

From Mary Walrand's and many other witnesses' accounts—however romanticized and self-serving they might be—it is clear that on Monday and Tuesday, slaves had gone in large groups to plantations other than those where they lived and worked, so they would not be recognized. They were armed with cutlasses, guns, and other weapons. They broke into plantation houses and took arms and ammunition, seized managers, owners, and visitors and put them in the stocks. A few slaves were left to stand guard while the crowd moved on to other plantations, where similar scenes took place.

There were cases of looting and property destruction. The slaves broke glasses, doors, windows, and furniture. They destroyed several bridges. On a few plantations, where whites barricaded themselves in the main house, slaves set fire to it. At others, they expressed their hostility to owners, managers, doctors, and overseers by cursing and beating then up. One slave, Kinsale, took the joint proprietor of plantation *Clonbrook,* Hugh Rogers, by the ear and told him he was "a very wicked fellow . . . a second Pharaoh," even worse than the overseers, and deserved to have his head off. At plantation *Enterprise,* when the resident doctor asked to be released from the stocks, arguing that he had to visit some sick people in the morning, Kinsale dismissed his request: "Who wants you for a doctor again?" On some plantations whites were wounded trying to resist. But only a few whites were killed, when they tried to shoot at the slaves and the slaves fired back.

Slaves told whites they were doing to them what the whites had always done to slaves. Sometimes they said if they were left to do what they wished they would cut the whites heads off, though they were never very explicit about just who or what was constraining them. Apparently, the leaders of the rebellion had given strict orders to the rebels not to harm any white. Some ring-leaders had also gone from one plantation to another asking owners and managers to sign a paper they intended to present to colonial authorities certifying that the slaves had done no harm. With it the slaves seem to have wanted not only to demonstrate their good behavior but also to show their understanding of the white men's rules.

The few slaves, mostly drivers and house servants, who sided with their masters were beaten and put in the stocks. Those who tried not to be involved were forced by the rebels to participate, as is clear in the story Thomas, the head boiler on plantation *Bee Hive,* later told the court. At seven in the evening Duke, a slave from *Clonbrook,* came to his door and started shouting "Thomas! Thomas! what do you mean by leaving us to fight your battles? We had you on the water side just now, and you have run away from us, and come to your house again!" Duke then went to another slave's door and chopped it down. He was bragging that he had been fighting all the way up from *Chateau Margo* to *Bee Hive.* Another witness added that Duke bragged that he had put the governor, the fiscal, and all "the great people" in the stocks. At *Bee Hive* he gathered all men next to the canal dam and ordered one of them to destroy the bridge. Apparently the slaves he forced out of their houses had returned home after he left. When Duke returned to the plantation later that day, he again ordered every man upon the dam, "then went to Edwin's house, broke open his door and beat him" and "threatened to set the negro-houses on fire."

Duke had been very active at other plantations too. Wherever he went he harassed blacks who had stayed home. Brutus of *Northbrook,* a few miles from *Bee Hive,* said that on Tuesday around seven in the morning, Duke had come to his house and reprimanded him. "Here is a great sergeant here, what is he doing?" He told Brutus that he had not slept the whole night and asked him, "Do you think we are going to prepare freedom for you, and you sitting down in your house all the while? You had better turn out and do your duty." He then threatened to chop off his head.[91]

It is obviously impossible to determine with any real certainty who told the truth and who lied during the trials. Slaves may have tried to protect themselves by inventing stories about how they had been forced to join the uprising. This seems to be the case with Barson of plantation *Paradise.* In his deposition, Barson said that successive groups of blacks had come to the plantation. Among the first group were Austin and Allick of plantation *Cove.* Later a second party came. The leaders of this party were Telemachus and Joseph of *Bachelor's Adventure,* Natty of *Enterprise,* Scipio of *Non Pareil,* and Hans of *Elizabeth Hall.* "They were the busiest among them giving orders." According to Barson, Natty heard him saying that the white people should not be put in the stocks and grabbed him by the collar and pulled him down, striking him with a cutlass. Sandy intervened: "Don't strike him," he said. "Take him with us." As Barson testified, he had been forced to go with them, but as soon as he had a chance he returned to *Paradise* and tried, without success, to free the whites from the stocks. When Natty returned he was told that Barson had been plaguing the guards for the key to let the white people out. Someone put a double-barrelled gun in Barson's hand and forced him to walk toward *Nabaclis.* On the road they met the driver Joseph, who had joined the rebels. Seeing Barson, Joseph said to him: "Look, there's a rascal like yourself, white people's servant." Barson's story may be true, but in Nelson's trial, Nero of *Annandale* said that he had tried to stop a large body of blacks that came on Monday night to the plantation. Barson stamped his foot violently on the ground and shouted defiantly. "Who dares to stop us?" Nero's testimony raises suspicion about Barson's story. But if in this case it is difficult to known whether Barson had joined the rebellion willingly or unwillingly, there are others in which the testimonies of both blacks and whites coincided, indicating that some slaves (about 10 percent) had indeed refused to follow the rebels.

The trial records show that only a small group of slaves had been privy to the conspiracy. Most slaves were ignorant of the plan until the day before the uprising. Some were taken by surprise when large numbers of blacks from other plantations came to their quarters on Monday night or Tuesday morning calling on them to rise up. Those who did not join spontaneously were scorned, accused of being cowards, and even beaten by the rebels. Forced to carry arms and to follow the rebels from one plantation to another, some slaves escaped and returned to their plantations, just to be dragged out again by another group of rebels. Others, like Quamina, stayed quietly in the "backs" of the plantations.

Except for those who had participated in the plot, most slaves did not have a clear idea what they wanted, or what they were supposed to do. Some had been told in a vague way that good news had come from England, but their masters and managers had hidden it from them. They were not sure what the news was. Some thought they were going to be free. Others expected to have two or three days a week for themselves, so that they could attend religious services on Sundays, cultivate their own gardens, and go to the market. They had hoped to force the governor to make such concessions. Some slaves had expected that the slaves in the neighboring colonies of Essequibo and Berbice would also rise. They also had nourished the idea that the "Bush negroes" would come and join them—a dream often present in rebellions throughout the Caribbean, where maroons had come to symbolize freedom.[92]

As with all popular risings—of slaves or others—it is easier to examine the conditions that created a rebellious situation and to identify the events that triggered the uprising than to say with any precision why some individuals were more involved than others. Among those who joined the conspiracy or later participated in the rebellion, many had quite personal motives. Aside from the usual complaints registered in the fiscals' records, some slaves had experienced particular types of hardships or were under some specific stress. At *Nabaclis,* seventy-eight slaves had been taken away by John Reed, the owner of plantation *Dochfour,* and others had been advertised to be sold just a week before the rebellion.[93] Plantation *Friendship* was also for sale, and Paul (who was later executed) had been desperately looking for someone to buy him, afraid he would be sold to some distant place.

Something similar was happening at *Clonbrook* and *Bachelor's Adventure.* One of the owners had died and several slaves were to be sold, among them Billy's wife Nanny and their twelve children. They had lived together for twenty years and were afraid of being separated. Telemachus, who belonged to *Clonbrook* but was living at *Bachelor's Adventure* with his wife and child, was facing a similar problem. He had gone to Whitlock, the manager of *Foulis,* to see if the manager would buy him so that he would not be sent away from his family. Telemachus had many other reasons to be resentful. He had several times been harassed for being a Christian, and about two years before the rebellion he had been sent to work twenty-four miles away from home because he was "too religious."

Susanna, who had enjoyed a privileged position as Hamilton's mistress, had been let down by him. Jacky Reed from *Dochfour* also had a specific reason to be bitter. He wanted his mother-in-law to be baptized—something she also desperately wanted—but the manager refused to give her a pass to go to the chapel. Jacky had complained several times to Smith that managers were constantly chastising Christian slaves. A few months before the rebellion, in November of 1822, Reed and Peter from *Hope* had gone to Smith in a very depressed state of mind. They told him that their manager accosted the "Christian negroes" with taunting jokes on the subject of religion in the presence of the "heathen negroes," insinuating that their profession was only hypocrisy and that a trifling consideration would cause them to abandon it, so they ought to be treated with scorn. Offended, some slaves had answered in a way the manager considered "disrespectful," and for this "insolence" they had been repeatedly flogged and confined to the stocks.

Immanuel, who lived at *Chateau Margo,* had been put in the stocks because he held prayer meetings (and also for beating a slave who had refused to make a contribution to the chapel). Sandy of *Non Pareil* had been persecuted because he held religious meetings. Prince, who (like Telemachus) belonged to *Bachelor's Adventure,* had also endured "much suffering on account of religion." Betsy had plenty of reason to be upset when she was raped by the manager. Gracy also had cause to be angry when she had been locked in the stocks and then sent to work in the task-gang, because of her affair with Jack. Rachel had good reason to feel resentment when she was flogged for taking a sick child to the fields after the manager had told her to leave the child with the old woman who took care of the slaves' children.[94]

But if some slaves had private grievances, they all shared the same fears. All lived with the fear of being sold. In Demerara, plantations were constantly being mortgaged and slaves were often given as payment for debts. Slaves were also sold because their masters needed cash, because they went bankrupt, because they had to "settle some pressing demands,"[95] because they returned to England or Holland, because they sold their plantations, or because they masters died and the slaves were auctioned or divided among the heirs. The threat of punishment was no less constant. All slaves at one time or another had either been punished or had seen their friends punished for frivolous reasons.[96]

The severity of treatment and frequency of punishment varied from one plantation to another, depending on the whims of masters and managers. Some, like Michael McTurk and James Spencer, did not allow slaves to go to the chapel, and often made them work on Sundays. Others, like John Pollard, who was responsible for the management of *Non Pareil, Bachelor's Adventure,* and several other plantations on the East Coast, were particularly violent and cruel and resorted with great frequency to the whip. Some planters, like Alexander Simpson, burdened by mortgages, might have driven their slaves more harshly than others.[97] But everywhere on the East Coast, the intensification of labor exploitation and the reduction of the time the slaves had always had for cultivating their gardens and provision grounds and for going to the market had had their effect. The infringement upon these traditional "rights" explains the slaves' demands for two or three free days a week.[98]

Although the situation varied from one plantation to another and from one individual slave to another, and although these differences may have worked to divide the slaves, there was one experience they all shared: that of being a slave. And being a slave at this point in history meant not only being in Demerara, on a particular plantation, at a particular place, under particular conditions of labor; it also meant to be part of a broader world, in which slavery was under attack. This was a changing world, in which slavery, once a necessity, was becoming a contingency; a world in which industries were gradually transforming the pace of life, and the traditional colonial system based on monopolies and privileges was collapsing while new opportunities for trade were opening in the international market and England was incorporating new areas into its empire. It was a world in which new interest groups were emerging, the consensus among the ruling groups regarding slavery was being broken, and slaves could expect to find powerful allies; a world in which the ongoing social processes were redefining what was fair, what was right, and what was possible, and rewriting the traditional codes of honor, rules of property, and notions of citizenship; a world in which new ideologies were undermining the system of sanctions and assertions that for centuries had maintained slavery, transforming what once had been an impossible dream of freedom into a tangible possibility, bringing hope where once was fear and despair.

The rebellion would bring to the surface the contradictions between resident and absentee planters, West Indian proprietors, and other groups that struggled for power in the British Parliament; between conservative members of the Established Church and evangelicals; between those who clung to traditional ideologies stressing monopolies and privileges, social inequality, hierarchy, and authority, and those who espoused new ideologies that postulated the universality of the rights of man, individual responsibility, and equality before the law. Like any other historical event, the Demerara Rebellion was the product of many contradictory historical forces. Jack Gladstone, Quamina, Susanna, John and Jane Smith, McTurk, Governor Murray, masters and managers, slaves and missionaries, all were trapped in such contradictions.

There was certainly chance in the lives of Jack Gladstone and John Smith, and in the lives of all those who survived and of those who were killed.

There was also choice. Yet, although individuals' options, personal motives, circumstantial networks, dreams of emancipation, all contributed to the making of the rebellion, they were subordinated to a larger historical process which transcended the consciousness of the participants and over which they had no control, a process which defined their limits, their possibilities, and even their dreams.

CHAPTER SIX

A Man Is Never Safe

"I could weep tears of blood." —*John Wray*

"It seems as if the Devil was on them, nay I am sure he is."
—*John Smith to John Wray, May 1823*

The rebellion caught the whites by surprise. They had lived so long with the fear that the slaves might rise that they had learned to dismiss it. To reassure themselves, they had created rituals of domination that spoke of total power. They had ignored the signs of tension and discontent among slaves and had convinced themselves that the slaves had no reason to rebel. Regarding their slaves as tools, masters had become blind to their humanity. Yet the threat of rebellion was always there. And the cutlasses that cut the cane could easily become weapons. No one captured better the masters' predicament than George Lamming, in a meditation on the Haitian Revolution:

> It must have been clear to the owner that the mournful silence of this property contained a danger which would last as long as their hands were alive. One day some change akin to mystery would reveal itself through these man-shaped ploughs. The mystery would assume the behavior of a plough which refused contact with a free hand. Imagine a plough, in the field. Ordinary as ever, prongs and spine unchanged, it is simply there, stuck to its post beside the cane shoot. Then some hand identical with the routine of its work, reaches to lift this familiar instrument. But the plough escapes contact. It refuses to surrender its present position. There is a change in the relation between this plough and one free hand. The crops wait and wonder what will happen next.

> More hands arrive to confirm the extraordinary conduct of this plough: but no one can explain the terror of those hands as they withdraw from the plough. Some new sight as well as some new sense of language is required to bear witness to the miracle of the plough which now talks. For as those hands in unison move forward, the plough achieves a somersault which reverses its traditional posture. Its head goes into the ground and the prongs, throat-near, stand erect in the air, ten points of steel announcing danger."[1]

When the threat became real in Demerara, the planters' first reaction was to muster all the military forces they had to put down the rebellion, the second was to look for culprits: someone or something to blame.

Early on Monday the 18th, hours before the rebellion broke out at *Success,* Captain Alexander Simpson of plantation *Le Reduit* was awakened by a bugle outside his window. His "boy" Joe, visibly troubled, told him that on Sunday at the chapel the slaves had decided to rise the next day. Simpson wasted no time. He saddled his horse and set off for Georgetown to warn the governor.[2] On his way he stopped at several plantations, alerting managers and proprietors of the imminent danger of a rebellion. At ten o'clock he saw the governor, who immediately ordered the cavalry to assemble. Captain Simpson was told to rush his fourteen cavalrymen back to *Le Reduit,* where the governor would join him later. He was to inform the other plantations along the road. By five o'clock, the governor had arrived at *Le Reduit.* He dispatched a sergeant and four troopers to the military post at Mahaica Creek, some twenty-five miles from Georgetown, where the lieutenant in charge had only one sergeant and sixteen privates under his command.[3] On their way, they were to caution burgher captains, particularly Michael McTurk and James Spencer.

McTurk, however, had already been warned. While preparations were taking place at *Le Reduit* on Monday morning, the rumor had been spreading that the slaves intended to rise that night. About four in the afternoon, McTurk was getting ready to dine with Van Waterschoodt and Lachlan Cuming at *Plaisance* when he was informed of the plot by William Cuming, a free black man who had heard it from Joe, Simpson's "boy." McTurk sent for Cato, whom Joe had named as one of those involved in the rebellion. Cato—who indeed had been privy to the conspiracy from the beginning—denied knowing anything about it, but was immediately arrested and sent to the stocks. McTurk then interrogated several slaves. From what they said he became convinced that Quamina and his son Jack Gladstone were behind the plotted rebellion.

Without further investigation, McTurk sent a message to John Stewart, the manager of *Success,* ordering him to arrest Quamina and Jack and send them to him.[4] Stewart immediately sent them under the custody of two overseers. On the road, the little party was surprised by a group of slaves (Ralph, Beffaney, and Windsor, among them) who managed to free the prisoners after threatening one of the overseers with cutlasses. Terrified, the man tried to escape by jumping into a nearby canal, but he was closely followed by the slaves. Jack's prompt intervention saved the overseer's life. Quamina and Jack returned to *Success,* followed by the slaves who had come to their rescue, taking the two overseers with them.

About five in the afternoon, John Stewart was standing in the gallery of the dwelling house at *Success,* when he saw a large body of blacks coming. They were led by the driver Richard, a slave who had been a fugitive for two or three months. When they came closer, Stewart ordered them to stop. But Richard and his party seemed determined to get hold of the manager. Several slaves from *Success* tried unsuccessfully to stop them. Stewart was quickly surrounded by blacks, some of whom he did not recognize since they did not belong to the plantation. One of them grabbed him by his feet, others by the collar of his coat, and they dragged him to the "sick-house" with the intention of putting him in the stocks. But, thanks to Jack Gladstone, Stewart and the overseers were allowed to return to the house, where they were locked up. The slaves searched the house for arms and ammunition and took away all they could find. The rebellion had started at *Success,* and was rapidly spreading to other plantations on the East Coast.

While the governor was assessing the situation and taking measures to protect the colony, Simpson and McTurk with small bands of the white militia were checking the neighboring plantations. Late in the afternoon, approaching the main building at *Success,* Simpson and his four cavalrymen saw blacks armed with cutlasses surrounding the manager's house. "Are there no white people here?" he shouted. From a window upstairs, Stewart answered visibly agitated. "Do not fire at them, they are doing no harm." But soon after, he and the two overseers escaped through a window and ran toward Simpson and his party hoping to be rescued. Realizing that he and his men would be overpowered by the slaves, who had become increasingly menacing, Simpson forgot all about solidarity among whites. He ordered Stewart, who was vainly trying to climb onto the back of his horse, to let go or he would cut him free with his sword. From the crowd, someone fired at Simpson, but missed. Stewart had just heard the shot when he was violently seized by Ralph, one of the slaves from *Success* who had helped to liberate Jack and Quamina. Holding a cutlass over Stewart's head, Ralph forced him to go to the "sick-house," where he put Stewart in the stocks. The two overseers who had tried to take Quamina and Jack to McTurk were also locked up in the stocks. This time Jack was not around to help. A slave guard was left to watch them while the others walked away. Simpson and his party rode away as fast as they could.[5]

At *Le Resouvenir* also the rebellion came as a surprise. Around six in the afternoon, John Smith was preparing to go for his usual after-dinner walk with his wife. As Jane went to pick up her bonnet, Guildford, a "boy" from plantation *Dochfour,* arrived with a note from Jacky Reed, a slave who belonged to Smith's congregation. The note was a remarkable blend of painstaking, conventional gentility and vague but ominous warnings.

> Dear Sir,
>
> Excuse the liberty I take in writing to you; I hope this letter may find yourself and Mrs. Smith well. Jack Gladstone has sent me a letter, which appears as if I had made an agreement upon some actions, which I never did; neither did I promise him any thing, and I hope that you will see to it, and inquire of members, whatever it is they may have in view, which I am ignorant of, and to inquire after it, and know what it is. The time is determined on for seven o'clock to night.

Enclosed was the letter Jack Gladstone had sent Jacky Reed. Like Reed's note to Smith, it was written in an awkwardly well-mannered and half-coded way. It was permeated not by warnings but by hope (the word itself appeared four times in the text):

> Dear Brother Jacky,
>
> I hope this letter will find you well, and I write to you concerning our discourse, and I hope you will do according to your promise; this letter is written by Jack Gladstone, and all the rest of the brothers in Bethel chapel, and I hope you will do according to our agreement; we shall begin at the Thomas, and hope, you will try your best up the coast.[6]

Smith asked the evasive and confused Guildford several questions. If the missionary did not understand clearly what either Jacky Reed or Jack Gladstone was talking about—as he later would claim during his trial—he must have suspected the worst. Perhaps fearing that Guildford might distort a word-of-mouth message, Smith wrote out a straightforward yet careful note to Reed:

> I am ignorant of the affair you allude to, and your note is too late for me to make any inquiry. I learnt yesterday, that some scheme was in agitation, but

> without asking questions on the subject; I begged them to be quiet; and trust they will. Hasty, violent, or unconcerted measures, are quite contrary to the religion we profess, and I hope you will have nothing to do with them. Yours, for Christ's sake.[7]

Jacky Reed's letter made the Smiths very anxious.[8] John and Jane were fretting together as they made their way toward the seashore, along the middle path of *Le Resouvenir,* when they heard a "great and unusual noise,"[9] and the plantation manager John Hamilton calling for help. They rushed to the main house, and found it besieged by forty or fifty men, "all naked, armed with cutlasses . . . and looking very fierce."

The slaves had forced the doors and gained the ground floor. When the missionary asked what they wanted, they brandished their cutlasses and shouted, "We want the guns and our rights." Infuriated and determined, they told Smith to go home, and as soon as they obtained the guns, they "gave a shout of triumph," fired into the air and blew their conch-shells.* Then, ringing the plantation bell, they disappeared into the darkness.[10]

The Smiths went home but could hardly sleep. In the middle of the night they heard noises outside their door. The missionary opened his bedroom window and saw four men. They turned out to be the servants of important figures in the colony: John Alves, Lieut.-Colonel Goodman's coachman; Cornelius, a servant of Charles Wray, president of the Court of Justice; John Bailey, a groom in the house of the merchant Richard Chapman; and a man who was later identified as a Mr. Robertson's servant. They had been driving along the coast road, transporting a detachment of six soldiers and an officer of the Twenty-first Regiment. When they found that the slaves had destroyed the bridge next to *Le Resouvenir,* the soldiers had forded the canal, but the coachmen returned with the carriages and were looking for a place to keep their horses for the night. In spite of the late hour, the missionary invited them to come in and offered them refreshments. The men talked excitedly about the slaves' revolt. When one of them commented on how terrible it was "the negroes rising in this manner," Smith—oblivious of the risks he was taking by sharing his thoughts with servants of such powerful men—said that he had been expecting something like this to happen for several weeks. He described how he had seen the rebellious slaves, armed with muskets, cutlasses, and other weapons, and remarked that he was not surprised to see them rise in view of the bad treatment they received. He knew the slaves were aware that new legislation had come from England that

*Conch-shells were usually used to call the slaves to work in the mornings and send them home in the afternoon.

would benefit them. But instead of implementing the new laws (which prohibited the flogging of slaves in the fields) the manager on that estate had given the driver a "cat-'o-nine-tails" and threatened to use it as long as he could.[11] Before they left, the men asked Smith if he were not afraid to stay in his house. Smith answered confidently that the "negroes would not trouble people like him."[12]

The Smiths spent the next day in conjectures about what was happening. Jane felt increasingly alarmed. She walked anxiously back and forth on the front gallery until it occurred to her to ask Ankey—a black woman living on the plantation—what the slaves were up to. But Ankey appeared to be as uninformed as she was. "I do not know, Ma'am, the people wish to get their liberty," was all Ankey said.[13] The two women shared their feelings of fear. Jane said she had not slept the whole night. Ankey confessed that she too had been afraid; she did not know whether to go to the "great house" or to the "negroes' house," a remark that captured her ambiguity about the events. Jane tried to comfort Ankey by saying that the slaves would do her no harm. She then asked her to send for Quamina or for Bristol. But she did not tell John of what she intended to do, probably afraid he might disapprove.

On Wednesday, Ankey came to tell her that Quamina had arrived. Jane hesitated. Sitting on the steps of her back door was a free black woman, Kitty Cumming, who had come from *Success*, saying that all the slaves had gone away. Afraid to remain there alone, she had asked to stay at the missionaries' house. Kitty was one of those free blacks who continued to live on plantations after being manumitted, and whose loyalties were always uncertain. Not knowing whether she could trust Kitty, Jane told Ankey to invite her to go to the slaves' quarters. Kitty was reluctant and mistrustful, and agreed only after Jane reassured her that she could come back before nightfall. A few minutes after the women left, Quamina came to the house. No one will ever know exactly what passed between the Smiths and Quamina. During his trial, John Smith would testify that he told Quamina he "was sorry and grieved to find that the people had been so foolish and wicked and mad, as to be guilty of revolting and hoped that he was not concerned in it." According to Smith, Quamina appeared "confounded and abashed," and left without a word.

After the conversation with Quamina, the missionaries felt even more distressed. John wondered what had brought Quamina there, and Jane finally confessed that she had sent for him. John reprimanded her. She had been very foolish. From the way Quamina had behaved, it appeared that he was involved in the revolt, and if that was true, he never wished to see him again.[14] The missionary may also have said that Quamina's presence in their house could compromise them. As it happened, the maid Elizabeth later testified that Jane had threatened to whip her if she told any one that Quamina had been there.[15]

Still very tense, John Smith started a letter to his superior, the Reverend George Burder, the secretary of the London Missionary Society. Smith was ready to absolve the slaves and condemn the masters. He wrote that the slaves of the East Coast had seized firearms at several plantations and had put managers and overseers in the stocks, "to prevent their escaping to give an alarm." But they had done no personal violence to anyone. They had not set fire to a single building or robbed any house, except of arms and ammunition. Nor had they attempted anything "like an outrage, either upon persons or property." The slaves had told him and his wife that they did not intend to injure anyone, "but their rights they would have." He thought "they were sincere in what they said, for they had the fairest opportunity of murdering every white person on the coast."

Smith noted that the colony was under martial law, and all adult males were called to join the militia. He then went on to speculate about the causes of the rebellion.

> Ever since I have been in the colony, the slaves have been most grievously oppressed. A most immoderate quantity of work has, very generally, been exacted of them, not excepting women far advanced in pregnancy. When sick they have been commonly neglected, ill-treated, or half-starved. Their punishments have been frequent and severe. Redress they have been so seldom able to obtain, that many of them have long discontinued to seek it, even when they have been notoriously wronged. Although the whip has been used with an unsparing hand, still, it seems the negroes have not been more frequently nor more severely flogged of late than formerly. But the planters do not appear to have considered that the increase of knowledge among the slaves, required that an alteration should be made in the mode of treating them. However intelligent a negro might be, still he must be ruled by terror, instead of reason. The most vexatious system of management has been generally adopted; and their religion has long rendered them obnoxious to most of the planters. On this account, many of them have suffered an almost uninterrupted series of contumely and persecution.[16]

The letter expressed admirably Smith's biases. His description and interpretations of the events were shaped by notions he had brought from England, notions that had been legitimized and confirmed by his day-to-day experience in Demerara. His conviction that oppression leads automatically to rebellions was a naive belief that Smith shared with many of his contemporaries. The connection had been established over and over again, in decades of revolutionary discourse. By the time Smith was writing his letter, the notion had become a powerful myth that friends of the French Revolution had helped to spread, and that the revolution itself had helped to give credibility. On the other hand, his life as a missionary preaching the Gospel to slaves had done nothing but reinforce his earlier commitments to emancipation. For him, slavery was evil, and Christianity and slavery were incompatible. Only free men could be good Christians. Slavery was a system that corrupted whites and blacks, masters and slaves, and even missionaries. There was nothing to be surprised at, if after so much "degradation," "violence," and "unfairness," the slaves had finally rebelled. This was the main thrust of Smith's analysis of the events.

He was still at work on his letter when Lieutenant Thomas Nurse of the Demerara militia arrived at the head of a company of infantry and demanded to see him. Nurse said he had been sent by Captain McTurk to find out why Smith had not obeyed the governor's proclamation of martial law, which required every man capable of bearing arms to enroll in the militia. He also insisted that Smith's wife be removed to McTurk's plantation *Felicity,* or some other place where she would be "safer."[17]

The interaction between Smith and Nurse was full of tension. Born out of two contradictory views of the world, their mutual distrust and hostility were impossible to disguise. Their encounter brought to light the conflict that had pitted missionaries against managers and masters from the moment the first missionary from the London Missionary Society had landed in the colony fifteen years earlier. To Smith, the order to enroll in the militia felt like one more harassment on the part of his neighbor McTurk. After many unpleasant encounters, a deep animosity had developed between Smith and McTurk. Both were unyielding, both were passionate, both were proud. But one was a missionary, the other a planter; one wanted to control the slaves' souls, the other their bodies. Their contradictory goals and ideologies had often driven one man against the other. Although *Felicity* was next to *Le Resouvenir,* McTurk had never allowed his slaves to attend religious services.[18] Smith thought McTurk was a godless man, always forcing his slaves to work on the Sabbath, a relentless opponent of the missionaries and an enemy of the Gospel, "a disgrace and a curse to the polity."[19] Because Nurse was speaking for McTurk, Smith was ready to dislike him too.

Nurse, who had close ties to planters, did not show much sympathy for the missionary either. To him, Smith was a troublemaker: one of those fanatic missionaries the *Guiana Chronicle* was constantly describing as "propagandizers of incendiary doctrines," "promulgators of liberty and equality,"[20] men always ready to challenge local authorities, and always sowing discontent and rebellion among the slaves. Not surprisingly, Nurse found Smith's manners "supercilious and offensive."

When Nurse told the missionary that McTurk had ordered Smith to join the militia, Smith answered that McTurk had no authority to issue such an order, and even if he had, he would not obey it. As a missionary he was entitled to a legal exemption. He also said—in a way that must have sounded arrogant—that his wife was as safe in their house as in any other place in the colony. Nurse insisted that under martial law all the inhabitants of the colony, without distinction, were obliged to serve in the militia. It was Smith's duty to comply with the order of Captain McTurk—or any other officer employed by the commander-in-chief—and his clerical vocation did not exempt him from military duties. Smith dismissed Nurse's argument altogether: "I differ, Sir, from you in opinion, and I do not intend to join any militia corps or company, or do duty with them."

Infuriated, Nurse told Smith that he had been ordered by McTurk to seal all his papers—which he proceeded to do. Smith asked permission to retain a few papers, in particular his class-books and a letter he had just received. Nurse insisted "on taking everything in manuscript." After some hesitation, Smith gave him the letter in question, saying that it was from a friend and brother-missionary in Berbice, "that it contained pleasing information as to the manner in which the inhabitants of that colony had met the views of Government and the people of England, for ameliorating and improving the condition of the slaves." If the people of Demerara had acted "with the same generous and liberal feeling, the revolt would never have taken place." Nurse dismissed the obvious provocation. He finished collecting Smith's papers and put them into a drawer, which he sealed, cautioning Smith against the violation of the seals. After that he left.[21]

Smith had not had time to recover when Nurse reappeared at his door. This time he was accompanied by a troop of cavalry under the command of Alexander Simpson, proprietor of *Le Reduit* and *Montrose,* burgher officer, and captain in the Demerara militia.[22] Menacingly, Simpson asked Smith how had he dared to disobey the orders of Captain McTurk. When the missionary tried to argue that he was exempt from military service, Simpson went into a fit of rage. He brandished his sabre and shouted "Damn your eyes, Sir, if you give me any of your logic, I'll sabre you in a minute; if you don't know what martial law is, I'll show you." He accused Smith of being the cause of the rebellion—if he were not connected with it why would he stay among the rebels? He ordered the soldiers to seize Smith's papers and to arrest him.[23]

Since Jane refused to leave her husband, she was also placed under arrest. The missionaries were not allowed to take anything with them. When Jane went upstairs to pack a few clothes, the soldiers rudely told the missionary to hurry her up: "If you don't fetch Mrs. Smith, by God, Sir, we will." The Smiths were marched to Georgetown where they were kept together in a small room, just under the roof of the Colony House—an old barn-like wooden building where the official business of the colony was transacted. No one was allowed to visit them, and they were prohibited to write to anyone. Two sentries were posted day and night outside the door. One of them, a young man who worked as a clerk in a local store when not on militia duty, described the pitiful conditions in which he had found the Smiths. (As a man of modest origins, imbued with antislavery notions and evangelical piety that had moved him to secretively teach slave children to read, he was ready to sympathize with the Smiths and even take personal risks to express his solidarity.)

> He is confined in a room, the door always open at night; if in bed the new sentry goes into his room and lifts the curtain up to see if he is there. You see how strict they are with him. Poor Mrs. Smith was taken away in such a hurry that she had no clothes but what was on her. What heart felt pleasure would it give me if I could be of any use to them, but I dare scarcely speak. I got some pieces of muslins out of the stores and sent them to Mrs. Smith. A. H. [the owner of the store in which he was working] must forgive me for parting with her thread case. I sent it over to Mrs. Smith with thread, silk, and needles . . . Mr. Smith is not allowed pen, ink, or paper; no person is allowed to speak to him, or to enter his room except a strange servant woman.[24]

While the missionary and his wife spent their first days in Colony House wondering what had happened to the people in their congregation and fearing for their own future, soldiers and militiamen carried on the brutal work of repression they had started on Monday night. Had John Smith and Jane known what was happening, they would have feared even more for themselves.

By Monday night, Governor Murray had no doubts that the slaves were in rebellion. He had seen it with his own eyes. On his way back to town from *Le Reduit,* where he had gone to discuss the situation with planters and managers, Murray had come upon a group of about forty rebels armed with cutlasses. When he asked what they wanted, they replied, "Our rights."[25] The governor explained that he had received instructions from England that would benefit them. The abolition of the flogging of females and the prohibition against carrying whips into the fields were but the first steps. "These things," the slaves said, "were no comfort to them; God had made them of the same flesh and blood as the whites; they were tired of being slaves; their good King had sent orders that they should be free, and they would not work any more."[26]

The governor told the rebels he would negotiate with them only if they laid down their arms. They refused. He ordered them to return peacefully to their plantations and promised to meet with them the next day at *Felicity*.[27] They did not move. By this time, their number had grown to more than two hundred. "A mounted negro persisted in sounding a shell," something that made Murray increasingly uncomfortable. He could hear some people in the back of the crowd insisting on firing upon his group. Probably fearing that the situation would get out of control, he decided to leave. He drove away under shouts of "Go! Go!" As he passed next to *Plaisance,* a shot or two was fired at his party. The governor moved hurriedly toward Georgetown. He had a serious problem on his hands and prompt action was required.

The situation was especially dangerous because whites were a tiny minority of the population and could easily be overpowered. Of the 80,000 people living in Essequibo and Demerara, about 75,000 were slaves. In the whole colony, there were only 2,500 whites, about half of them living in Georgetown.[28] The area known as the East Coast, where the rebellion had started, had perhaps one of the highest concentrations of slaves in the West Indies.[29] Within the twenty-five-mile-long strip from Georgetown to Mahaica Creek, the rebels could easily gather from 10,000 to 12,000 slaves.[30] There was also the danger that the rebellion would spread to the West Coast and to the neighboring colonies of Berbice and Essequibo, and (even worse) that the maroons living in the interior would join in. In such circumstances, a slave rebellion could be a total disaster for the whites.

When he got back to Georgetown after his encounter with the rebels, Murray convened a meeting of the Court of Policy and summoned the militia. He dispatched to the East Coast a detachment of the Twenty-first North British Fusileers and the First West India Regiment.[31] A second detachment of the Twenty-first Fusileers followed. The next day the colony woke up in a state of war. Governor Murray declared martial law. All stores in Georgetown were closed. Ladies sought refuge on the vessels in the harbor. Only soldiers and militiamen were seen in the streets. The Scottish Presbyterian Church became the gathering point for the militia. Lieut.-Colonel Goodman, the vendue master, was appointed head of the Georgetown militia. By four in the afternoon, a battalion consisting of 41 sergeants, 30 corporals, and 507 privates had been formed. A marine battalion of about 400 men—some seamen from the ships anchored in the river and some townspeople—was also ready. Their commander was Captain Muddle of the Royal Navy.[32] Two pieces of artillery were placed at each of the two principal entrances into town, and a heavy guard was mounted. New reinforcements were sent to Mahaica: a contingent of artillery-men, sailors and infantrymen, with two three-pounders and a large quantity of ammunition.

At this point, the governor still had no clear picture of events. Everything seemed confusing and unpredictable. It was clear that the slaves had stopped working and some had arms. The rebels had set some buildings and cane fields on fire and had destroyed bridges and ransacked houses, taking any weapons or ammunition they found. Wherever they went they put masters, managers, and overseers in the stocks. On a few plantations where whites resisted, there were more serious confrontations, but for the most part the slaves did not seem inclined to commit acts of violence against the whites. They talked about their "rights" and insisted on presenting their grievances to the governor. In spite of his appeals, they did not seem willing to return to work, and the rebellion was rapidly spreading.

It was impossible to anticipate what would happen next. Would the slaves attack Georgetown? Would they be intimidated by the garrison and the militia and return to work? Would it be possible—or necessary—to make a bargain with them? What did they really want? Why had they rebelled? On Monday, Murray may not have had a ready answer to such questions. But he soon would find one. On the 24th, three days after John Smith was arrested, the governor wrote to Earl Bathurst: "It is evident that this mischief was plotted at the Bethel Chapel, on Mr. Post's estate (*Le Resouvenir*); that the leaders are the chief men of that chapel; that the parson could not have been ignorant of some such project under these circumstances; and, in consequence of his having refused to arm in opposition to them, he was sent in a prisoner by Captain McTurk; and as our situation was the most critical, I detained him on his parole at the Colony House." After mentioning that he had also been forced to arrest another missionary from the London Missionary Society (Richard Elliot), Murray concluded: "I hope this rebellion will soon be put down, but should it become more general, as there is great cause to apprehend, it will not be possible for me to give protection to the country, unless the commander of the forces can spare me very strong reinforcements, for which I have applied."[33] The military and naval power of the mother country was crucial to maintain the colonial social "order."

The troops from the garrison left Georgetown for the East Coast Monday evening. While the regulars were moving slowly from one plantation to another, the militia had been gathering in town. John Cheveley, a young man from Liverpool, who clerked in the merchant house of John and William Pattinson, went with them. He described in detail in his diary the fear and ambiguous feelings of a youth brought up in England who, like others, had come to the colony to make a living and had suddenly found himself in the midst of a slave rebellion. Monday night, when he was already in bed, he had been roused by the sound of a bugle, which he immediately identified as that of his rifle corps, "calling to muster." Before he could figure out what was happening, someone had called him "to make all haste and join the Corps, as the negroes were all in rebellion."[34]

"A pretty piece of business this, to be called out of bed at this time at night, and go, I don't know where, for the pleasant chance of having my throat cut by these same savages." With such thoughts and a confusing mixture of fear and pride, Cheveley picked up his accoutrements, buckled on his sword, slung his rifle over his shoulder, and found his way into the street. There he walked toward the parade ground, "where the bugle was sounding fast and furious and the whole town turning out to see what was the matter."

He found that the first detachment of the troop had gathered, and he was soon jogging along with the second detachment toward the East Coast. No one seemed to know exactly what was going on. During the day, rumors of something wrong up the coast had reached the town. But initially the affair had been kept secret for fear of exciting the slaves in town and on the West Coast. The militiamen marched in gloomy silence, "not knowing if the danger was far or near," or how soon they "might be set upon by a ferocious band." They went "along the road, which ran parallel with the sea shore, and in front of the various Estates, mostly sugar plantations, which lined this part of the coast, for about four miles." The night was quiet, "everything perfectly still, save the croakings of all the varieties of frogs" and the sound of countless crickets. Suddenly the commander called a halt. Something lay by the roadside. It was the dead body of a black, recently shot—a sign that something serious was happening in the neighborhood. "Keep a good look out. . . . Don't fire without orders," warned the commander.

They continued warily for another three miles until they arrived at *Vryheid's Lust,* Van Cooten's estate. On the bridge they saw a black with a musket and bayonet, who told them to stop. The next moment two militiamen had the man by his collar. He protested he had been sent by his master to invite the soldiers to come up to his house. Not trusting the slave, they followed him with a loaded rifle held to each of his ears. After a short but fearful walk, they arrived at the main building of the plantation and found another militia detachment and the "negroes of the estate quite peaceably disposed." There they spent the night.

The next day they continued their march forward, anxious to join the regular troops. Wherever they went they heard that the blacks had stopped working and had retired to the "backs" of the estates—a bushy area, left uncultivated, except for some slaves' gardens here and there. A large group of insurgents reportedly had gathered in the neighborhood of plantations *Elizabeth Hall* and *Bachelor's Adventure.*

At plantation *Chateau Margo* the militia found the proprietor Lachlan Cuming in an "awful temper," with a broken nose he had gotten while struggling with slaves who had tried to take his musket from him. According to Cheveley, the blacks had put Cuming, the manager, and the overseers in the stocks, "to the infinite diversion of the negro women, whom the manager had treated with great severity, and who now took it out on him, by each saluting him . . . with a slap on his face." After this ritual was performed an old slave, "who had been in the habit of very frequently going on the sick list, instead of going to work, for which the manager invariably treated him with a copious dose of salts every morning, thought this no bad opportunity to pay him off in his own coin, so accordingly mixing up the dose, which (had all gone on as usual) would have been his morning's portion, he presented it to the manager. . . . 'Here Massa, here something for do you good'." The manager tried to resist, but the slave insisted. " 'Drink um, Massa, drink um arl up, pose be good for me, be bery good for Massa Buckra, so drink um up 'rectly, I say.' " After that most of the blacks had gone, leaving the manager and overseers in the stocks.

After liberating them, the militia rested for a while at *Chateau Margo* before continuing its march in the afternoon. They soon joined the regular troops with Colonel Leahy at their head. Together they numbered 400 or 500 men. According to Cheveley, they found the Colonel "quite disposed to exercise the privileges conferred upon him by Martial law." The first thing the Colonel did was to reprimand Captain Croal for not preserving discipline. Croal once had been a slave trader and now worked at Troughton and Co., an import-export firm in Georgetown. As a career officer, Leahy did not trust the military skills of civilians. With a mixture of military zeal and native rudeness, he swore with an oath his determination to hang the first offender from the nearest tree, "a measure which the whole aspect of the man seemed to bespeak him quite capable of carrying into effect should occasion offer."

The militiamen marched gloomily under this new command toward *Bachelor's Adventure.* As they marched along, through the darkness, firing at intervals, they began to spot heads poking up in the cotton fields. They heard women and children's voices calling for mercy, coming from the fields: "Oh Massa, Oh Massa, spare arl we, do Massa, spare we pickny." "Go and tell the men to go to work then, or we'll shoot them and you too," the Colonel replied.

The detachments moved on till they arrived at *Bachelor's Adventure.* There they found themselves suddenly surrounded by an enormous crowd of slaves, which the terrified Cheveley estimated to be around 3,000 or 4,000. The slaves were all armed, some with muskets, others with cutlasses. "It was an awful moment of suspense." With a keen perception of the nature of the confrontation that was taking place, Cheveley observed that "Everyone felt that the crisis had arrived when it was decided who should be master." The officers held a conference and resolved that a deputation with a flag of truce should be sent to the insurgents to persuade them to lay down their arms and disperse. Colonel Leahy, Captain Croal, and one or two captains of the regular troops went on the mission. When they asked the blacks what they wanted, they replied: "Massa treat arl too bad, make we work Sundays, no let we go Chapel, no give time for work in we garden, lick arl we too much. We hear for true great Buckra give we free, and Massa no let we hab nothing." From among the crowd, Jack Gladstone stepped forward and handed to Leahy a paper signed by several managers saying that they had been well-treated by the rebels. Leahy responded by giving Jack a copy of the governor's proclamation of martial law. He then ordered the slaves to lay down their arms and return to their work. But the rebels did not move. Soldiers and slaves stood watching each other, the tension growing. To Cheveley, it seemed that an hour had passed when people in the crowd started calling out, "Catch the big Buckra, tie um, tie um." Hearing this, Leahy spurred his horse, galloped back toward the plantation buildings, and ordered the soldiers to attack.[35]

The soldiers poured in volley after volley. The slaves returned the fire but soon began to run, "leaping the trenches, into which many tumbled lifeless." Many were shot down on the road and in the cotton fields. By noon, the roadside was littered with dead bodies. About two hundred slaves had been killed. On the side of the whites only the bugler (who had been shot accidentally by one of his fellow soldiers) was dead, and one or two soldiers injured.

From *Bachelor's Adventure* the troops continued toward Mahaica going from one plantation to another, liberating managers and overseers from the stocks. Here and there they met groups of slaves. But after small skirmishes the slaves would withdraw to the bushy areas at the backs of the plantations.

While the regulars and the militia did their bloody "duty," the governor, in an attempt to avoid the spread of the rebellion, issued a proclamation. He announced that various measures were being contemplated by His Majesty's Government for ameliorating the conditions of the slave population in the colonies and for progressively qualifying them for an extension of privileges. But any misconduct or acts of insubordination on the part of the slaves would nullify such measures. It would "forfeit all their claims upon the liberality of the British Government, and utterly disqualify them from benefiting in any manner by its favourable disposition towards them."[36]

The proclamation was nominally addressed to the slaves who had "continued faithful and obedient to their masters," though Murray must have had other audiences in mind, since he also instructed his subordinates that whenever they met rebellious "negroes" they should announce that the Court of Policy and their masters had agreed to issue certain regulations that favored the slaves. The first step would have been to abolish the flogging of women and the carrying of the whip or any other instrument of punishment into the fields. Other improvements were to have followed. But now they had forfeited all claim to favor and the "only hope that the measures intended" would not be stopped forever depended on their immediate and unconditional return to their duties. "You must lay down your arms, and come in within twenty-four hours, and your Governor will extend what mercy is possible to you."

The governor continued his psychological warfare two days later. He issued another proclamation, hoping to sow confusion among the slaves and make them believe that if they gave up resistance and returned to their work, they would be spared and forgiven. While in his first proclamation Murray had threatened the slaves, now he used a different strategy. After making a distinction between "good" and "bad" slaves, he offered pardon to the first, and punishment to the second. He said that many "faithful and well-disposed slaves had been forced by the more evil-minded among them to join their revolt" and had continued among them "either by force or by apprehension of the consequences of them having appeared in arms." He promised a "full free pardon to all slaves" (provided they had not been ringleaders or committed excesses) who would within forty-eight hours "deliver themselves up to any burgher captain or officers commanding detachments or parties of troops, giving up their arms, accoutrements and ammunition, as pledges of their sincerity." Those who did not respond to this appeal would be dealt with as rebels and "could place little hope in mercy."[37]

Murray's proclamations, combined with arrests, killings, and executions, would in the long run have the desired effect. They would deepen the divisions existing among the slaves from the beginning, and help to put down the rebellion. But it would take several weeks for "order" to be completely restored.

According to the governor's own bulletin, 255 slaves had been killed or wounded in the skirmishes between the troops and the rebels during the first three days of the rebellion. He calculated that around 9,000 slaves had risen more or less at the same time, in different parts of the East Coast. But although they were armed with cutlasses and other weapons they had taken from the plantations, most lacked military training, and when they fired they often missed their targets. More important, most slaves had opposed killing the whites. Initially, many slaves believed that the soldiers would not attack them, but still took some precautions, in case the soldiers did. When the soldiers started shooting at them, some slaves fired back. But most ran away in confusion. (In fact, one of the remarkable things about the rebellion was that so few whites were killed, and so few were wounded.) The whites, however, were in a different mood. Managers and proprietors may have considered the financial loss they might face if slaves were killed (although they could always expect to get government compensation). But this concern was not enough to counteract their fear of the slaves. Except for a few militiamen like Cheveley, most whites were more than ready to shoot, particularly when they felt personally threatened. The professional soldiers were even more ready to kill. Their business was war. They were there to suppress the rebellion, at any cost. Whenever an inexperienced militia recruit revealed any scruple or fear the soldiers treated him as a traitor. Officers maintained an iron discipline over their troops and resorted to whatever display of violence and brutality necessary to intimidate the slaves and show them who the real masters of the colony were.

Repression fell violently upon the rebels. The moment called for ritual. There were "truths to be played out." It was necessary to show who had power and who must obey, and to make fear stronger than hope.[38] Under these circumstances, some officers, moved either by an over-zealous sense of duty or by natural bad instincts, committed all sorts of atrocities in the name of "wisdom," "necessity," or "duty." The most outstanding was Colonel Leahy. After his victory at *Bachelor's Adventure,* he led his men to *John* and *Cove,* the property of John Hopkinson. The proprietor and several other planters were there. When the soldiers arrived they saw the space before the plantation house crowded with slaves—men, women and children—who Cheveley thought looked "dispirited and alarmed." As the troops were forming in regular order around the front of the house, orders were given to prevent any black from leaving the place. One man was spotted "making off at a quick walk, along the side line of the estate, towards the back." Colonel Leahy dispatched a corporal and two privates to bring him back. The man was easily overtaken. He was trembling all over. "One Corporal said, 'Come boy, you must go back with us.' 'Oh Massa,' he said, 'you go for kill me. Oh, Massa, me innocent, we been quiet arl the time for true.' 'Well, well,' said the Corporal, a Scotsman who had a store in George Town. 'Well, well, my lad, gyn ye be innocent ther is na hamm.' 'Deed, Massa, me innocent. What for Massa want me, Oh, Massa, feel in myself they go for kill me, peak for me, do, Massa.' " Cheveley, who knew the slave, tried to reassure him that nothing would happen: "If you are innocent, Allick, you'll come to no harm, you tell true that's all." The slave was then taken to the plantation house.[39]

Meanwhile, Colonel Leahy, Captain Croal, two regular army officers, Hopkinson, and several planters had been trying Dublin, one of the blacks who supposedly had been a ring-leader in the rebellion. The "trial" took only a few minutes and they came out bringing the "culprit, his hands bound behind him." "Bring that fellow to the front," the old Colonel shouted. "There was a dead silence. The slaves looked on silently and apprehensively. 'You rascal, what have you to say?' " asked Leahy. "The man protested his innocence, calling God to witness." Cheveley was deeply moved by the prisoner's appeal, "which was made in a manly spirit of candour, betraying no sense of guilt or fear."

Hopkinson, the plantation owner (and himself a mulatto), stepped forward to plead for the slave: "Colonel Leahy, I must beg to intercede for this man. I have always found him a most faithful servant. I cannot believe he is guilty, let me entreat you to give his case further consideration." Indeed, two years and a half earlier Hopkinson had filed a petition for the manumission of Dublin.[40] But nothing would move the Colonel. "Who are you, Mister?" Leahy replied with apparent indignation. "Go back to your business, I am here to punish these fellows, and by God, they shall receive their deserts. Tie that fellow up." Dublin pleaded his cause over and over again while the soldiers tied him to a tree. His master once again begged Leahy not to act so precipitately, and to spare the slave till his case could be more fully investigated.

"The old colonel glared round upon him, and said, 'I'll tell you what it is Sir, it's of no use your talking to me, you are acting from interested motives, and by God, if you talk to me any longer I'll put you under arms and send you down to the governor. If you are afraid of losing your negroes, I am not coming up here to be humbugged by you, and have all this trouble for nothing.' " Let me do my duty, he continued with an unmistakable military logic, "and you may all sleep quiet, in your beds for years to come, but if I am interfered with, you'll all have your throats cut, before you're twelve months older." Turning to the slave he said: "Pray to God, daddy, Pray to God." Then he ordered the soldiers to fire.

Allick, who had been watching the whole scene with terror was then called. " 'What have you to say for yourself?' 'Mister Buckra Massa, me innocent for true.' 'You lie, you rascal, haven't we the clearest evidence that you were one of the ringleaders?' " Allick pleaded his innocence again. "You scoundrel," roared the old colonel, in a fury, "Do you persist in telling such a lies? Do you know you will be in hell in five minutes?" On Leahy's orders the soldiers shot and killed Allick.

Troops reenacted the same bloody ritual on other plantations. After summary trials, alleged rebel leaders were executed in front of paralyzed gangs, their bodies were laid out side by side on the grass. The slaves' actual degree of involvement in the rebellion mattered little to whites. The officers were interested only in reasserting their authority and terrifying the slaves. The mock trials they conducted were not born out of an abstract commitment to justice, but rather out of their desire to reestablish the "order" the slaves had temporarily inverted. As in most political trials, the circumstances of each case were irrelevant. It did not make any difference whether those who were condemned to death had been ring-leaders or not, whether they had spared their masters' and managers' lives, or had treated them roughly. Slaves had challenged their masters' power, violated their masters' rights, and some would have to pay with their own lives for these outrages. Since it did not make sense to kill them all (after all, they were valuable property), slaves would be randomly chosen as examples. And the rituals of punishment would have to be terrifying so that no slave would ever dare try it again.

The killings continued for several days. Between Friday the 22nd and Wednesday the 27th, more than twenty slaves were executed on different plantations. The circumstances were often as absurd and pathetic as at the execution of Allick and Dublin. On the morning of the 22nd, one of the prisoners, in the hope that he would be recompensed, offered to give Colonel Leahy the names of the leaders of the rebellion. When the man was brought back to where the other prisoners were staying, one of them, Beard, told him that he had been a "damned coward and a fool" for betraying his friends. He would have preferred to be shot than to give up their names. Beard's remark was immediately reported to Colonel Leahy. As a result, Beard was marched to plantation *Clonbrook* and shot "by two file of the 21st Fusileers and Rifles."[41] On the same day and place, three other slaves, January, Edward, and Primo came to the same end. At *Le Resouvenir* and *Success,* where the slaves had hoisted a white flag, Captain McTurk (disregarding the full pardon promised by the governor to those who surrendered) arrested Toby, Jim, and Hill, who were executed together with three others on the following Tuesday, the 26th, at plantation *Beter Verwagting* in the presence of the assembled gangs of the neighboring plantations.[42]

The slaves' official trials had already started in Georgetown on the 25th, but summary executions on the plantations continued for several days. At *Nabaclis,* the same detachment that had been executing slaves on other estates arrested Caleb and Sloane, one for the murder of an overseer and the other for "maltreatment" of Mary Walrand, wife of Francis Alexander Walrand, part-owner of *Nabaclis.* The prisoners were shot and then decapitated by Joseph, head-driver of the plantation (who had also been arrested). Their heads were affixed to poles on the road in front of the estate.

Joshua Bryant, who lived in Demerara for fifteen years as an artist, and served in one of the regiments during repression, wrote the first history of the rebellion. In reporting the summary executions, he carefully registered that the slaves had been executed "after full proof had been received of their guilt." But to Cheveley, the whole thing looked like butchery, an "odious, painful and sickening" experience, "a dreadful affair," alien to any notion of justice.[43]

During the mock trials on the plantations, no one was concerned with keeping records. But indirect evidence that emerged later in the official trials suggests that Cheveley was right and Bryant wrong. Slaves were singled out more or less at random for execution. The sentences were less a result of fair deliberation and legal procedure than of hearsay and hasty decisions by officers, managers, and overseers. The purpose was terror, not justice. Managers and plantation owners had not recovered from the impact of the fear and humiliation the slaves had imposed on them. For a few days, slaves had inverted the social order and had treated masters the way masters had always treated slaves. Slaves had been whipped, so they whipped. They had been put in the stocks, so they put masters and managers in the stocks. They had been verbally abused, so they called their masters "scoundrels" and "rascals." For a few days the slaves had played masters. Now they had to be severely punished, to learn their place.

Masters, managers, and officers forgot their boastful admiration for the "great tradition" of British law. They even forgot to keep up the appearance of civility, decorum, and legitimacy in the eyes of the mother country—something that seems to have been one of their greatest concerns in normal times. It was only retrospectively, when things returned to normal, that they felt the need to advertise to the world that everything had been done according to the rules of "civilized men" (which for them meant the rules of Great Britain). This explains Bryant's insistence that everything was done according to the law.

But in reality things were quite different. The slave January, for example, one of the first to be executed, was described in a later trial as the captain of the schooner of plantation *Clonbrook*, a man who before the rebellion had earned the confidence of his master because of his exemplary behavior. His main crime during the uprising was to have beaten Jack Adams, a black man working as an overseer. Jack Adams had tried to prevent his master's ammunition from falling into the rebels' hands by hiding it on top of a thatched building in front of the dwelling house of the estate. January ordered the house set on fire and when Jack Adams tried to escape, January knocked him down and dragged him to the stocks.[44] Apparently, that was enough to justify the execution of January.

Equally arbitrary was Allick's execution. His participation in the rebellion was undeniable, but even Bryant, who was not a sympathetic witness, reported that during the rebellion Allick had interfered in favor of Gainsfort, the manager of *Golden Grove*. Bryant's testimony was confirmed by Gainsfort himself in one of the official trials. According to Bryant, at four a.m. on August 19, about three hundred slaves surrounded the house. Gainsfort and others who were staying overnight at the plantation tried to resist. They fired upon the slaves to keep them away and apparently killed two. Irritated, the rebels set the house afire. In an escape attempt, Gainsfort was wounded in the jaw and the back of his head by musket fire and taken prisoner. The rebels debated whether to kill him or to put him in the stocks. They stripped him naked and dragged him by the feet over the rough shells that covered the ground around the house. Allick intervened and Gainsfort was allowed to stand up and walk to the "sick-house," where he was placed in the stocks. On the way Gainsfort was severely horse-whipped by "several insurgents, wounded in the arm with a bayonet, and sadly bruised in various parts of his body."[45] Allick was not among those who attacked him.

After Allick's execution, scattered information gathered in later trials did confirm his participation in the rebellion, but there is nothing to suggest that he played a leading role. Much of his violence was directed against slaves who stood by their masters. On one plantation, he had beaten two slaves, Cuffy and Ned, to force them to reveal their master's location. Ned testified that he was guarding the counting-house door when Allick and other slaves came looking for his master. According to Ned, Allick said: "Here is one of Mr. Spencer's servants, he must know where he is," and they started beating him.[46] Neither Cuffy nor Ned gave any evidence that could have justified the death penalty summarily imposed on Allick. He had done nothing worse than many others. On the contrary, he had even interceded in favor of the manager of *Golden Grove*. Later, when things calmed down, others would be acquitted for similar behavior. Allick had the bad luck to be one of the first to be "tried."

After several days of indiscriminate killings, the executions on plantations stopped. Regulars had done most of the work, but the militia had also participated in the ghastly rituals, to the disgust of some young men like Cheveley, who were new in the colony and had felt horrified at the sight of so much brutality. Many of them had shown "the greatest repugnance," and some had flatly refused to participate in the executions. But whatever their feelings, they had had little choice, and as Cheveley put it, "file after file of these young unpracticed hands were turned out to fire at those poor wretches." For Cheveley, however, the worst was having to stand by while the slaves were forced to cut the heads off the dead bodies and stick them on poles.

As days passed, whites must have begun to feel that they had gone too far. The cells were full of prisoners and dead bodies in chains and decapitated heads were on poles along the colony's roads. The court-martial had convened in Georgetown. The rebellion was effectively over. Most slaves had returned to work; only about a hundred were still loose, among them Jack

Gladstone and his father Quamina. The important task now was to try to catch the fugitives before they started a new rebellion.

As late as September 3, the *Royal Gazette* was offering a reward of 1,000 guilders for the apprehension of Quamina, Jack Gladstone, and eight other men and ten women.[47] An expedition composed of militiamen and Indians was organized. For four days whites and Indians searched the bush without success. The tireless McTurk went with one group. Despite his obsessive determination to find the fugitives—a determination that led him one day to force the men under his command to such a long march that some of them fainted from fatigue—they found only abandoned camps. One by one the expeditions returned to town empty-handed.

On September 6, a black man informed Captain McTurk that Jack Gladstone had been seen at *Chateau Margo* the night before.[48] McTurk "sent a boy to ascertain the truth." The "boy" came back confirming the information. McTurk did not waste time. With a detachment of the militia he set out for *Chateau Margo* at one o'clock in the morning. They surrounded the slave quarters and searched one house after another. In one they found Jack Gladstone. They also discovered a woman on the roof "astride across the rafters." It was Jack's wife, who was also taken prisoner. But they searched in vain for Quamina.

One expedition after another went to the bush in search of Quamina, but to no avail. Whites took slaves as their guides, promising rewards for his capture, but one slave after another defeated their intent. Not only did they not lead the expeditions to Quamina's hideout, but they seemed to have a system of communication with the local maroon groups because often when an expedition arrived at a bush camp no one was in sight. In one place they found huts with plenty of corn and rice in cultivation—evidence that maroons had been living there—but found no one. In the houses there were several books (two Bibles, one Testament, one Wesley hymn book, one Watts hymn book, a spelling book, and Sunday school tracts). There was also a case with a compass and other instruments, hammocks, saucepans, blankets, a packall, trunks containing musket-balls, powder, a white muslin dress, a chintz petticoat, one pair of trousers, shirts and a shift, a flannel nightgown, a brown surtout, a black silk spencer, remnants of osnaburg,[49] a pair of shoes, a piece of salempores, a shift, a hat, some razors, a cartridge box, and a hammer.

Disappointed, they continued their search, canvassing the backs of the plantations. They scoured plantain walks, coffee, cotton, and sugar cane fields, and inspected watch-houses. Finally, on September 16, after several fruitless attempts, they stumbled on three blacks: Quaco, Primo, and Jack. Quaco was the first to be taken prisoner. McTurk promised to spare his life, if he would lead them to the rest of the party. Quaco agreed, but apparently, like other slaves who had been taken as guides before, he led them nowhere. Primo was the second slave to be arrested. He said that Quamina had told them they were entitled to their freedom; that if they wished, they could leave him in the bush and return to the plantations as some others had, but they would be fools to do so. He intended to remain in the bush. No white man would take him alive. And if they took him by surprise, he would kill himself. Primo also said that Quamina and his party had been in the bush for days. Short of food, they had given Jack Gladstone money to buy fish and other provisions, but he had not returned. (They had no way of knowing he had already been captured.)[50] Suspecting that Quamina would not be far off, McTurk ordered his group to continue the search. They finally spotted him, in an area of heavy bushes where he could not be easily seized. The Indian who saw him first ordered him to stop, but Quamina neither stopped nor ran. Instead he went on walking without looking back, as if he had not heard the order. He seemed to be determined to be killed rather than arrested.[51] As he was about to get out of sight, the Indian shot him through the arm and temple. Quamina had kept his promise: no white man would take him alive. He carried no arms, only a knife and a Bible were found in his pockets. His body was carried back to *Success* by slaves who had been caught during the raids. On September 17, a gibbet was erected on the road in front of *Success*. Surrounded by the assembled slaves of the plantation, a party of militia under arms, and Indians, Quamina's body was hung up in chains.

The bush expeditions continued for several weeks. The colonists still feared that runaway slaves might try to come back and instigate other slaves to rebel. So with the help of Indians they went on searching the bush in many directions and combing the estates. They located several camps and arrested several runaways, "mostly very old absentees."[52] Many camps seemed to be only temporary settlements with little or no provision grounds. The runaways apparently had continued to rely on the plantations for survival.

The focus of the action was now in Georgetown, where the trials had started and new executions were taking place. Informal terror had subsided and the official rituals of exorcism and excommunication had begun. On Saturday, August 23, the militia escorted the prisoners to town. The militiamen arrived just before sunset, with soiled garments, unwashed and unshaven faces as badges of their hard service. Cheveley was among them. He described how they marched between crowded rows of people up the main street, cheered and hurrahed by the multitude "as heroes returning from a splendid triumph. 'Bravo rifles,' resounded from all sides, whilst execrations were heeped upon the unhappy prisoners," even by the town blacks, something that Cheveley had difficulty in understanding.[53] According to Cheveley, as the militiamen walked along what black girls "cheered" was "Oh, me dear Massa. Oh, me poor Buckra. Look he clothes, look he face; we really pity poor Buckra now." And the words would be passed from one to the other—"We really pity poor Buckra now." The prisoners were crammed into the local jail. The militia spent the next day drilling and marching, as in a ritual confirmation of their triumphs on the field of "battle."[54]

Like Cheveley, many of the militiamen were young and worked as clerks in towns. They had come to the colony from England, Scotland, Ireland, or islands in the Caribbean in search of positions. Some had brought antislavery notions, others a deep evangelical piety that compelled them to recognize the slaves' humanity, even though this recognition might be qualified by notions of racial and cultural superiority. They all shared a commitment to "British liberties," a naive belief in the "civilizing" mission of the empire, and ideas of fairness and justice that were obviously contradicted by the events. They were horrified by what they saw, but had little power to change things. However reluctant they may have been, they performed the tasks they were told to perform and wrote pathetic letters to their families describing the horrors of repression, condemning the atrocities, and marveling at the slaves' extraordinary endurance.

A young militiaman, a sentry in the jail where blacks waited for their trials, described what he saw:

> About one hundred men are pent up in this close place, but what is still worse, the poor negroes that are taken are in the same place with several of the wounded; their wounds not dressed or washed; the stench is dreadful. Sometimes we hear groans, when these poor fellows are in our part of the room. With their hands tied behind their backs. I am astonished with what patience they bear the pains of their wounds, without a friend to say one comforting word. When they ask for a little water to cool their fever, perhaps some sentry will curse them for damned black dogs; what would they do with water? A halter around their necks would fit them better. Often could I with all my heart, have taken my bayonet, and leveled the cruel hardhearted rascals to the ground. I have certainly experienced more of human life during these last six weeks than I have done during the last nineteen years.

The sentry could not disguise his antislavery feelings and his sympathies for the prisoners. His letter was couched in the unmistakable pious language of the new evangelicalism, mixing feelings of guilt and sin with a profound hatred of privilege.

> It is a strange world that we live in. What a meddley. Providence allows such outrages. Surely slavery must be the oldest form of hell. The prisoners generally ask me for any thing they want, although I never speak to them. Often, when I was receiving every luxury of meat and drink could I have wished that these poor fellows had it instead of myself. Why does God give me every thing that I can wish for, who am a sturdy rebel and daily sinning against him and yet these poor fellows who are wounded, have not the slightest comfort?[55]

During August and September, several slaves were tried and condemned to death. The court seemed eager to pick a few as examples. To have participated in the rebellion at all was enough to condemn any slave to death. The trials were pro forma. The verdict guilty was determined before they even started. What is puzzling is that the government felt the need to maintain the appearance of due process, particularly since the militia and the soldiers had already executed after summary trials many blacks whose only fault was to have gathered in what seemed to the whites a "menacing manner" to demand "their rights." The colonists could have continued their sinister task instead of bringing prisoners to town for public trial. On the surface it would seem that nothing but the need to keep a favorable image in the mother country could explain this legalistic ritual. But, ironically, behind the decision to bring the slaves to trial was the colonists' reverence for what they themselves called "rights."

When slaves and colonists talked about rights, they used the same word, but meant contradictory things. Slaves were claiming their right to freedom and equality, masters were asserting their right to property. In a slave society, the universal rights of men, equality before the law, and other such notions had nothing to do with slaves. Justice applied to slaves was the justice of the master—a type of justice which implied the power relations characteristic of slave societies. Like medieval justice, it made crime and punishment dependent on social status. It was an arbitrary, haphazard, and discriminatory justice, whose primary goal was to display and reinforce the absolute power of master over slave. Yet in Demerara, as in other slave societies in the nineteenth century, side by side with this traditional concept of justice, there was another, more consonant with a society of "free laborers."[56] It claimed to be universal and aimed at subordinating whim to carefully prescribed procedures of investigation, and at eliminating both legal privileges and patronage—a new concept of justice asserting the supremacy of the law and the equality of all citizens before the law.[57] This was the concept of justice that men like Smith and Cheveley had in mind. The two notions of justice were fundamentally contradictory. And in the day-to-day life of the colony a constant battle was waged between the masters' and managers' claims to absolute rights over the slaves and the British government's attempts to assert the supremacy of the law over both masters and slaves. The slaves understood this tension well and were ready to use it for their own benefit. The amelioration laws of 1823 had brought this conflict to the surface.

In Britain the concepts of justice and citizenship were changing and so was the system of punishment. In the 1820s and 1830s the bloody eighteenth-century code was substantially repealed.[58] These new trends shook the foundations of the slave system in the colonies. Demerara's masters and managers correctly perceived the new laws, abolishing the use of the whip in the fields and the flogging of females, as infringements upon their own "rights." Their authority lay symbolically in the whip. As one West Indian spokesman stated in a speech to Parliament, the whip was placed in the hands of the driver as a "badge of authority . . . a symbol of office." But as one of his political adversaries pointed out, the whip was more than a symbol, it was also a form of exertion *necessary* to the maintenance of slavery.[59]

As in the conflict between those who favored teaching slaves to read and those who were against it, the conflict between those who supported the law prohibiting the use of the whip in the fields and those who opposed it originated in two different power strategies and forms of social control. One was based on physical punishment, the other on moral persuasion and economic coercion; one was addressed primarily to the body, the other to the mind. An article in the *Royal Gazette* in 1808 exemplified well the point of view of masters and managers: slaves were not fit for a "regular" government based on moral persuasion, and even the most humane and Christian masters were necessarily obliged to resort to corporal punishment.[60] As another spokesman for the colonists put it: "The authority of the master cannot be limited. European workers can be fired, West Indies slaves have to be punished." Demerara planters and managers would gladly have endorsed a Trinidad planter's statement, reproduced in the *Guiana Chronicle,* which forcefully argued that "to deprive the masters of the power of inflicting corporal chastisement on males or females is virtually a deprivation of property."[61] And there was nothing more sacred than the right to property. The colonists were in a quandary. How could they obey the law if it took away their power and threatened their property? Yet, how could they do without the law if they needed it to guarantee their property and to protect themselves?

Masters and managers were confronting a concept of justice that, because it purported to transcend class by considering all people equal before the law, was profoundly subversive of the slave system's protocols and forms of social control. Nothing could express better this trend than a speech made by Lord Combermere, a spokesman for the crown in Barbados: "All classes of his Majesty's subjects in this Community enjoy to the fullest extent the privileges and blessings of our glorious Constitution, and they know that upon no occasion will their just grievances . . . be unattended to, provided the proper and legitimate mode of seeking redress be adopted." This was a concept of citizenship, justice, and law alien to the experience of societies grounded on privilege and discrimination based on status, gender, and race. How to accommodate this new trend to a slave society was the dilemma the colonists faced.[62]

The dilemma was common to all nineteenth-century slave societies, but in Demerara the situation was even more complicated because its colonial status made it dependent on decisions made in Great Britain. The colonists were expected to obey the orders of the King-in-Council. Even when the "recommendations" of the British government were cast in diplomatic formulas stressing the relative autonomy of the local assembly—as the recent amelioration laws had been—they still had a coercive power. This was true not so much because of English military and naval superiority (although that counted) or because the colonists depended on British markets, capital, and commodities (although that was also important), but because most of the colonists—except perhaps for the Dutch—were fascinated by British institutions and fashions, which they identified with "progress" and "civilization." Most of all they did not like to be called "backward" and "unenlightened." They did not want to be seen as brutes.

Like other members of the empire, they took pride in "the high moral feelings of the British people." Typical was an article in the *Royal Gazette* in 1821, entitled "Political Influence of England," which reminded the colonists that there was "yet no other country in Europe where the principles of liberty, and the rights and duties of nations, are so well understood as with us, or in which so great a number of men, qualified to write, speak, and act with authority are at all times ready to take a reasonable, liberal, and practical view of those principles and duties."[63] How could the colonists deny such traditions? On the reverse side of this ideology stood Tyranny, Corruption, Ignorance, Superstition, Anarchy, and most of all Treason. They had to give the slaves at least a semblance of a trial.[64]

Alone, such ideological and cultural habits might not have been enough to compel the colonists to adopt the rituals of due process. But there were also other reasons of a more practical nature that explain their behavior. If the law and legal procedures might sometimes appear cumbersome to slaveowners or plantation managers, there were times when the law was needed to protect the rights of the colonists themselves, to settle their disputes and satisfy their claims. More important, for the colonial authorities the law was the medium that helped them to establish their own authority and to consolidate the empire. To enforce the supremacy of the empire was to enforce its laws. To disregard its laws was treason and rebellion. The arbitrary power of the masters and managers had to be curbed by the authorities and subordinated to law. How to solve the many contradictions which issued from the need to impose legal procedures from a country of free citizens on a slave society was one of the tasks the authorities in Demerara had to face. Their solution was to keep the appearance of legality, but bend the procedures in such a way that they managed to transform the trials into a show of force rather than a search for truth.[65]

The slaves' trials were brief. With few exceptions they lasted only one or two days. The same charges were monotonously repeated in every trial; so were the sentences, whatever the crimes. Adonis, from *Plaisance,* a slave who had had only a minor role in the rebellion, was one of the first to be tried. On September 1, he was charged with "having, on or about the night of Monday the 18th of August last, been in open Revolt and Rebellion, and actively engaged therein, against the peace of our Sovereign Lord the King, and the laws in force within this colony, and also for aiding and assisting others in such Rebellion." This was to be the charge against many others tried after him.[66] All the slaves pleaded not guilty. All were condemned to death, except for one woman, Kate, tried and sentenced to solitary confinement for two months. For all the others the sentence read: "The Court having most maturely and deliberately weighed and considered the evidence adduced in support of the charge preferred against the prisoner . . . as well as the statement made by him, in his defense, is of opinion, that he, the prisoner . . . is guilty of the charge preferred against him, and does therefore sentence him, the prisoner . . . to be hanged by the neck until he be dead, at such time and place as his Excellency the commander in chief may deem fit." Occasionally there was a variant, instead of "hanged by the neck," the prisoner was sentenced "to suffer death" (although it is not clear whether this meant anything other than hanging). There was, however, one exception: Jack Gladstone. In his case, to the usual charges the judge-advocate added the words "and further for acting as a chief or leader or headman in such Revolt and Rebellion." But, like the others, Jack was sentenced to death. His trial was apparently handled with somewhat more care. It took several days and he was allowed to bring in several witnesses to testify for him.[67]

One of the most striking things about the trials is that among the witnesses there were slaves. In Demerara the criminal court had heard slaves' testimony in a few cases, but usually they were not allowed to testify in court. As late as 1819, the editor of the *Royal Gazette,* reporting on a trial in Dominica against a manager for "an alleged ill treatment of the slaves," commented on the "prevarication if not to say perjury of most of the slaves witnesses for the crown, what little knowledge they had of, or regard they paid to, the sacred obligation of an oath."[68] Like the editor of the *Royal Gazette,* most whites in Demerara considered slave testimony unreliable. In recent years, however, there had been much discussion about this issue. There was a growing tendency to accept slaves as witnesses, particularly when other slaves were on trial. In Antigua, an act of 1821 regulating the trial of criminal slaves established that "in all trials for felonies, or other offences . . . the testimony of slaves for and against one and another shall be valid and admissible in law; and such testimony shall be taken (as has always heretofore been usual and customary in this island) without oath." It also stipulated that slaves charged with offenses which might be punishable with death should be tried "upon a regular indictment, and with every other legal formality which is essential to the trial of free persons under a similar charge."[69] In Montserrat, an Act for the Trial of Slaves Accused of Criminal Matters or Offences by Jury, enacted in 1822, established that since in cases of criminal offenses committed by slaves it was not always possible to procure the testimony of free persons, and "as it may therefore become indispensably necessary, in order that the ends and purposes of justice may not be defeated," slaves as well as persons "of whatever colour or condition, class or denomination" should be accepted as witnesses. The same act also stipulated that whenever the slaveowner failed to provide counsel to defend a slave, judges would nominate one of the barristers of the Court of King's Bench and Common Pleas to act as counsel.[70]

All this reflected a desire to subordinate slaves' trials to systematic procedures. Yet there was still much ambiguity left. Slaves could testify against slaves, but could they testify against whites? And what about free blacks? Could they testify against whites? In two cases tried in Dominica and amply publicized in Demerara newspapers, the rulings revealed a double standard. In one case a white person had "maliciously" shot at a black soldier. The court decided that blacks, even if soldiers, could not testify against whites, so the defendant was acquitted for want of evidence. In the other case a free black man was indicted for buying coffee from a slave. The court decided to accept the testimony of slaves against the defendant but not in his favor. The man was condemned.[71]

The procedure chosen in the slaves' trials in Demerara was to accept slaves' testimony under oath. Whites and blacks were all "duly sworn." All witnesses were asked if they understood the nature of an oath. And in one case at least, a slave was authorized to use a Muslim oath. But in the identification of the witnesses there was already obvious discrimination. In the trials the slaves were usually identified only by their first name and the name of the plantation to which they belonged. Their color was also often mentioned—a reflection of the complex system of social stratification and of the colonists' preoccupation with differences among blacks, mulattos, and whites. There would be "Sam, a negro from plantation *Mon Repos*," "the mulatta [*sic*] boy Isaac, belonging to *Storeck*," "Harry, a negro of plantation *Dochfour*." Only exceptionally did slaves have a name and a surname: Jacky Reed, an important witness in Jack Gladstone's trial and later at Smith's trial, carried the surname of his master. The same was true of Jack Gladstone, who lived at *Success,* one of the plantations owned by John Gladstone. Joe Simpson, the first to uncover the plot, was a slave belonging to Alexander Simpson. Joe carried his master's surname, although he was also referred to as Joe Packwood. Some slaves—probably Africans—had two names: one African and one English.[72] Quamine from plantation *Rome* was also called Morris, and Quashy was called Laurence.[73] There were also slaves like Goodluck who had several names.[74] Free black men were identified by one name if they were defendants, or by a first and last name if they were witnesses for the prosecution.

No slave woman was given a surname. This seems to indicate that they occupied the lowest position in the ranking system. Susanna, an important witness in Jack Gladstone's trial, was listed as "a negress of *Le Resouvenir.*" Free women were usually given both a first and a last name, and their color and residence were also mentioned: "Kitty Cummings, black, lives at *Success,*" "Jenny Grant, black, lives in town," or "Mary Chisholm, free woman living at *Success.*"

Slaves' crafts or jobs were never mentioned, although some were drivers, others were artisans, boatmen, servants, and field workers, serving a great variety of functions in complex hierarchies. Whites, by contrast, were always identified by their names and their profession or standing. This was true whether they were managers, overseers, plantation owners, merchants, or professionals. John Bowerbank, a white man, declared: "I am an overseer on plantation *Bachelor's Adventure.*" Donald Martin said: "I reside on plantation *Enterprise,* I am a medical practitioner there." The same formula applied to mulattos who occupied high positions on the social ladder. In this case color was omitted: Hugh Rogers, who belonged to a family described in other documents as "coloured" (although this was not mentioned in the trial), identified himself as "a joint proprietor of plantation *Clonbrook.*"[75] In this case, status and wealth apparently superseded color.

The bulk of the testimony incriminating slaves came from managers and overseers, mostly white, although some also came from black overseers, drivers, servants, or other slaves. Most of those who testified were men who had been put in the stocks or roughly treated by the rebels. Sometimes the same person appeared first as a planter, then as a manager or attorney. This apparent confusion derived from the fact that some planters were asked to be attorneys on plantations whose proprietors lived abroad. Henry Van Cooten, for example, was owner of *Vryheid's Lust* and the attorney of *Le Resouvenir.* But it is also possible that in a colony in which managers and attorneys constituted the great majority of white men, they had come to see themselves and to be seen by others as masters.

Whatever the witnesses' status, color, or gender—male or female, slave or free, black or white, manager, master, attorney, overseer—their testimony was always sufficient to convict the accused. Usually, one or two witnesses were enough, even when what the prisoner said in his defense made the witnesses' testimony seem quite implausible. The court's assumption was that the defendants always lied and the witnesses against them always told the truth. With the exception of Jack Gladstone, the prisoners had no counsel. And no witnesses for the defense were ever called. The "defense" consisted merely of the prisoner stating *his* version of the events. A few tried confessing and throwing themselves on the mercy of the court. This strategy did not help. Others vehemently denied the accusations. But that did not help either.

Some slaves attempted to establish a good record by describing the way they had treated their prisoners, how they had brought them water, given them a pillow, or tied only one foot to the stock—details that mattered to slaves but were not likely to impress the court. When they were allowed to question witnesses some slaves asked their managers whether they had ever given them any motive for complaint. Surprisingly, managers often answered: "No, you were always a good worker." But such admissions made no difference. Occasionally a manager would say that he had always considered the prisoner a dangerous man. James Allan, a manager testifying about Gilles, said he really believed Gilles was "a character bad enough to be guilty of any thing."[76] Often, in a vain attempt to prove his innocence, a prisoner asked witnesses whether they were sure he had actually done the things they had accused him of. Had he really been at such time and place? With few exceptions, witnesses confirmed their earlier testimony.

Probably because the defendants were ignorant of procedure, intimidated by legal rituals, or simply paralyzed by fear, they seldom challenged witnesses—except for fellow slaves. In such cases they were at times so outraged by their fellows' testimony that, rather than trying to refute accusations, they attempted to incriminate the witnesses. At a certain point in his trial Jack Gladstone seemed more intent on proving Jacky Reed's involvement in the plot than arguing his own innocence. Apparently, he had not forgiven Reed for betraying the plot to his master and to Smith the day the uprising was to start.

The witnesses' testimony was not very reliable. Blacks were under pressure from their masters, and most whites were angry and resentful. Many had been awakened in the middle of the night and must have been terrified by the sight of a threatening crowd of blacks—most of whom they did not even know since the rebels had taken the precaution of sending slaves from one plantation to another, precisely so that they would not be recognized. In the dark, in the midst of all the confusion and fear, how could they tell the difference between one slave and another? Only a few witnesses were honest enough to admit that they could not. They were not sure whether the prisoner was among those who had dragged them to the stocks; they did not remember whether the prisoner was armed, and if armed, whether they carried guns or cutlasses. But when managers did not certify that the prisoner had been involved in the rebellion, the court would simply accept the testimony of slaves who said he had. The truth was always with the accuser.

From the time the first slaves had been arrested, the colony's authorities had been actively investigating the "causes" of the rebellion. The governor created a committee to gather information, and during the preliminary inquiries a picture that incriminated Smith took shape. Jack Gladstone's trial was particularly valuable to the authorities, not only because he had played an important role in the organization of the uprising, but also because his testimony incriminated Smith. And he was not the only one to do so. Apparently, slaves had been told by the court that they could expect mercy if they revealed the parson's participation in the conspiracy.[77] And a few of them did. Later, at their executions, some of them confessed they had lied.

It is impossible to know how much of what appears in the record as the slaves' own recollections of what had happened was really their own version of the truth; how much was their attempt to say what they thought would help them escape; and how much was a distortion intentionally created by those who took depositions. Because many of the slaves spoke only Dutch, some African language, or some sort of creole, William Young Playter, Esq., and Robert Edmonstone, Esq., acted as interpreters, and they were accused at least in the case of Jack Gladstone of altering his testimony. But even such contrived and distorted legal records can be made to cast much light on what really occurred during the rebellion if read closely and carefully enough. During the trials a few names appeared time and again as the leaders of the rebellion: Telemachus, Sandy, Paris, Joseph, and Attila. But the slave who seemed to be at the center of the conspiracy was Jack Gladstone, from plantation *Success*.

Jack Gladstone was interrogated repeatedly, and in his depositions he provided many details about the conspiracy and the rebellion. At first, he left Smith out. But on the last day of his trial, at the end of a long written statement he presented in his defense,[78] Jack claimed that not only was every deacon and member of the chapel acquainted with the rebellion before it broke out, but that Smith knew about the whole plan. It had been revealed to him by Quamina and Bristol on Sunday, August 17. "Parson Smith wanted us to wait," he said. And when asked again whether he was sure Smith knew about the plot, Jack said, "If he did not know what we were going to do, would he have told us to wait?" And, after throwing himself "humbly upon the mercy of the court," Jack made an astonishing declaration, reminiscent not only of the colonists' rhetoric about missionaries, but even of their habitual inability to distinguish the missionaries of the London Missionary Society from the Methodists.

> Before this court, I solemnly avow, that many of the lessons and discourses taught, and the parts of scripture selected for us in chapel, tended to make us dissatisfied with our situation as slaves, and, had there been no Methodists on the east coast, there would have been no revolt, as you must have discovered by the evidence before you; the deepest concerned in the revolt were the negroes most in Parson Smith's confidence; the half sort of instruction we received, I now see, was highly improper; it put those who could read on examining the Bible, and selecting passages applicable to our situation as slaves; and the promises held out therein were, as we imagined, fit to be applied to our situation and served to make us dissatisfied and irritated against our owners, as we were not always able to make out the real meaning of these passages; for this I refer to my brother-in-law Bristol, if I am speaking the truth or not. I would not have avowed this to you now, were I not sensible that I ought to make every atonement for my past conduct and put you on your guard in future.[79]

This last part of Jack Gladstone's statement had obviously been added to his defense at the last minute. Indeed, Charles Herbert, barrister of the Middle Temple who had been practicing in the colony, admitted later that he had written Jack's defense only down to the words "I humbly throw myself on the mercy of the Court." The next day he returned to talk to the prisoner, with the idea of including some other statements, but found that Robert Edmonstone had already gathered the information he needed. Herbert swore he had only "endeavoured to express the prisoner's meaning." Edmonstone, a merchant in Georgetown who was serving as an interpreter in the trials, explained that he had the habit of going round the jail every morning.[80] He had seen what Herbert had written for Jack Gladstone and thought that many things the prisoner had previously told him personally had not been said in his defense. So he had gone to see Jack Gladstone, who had agreed to include them. Edmonstone's tale was quite transparent: he had added the final paragraph to Jack Gladstone's statement with the obvious intention of incriminating the missionary.

Telemachus's testimony would also be used to incriminate John Smith. Telemachus was a member of Smith's congregation and a communicant. He told the "Court of Enquiry" that although the parson had never preached that they should "take their freedom," they always understood he meant they should do so. He also said that Parson Smith "counted Quamina and Bristol . . . more than white men." A year before the rebellion he had heard

Smith say in the chapel and again in his house that the slaves "were fools for obeying the managers" and that the slaves should not, since the King did not wish it. About the time Governor Murray ordered the slaves not to go to the chapel without passes, Smith had told the slaves that something good "had come out for the slaves and that he hoped soon to see all slavery abolished." Telemachus claimed that Quamina had consulted with Smith about whether Jack Gladstone should make peace with Colonel Leahy, and that Smith had told him they should go on. According to Telemachus, if the rebellion succeeded, Quamina was to be King, Jack Gladstone, governor, Mr. Hamilton, "a great man," Paris, "an officer." The "white ladies" were to be allowed to leave the colony, but the white men would be put to work in the fields.

Paris presented an even more extreme picture of what was supposed to happen. All the white men were to be killed and the white women to be taken as wives. If the slaves failed, the ships in the river were to be burnt. All the doctors the slaves were fond of were to be saved, the others murdered. The parsons were to be spared. Quamina insisted on being the King and Jack the governor. Smith was to be the "emperor" and to rule over everything. Hamilton, the manager of *Le Resouvenir,* was to be the general. He had told the slaves to destroy the bridges "to prevent the great guns from advancing." Paris insisted that Hamilton and Smith had often said that the slaves were to be free and that they ought to "take their freedom." On the day before the rebellion, Smith had administered the sacraments and had exhorted the slaves "to go on with the business now, or die."

Paris's "confession" was dated August 18—the first day of the rebellion. He was called again on September 12, and confirmed his deposition, adding some new and colorful details. He said that Hamilton was to have taken the court president's wife for himself, and that Jack had said the governor's wife should be for his father Quamina, and that he would take some young woman for himself. Paris, who was not a member of the congregation and could not have been a communicant, also said that he had gone to the chapel on Sunday the 17th, hoping that Smith would tell whether the "story" was true. But when they all took the sacrament and oath, he had felt obliged to take it too, lest he should be suspected. He went on to say that "The hymn the Sunday before the fight was all about war: 'The Lord will help us in the fight, he is able, he is able, he is able.' " Called again on September 29, Paris added still more detail.[81] Sunday, after giving the sacrament, Smith had asked the slaves if they intended to do him any harm if they took the country. He then brought the Bible, and everybody put their hands on it, and the parson repeated the words: "We are servants of Jesus Christ, as we begin with Christ so will we end with Christ, dead or alive." Then they all bowed their heads, and it was agreed that whatever happened, "we were to call no names." The "parson" also said of the thing they intended, that if they did not seek it they should not receive it any more, neither them, nor their children, and children's children.

Sandy of *Non Pareil* also implicated Smith. He claimed that three weeks before the insurrection he, Allick of *Dochfour*, and Quaw (Quaco) of *Northbrook* had gone to see Smith about a Bible, and Smith had told them that they were "going to lose a good thing" if they did not seek it; "they [the whites] would press [meaning oppress] or trample on our sons, and our sons' sons, and as long as the world remains, we would then receive no rest."[82]

It is impossible to determine whether Sandy, Paris, and Telemachus had said such things just to incriminate Smith and to please the whites or whether they were telling what they believed to be true. No one else spoke of Smith becoming an emperor, or about slaves killing whites and taking their wives. More important, the conduct of the slaves during the rebellion does not give credence to such statements, and they later confessed they had lied. It is also possible that Sandy and Paris had misinterpreted Smith's words directed to a group of slaves who had been sold away. He had told them to hold onto religion. The "good thing" they would lose if they did not seek for it was God's guidance and protection. The fight he spoke about was against Satan, not the white men. Smith had talked about religion, not rebellion. But Sandy and Telemachus had invested Smith's metaphorical language with new meanings, even the sacrament had appeared to them as a ritualistic oath before the war (a usual practice in some African societies).

While Jack Gladstone, Telemachus, Paris, and Sandy—later assumed to be ring-leaders—implicated Smith in the preliminary inquiries, most slaves seldom mentioned him and some explicitly exempted him from any responsibility, although some of their words could be used by the prosecution to incriminate Smith. This was particularly true of Bristol, from *Chateau Margo,* a deacon in Bethel Chapel. In his deposition Bristol insisted that Smith had never told them that they were to be free; he had said instead that the King could not give them freedom, since they did not belong to him. Bristol mentioned that when he heard about the plot, he had gone to Smith and told him the people were all dissatisfied. Smith cautioned them to wait three weeks or a month longer, and not commit any violence. If he saw that "the people kept on," he would have to tell the authorities. Bristol also confessed that his brother-in-law, Jack Gladstone, had been displeased with him for telling Smith about the rebellion. All this pointed to Smith's innocence. But Bristol also mentioned that one Sunday he heard Smith say that the children must all learn to read. With the help of God they would some day be free. He had also told the slaves that since there was no point in going to the fiscal, they were fools for not going to the governor instead, and if the governor did not give them their rights, they could go into the bush. Such remarks would compromise the missionary in the eyes of the prosecution. All along Bristol insisted that Smith had not told the slaves to rise up. In fact, Smith had said that the Christian slaves should not have anything to do with it. But when cross-examined, Bristol admitted that Smith had not told them that it was "wrong."

Bristol was recalled as a witness during Smith's trial, but curiously enough the prosecution did not call Sandy, Telemachus, or Paris (who had made the more compromising statements). At the time of their executions they all recanted their stories, and admitted they had lied about Smith. Later, the Reverend Wiltshire S. Austin, to whom Sandy had initially surrendered, declared that when he asked Sandy why he had joined the rebellion, Sandy answered, "I think I have been a slave long enough." According to Austin, Sandy had complained to him that he had been cruelly treated by both the attorney and the manager of the estate, who had taken away his Bible. Sandy was convinced that there was no use to appeal to the burgher officers because they never gave justice. (To prove his point he mentioned the time he had gone to complain to the fiscal and had been locked up for several days in a dungeon.)[83]

Before the slaves' trials ended the executions started, with all the pomp and ceremony of a public spectacle.[84] The first to be tried, on August 26, were Natty of *Enterprise* and Louis of *Plaisance.* Both were sentenced to die the same evening. Since these were the first public executions, they were carried out with great solemnity. A procession was formed to conduct the prisoners to the gallows that had been erected on the Parade Ground at Cumingsburg. First came an advance guard, followed by blacks bearing empty coffins. Then came the prisoners between guards, the garrison chaplain, and the band of the First Battalion, Demerara militia. They were followed by Lieut. Colonel Goodman, attended by numerous field officers, and militia detachments.

The procession moved slowly through the streets, the band playing a funeral march.[85] As the procession passed up the main street of Cumingsburg, the whole of the Marine Battalion turned out and presented arms, until the procession had passed. When the prisoners were executed, a gun shot announced their deaths.

The next day several more slaves were tried and executed. Murphy, Daniel, and Philip of *Foulis;* Harry and Evan of *Good Hope;* and Damas from *Plaisance.* On September 6, six more rebels were executed.[86] This time, the rifle corps and units of the cavalry and "a body of Indians" joined the procession. On Friday, September 12, nine were executed.[87] The condemned men were accompanied by sixty of their fellow prisoners who were marched under a strong guard to witness the execution. Of the nine rebels' dead bodies, four were hung in chains on the East Coast by the side of the public road.[88] Some of the others were decapitated and their heads stuck up on poles within the colony fort.[89]

The public executions had a juridical and political function: to restore sovereignty by manifesting it at its most spectacular. They were an emphatic affirmation of power, an exercise in terror intended to make everyone aware of the unrestrained power of the masters. As Michel Foucault perceptively remarked in *Discipline and Punish,* the public execution did not reestablish justice, it reactivated power. "Its ruthlessness, its spectacle, its physical violence, its meticulous ceremonial, its entire apparatus was inscribed in the political functioning of the penal system. . . . More than an act of justice, it was a manifestation of force."[90]

It was against this form of punishment that reformers in Europe had risen in the last decades of the eighteenth century. But although public executions would eventually be forbidden in England, they were still taking place. The Demerara executions echoed those of Arthur Thistlewood and his friends, the Cato Street conspirators, who were tried for treason in 1820 in England and were hanged and decapitated, their heads shown to the spectators with the usual exclamation, behold the head of so-and-so, a traitor. There was, however, one fundamental difference. In London, the prisoners were cheered by the people. They called themselves "friends of liberty" and "enemies to all tyrants," and until the last minute they were convinced that they were rendering their "starving fellow men, women, and children, a service."[91] They addressed the crowd in the name of abstract principles they all shared, converting their executions into a conscious political statement. In Demerara, silence and gloom surrounded the prisoners' deaths. Those who dared to speak said they were dying for the sake of religion.

Elaborate rituals were also followed in the cases of slaves condemned to be flogged.[92] On November 6, at plantation *Success,* two blacks were flogged. One received 500 lashes, the other 350. There were present, under the command of Colonel Leahy, a detachment of the First West India, and a great part of the militia's troop of cavalry. After the punishments were completed, an escort of officers was sent to an adjoining plantation, to bring the governor. When Murray arrived, the drivers of the gangs of neighboring estates, with the overseers and the whole of the gang of slaves from *Success,* formed a semicircle before him. The governor addressed them in what seemed to Bryant "an appropriate and admonitory manner." An analogous scene took place the following day. These elaborate rituals were intended both to terrify the blacks and to placate the whites' thirst for vengeance.

Seventy-two slaves were tried between August 1823 and January 1824.[93] Fifty-one were condemned to death. Thirty-three of these were executed, of which ten were decapitated and their heads stuck on poles on the roadside.

Sixteen were spared capital punishment and were flogged with the cat-o'-nine-tails. The others were acquitted.[94] The governor requested mercy for Jack Gladstone and fourteen slaves who had been condemned to die.[95] He explained that in the case of Jack Gladstone his motives were entirely political.

> This man is clearly proved to have been a most active agent in promoting the revolt of which his father Quamina was undoubtedly the principal ringleader, and Jack appears to have had quite sufficient influence of himself over the minds of the other negroes, to have enabled him to guide them almost at his own will in the progress of their revolt, although certainly not sufficient to have checked its actual progress; there is no doubt of his having been in arms, and among some very desperate parties on the night of Monday the 18th of August, . . . but I look upon it to be good policy in the event of a repetition of such struggles on the part of the slaves, to show them that any benefit they bestow on the whites, even though in the act of rebellion, will not be lost sight of in awarding a punishment for their crimes. Jack, whether from politic cunning, or real good feeling, saved the lives of several white persons, amongst whom was the individual who had him in charge as a prisoner on the 18th of August, and from whom he and his father escaped by the assistance of their companions. . . . He is an athletic young man, and of that open and manly disposition which would naturally lead him to enter with heart and hand upon all his undertakings . . . in the midst of these proceedings he is found to have screened a party of whites from his companions, who were greatly exasperated against them. . . . [H]e prevented their being taken until the arrival of the troops secured them from danger, and I think it my duty to urge this circumstance in his favour upon the grounds already stated, as well as from an opinion that his escape would tend greatly to lessen the general confidence in any one who might hereafter attempt to lead them from their duty, by showing them that he has led them into the way of danger, and kept himself out of it.[96]

Murray knew that executing Jack would only make him a hero. It would be much better to banish him, so he suggested that Jack and several other prisoners be removed to Bermuda, where they could live as convicts. The Court of Policy, bowing to pressures from important colonists, opposed clemency. The King, however, eventually agreed with Murray, and Jack Gladstone was banished to Saint Lucia. A letter his owner, the powerful and prestigious John Gladstone, sent on his behalf may explain this decision that saved his life.[97] Cato, a free man, was also banished. The others were sent back to the plantations, where some would have to work in chains for a number of years.[98]

Punishment fell not so much upon those who had plotted the uprising as upon those who had behaved in an openly aggressive way during the rebellion.[99] Most slaves sentenced by the court lived on plantations contiguous either to *Success*[100] or to *Bachelor's Adventure*.[101] No slave living beyond *Orange Nassau,* about eighteen miles from Bethel Chapel, suffered punishment. This seems to confirm that the rebellion had been circumscribed to plantations within walking distance of *Success*. Surprisingly, no one from *Le Resouvenir,* where John Smith lived and where his chapel was located, was sentenced to die.

The slaves' trials and executions only increased the rage of the white community against the LMS missionaries. Davies and Elliot became the target of violent attacks from the local press. The colonists gave free expression to their long-standing suspicion that the missionaries were poisoning the minds of the slaves and working as spies for Wilberforce, the African Institution, and antislavery groups in England. Not even John Wray, who was living in the neighboring colony of Berbice, escaped persecution. But, most of all, everyone waited anxiously for the trial of John Smith.

Seventy miles away, in Berbice, where the slaves had remained quiet throughout the Demerara rebellion, John Wray lived through days of anxiety and terror. In late June, about a month and a half before the rebellion, he had received a letter from Smith reporting that Governor Murray prohibited the slaves from going to chapel without passes from their owners, and as a result many vexatious things had already come to his knowledge. Referring to the behavior of the colonists, Smith had written, "It seems as if the Devil was in them, nay, I am sure he is. Scarcely does a *Guiana Chronicle* appear now, at least not for this fortnight, without holding the missionaries to reproach. . . . I rejoice at these things on two accounts. Such conduct will accelerate the end of slavery, and it will prove our faith and patience, teach us experience and raise a hope that will not make us ashamed. . . . We have no cause to complain of a want of heaven, but of a deep laid scheme to render our efforts for the benefit of the negroes abortive."[102]

For Wray, the attacks on missionaries were no surprise. He had known for fifteen years how hard a thing it was to be a missionary among slaves. "No one knows the difficulty of living on an estate as a missionary, but those who have experienced it," he once wrote. "A person may live many years in Georgetown, Mahaica and New Amsterdam and in any other West India Town . . . but on an estate it is impossible to shut your eyes and ears against what daily passes before you, and the jealousy of the managers and overseers is very great. They look upon you as spies."[103] Wray knew how difficult it was for a missionary to get the confidence of both the master and the slave,[104] and to remain aloof when "slavery with all its evils" was exposed before his eyes in the daily management of a plantation.[105] Wray had been persecuted by governors, attacked by planters and managers, and criticized by the press. He had felt discouraged many times. In fact, it was precisely his dismay at the obstacles he had encountered to his work that had led him to quit his mission in Demerara in 1813 and move to Berbice. Life there had not been easy either, but lately he had been feeling pretty good.

Unlike Governor Murray of Demerara, with whom Wray had always felt at odds, Henry Beard, the governor of Berbice, had not denied him patronage and had been supportive of the mission. There were also other reasons to be hopeful. The British government was making progress toward gradual emancipation, and the government in Berbice (a crown colony) had created no obstacles to the implementation of the new laws. It even had held a public meeting to discuss the subject, with many slaves present. As in Demerara, some slaves had initially misunderstood the new legislation. They too had been under the impression that they were to be free, but once they were informed of their new "rights" they seemed satisfied. Or so it appeared to Wray.

Such things had put Wray in an optimistic mood: "I rejoice that I have lived to see the beginning of the time I have long anticipated when whips and drivers shall not be known, but when these people shall be governed by reason and religion," he wrote just a month before the rebellion broke out in Demerara. "The whipping of the females is also done away with in the colony, in consequence of the late discussions in Parliament, and other improvements will be adopted. . . . I understand that several planters and merchants have written out to request their agents to forward the religious instructions of the people as much as possible."[106] Everything seemed to be improving. But then came the terrible news.

Communication between Berbice and Demerara was difficult and it took a while for news to arrive. But a few days after the rebellion broke out in Demerara, Wray started hearing confusing rumors. Slaves had risen. Five or six hundred slaves had been shot. Both Smith and Elliot had been taken into custody on suspicion of instigating the slaves to rebel.[107] It did not take long before people around Wray started accusing him of all sorts of things. He received a note from the fiscal, about a representation to the governor accusing Wray of having told his congregation that he had received a letter of importance, that he would communicate to them at a private meeting to be held in his house. On that occasion, he supposedly had called the slaves "my degraded brethren and heroes," an expression which would have been particularly irritating to the colonists.[108] Although the inquiries ordered by the governor proved Wray innocent of the accusations, they made him terribly frightened and upset.

Wray thought such accusations were part of a plot to render his labors useless or abortive. What bothered him most was that the attack was coming from people of high standing in the community, so high that they could make the false sound true. "Had this false report originated with people in the colony in an inferior situation of life, without influence either here or in England," Wray wrote to the directors, "I should have passed it by in silent contempt, but unfortunately, Messrs. Atkinson and Watson are among the people of influence and respectability, the former being a planter and lately one of the first merchants in Berbice, at present holding important and serious office of a vestry man of the Church of England, and also some years since a Lieutenant in the Berbice Militia; the latter at the head of a large mercantile house in the colony." Wray feared their high status would give their testimony great weight in England and in other places where the two men had connections. Indeed, he already had evidence that was happening. One of the members of the Berbice council had asked him whether he thought it possible that those "gentlemen" would tell a falsehood. In times of troubles, Wray feared, accusations like these might even cost him his life.[109]

But these were small worries compared with what he continued to hear about Demerara. It was very difficult to find out what was really going on, so many contradictory things were said every day. At the beginning of September, people were saying that Smith had been tried and found guilty and had been shot.[110] (This was a month and a half before Smith's trial started.) Wray wrote a frantic note to the governor, but was told that no official information had been received. The next day, Wray heard that Smith and Elliot had been arrested and were kept in the tower of the Scottish Presbyterian Church. He also heard that the two Methodist missionaries had been arrested too. (This was not true either.) The newspaper reported that the rebellion had been chiefly on the plantations from which Smith's "hearers" came. But they also were saying that two of the ring-leaders were Governor Murray's servants, one of whom, Wray believed, was a member of the Methodist Church. How could he make sense of all this confusing welter of gossip and rumors?

"This is a severe trial to us and should Mr. Smith's life be taken, it is probable we shall never know the real cause of the insurrection," he wrote in a letter to George Burder, the secretary of the London Missionary Society. "Quamina is a member of Mr. Smith's Church. I have known him for fourteen years, a humble, quiet, peaceable man and always a peacemaker and should as soon suspect Mr. Burder of exciting the negroes to rebel. He must either have had insupportable provocations or it is a plot of the greatest Envy against his life. I am confounded when I read it, and could weep tears of blood. Jack is Quamina's son. He can read well but was a wild youth. He was married to a young woman on a neighbouring estate some years ago, by whom he had two children, but a white man, her master, took her to be his *wife.* This is all I know of the history of Jack."

Wray went on to say that every time Parliament adopted any measure for the benefit of the slaves "all the wrath of the West Indies poured upon the missionaries"—an interpretation that was very much to the point. "I do wish the Government would do at once what they intend to do and not tamper with them in this way. Something is absolutely necessary to be done to pressure the West Indies. . . . I have now been nearly sixteen years in this fiery furnace. . . . I feel quite weary of contentions and persecutions. I am just informed that a white merchant declared publicly the day after the news arrived about Mr. Smith being shot, that if any thing happened in Berbice, the first thing he would do, would be to set on fire the chapel and the house and burn my wife and children in it."[111]

Under such terrible pressure, Wray concluded that he had no heart to carry on and wished the directors to appoint some one better able to contend with opposition. So terrified was he that in a postscript he warned the directors that it was not safe to send letters by ship and asked them to acknowledge having received his letters. His fears were not unfounded. Letters sent by regular post were indeed opened, and the only safe way was to confide them to some friendly boat captain or traveler.

The worst, however, was still to come. If Smith were found guilty, no one could tell what the enraged population would do to the other missionaries. That became obvious even before Smith's trial started. On September 29, two weeks before the trial opened, Wray's chapel burned down. He had gone to town on some errands and had not been out of the house for more than half an hour when he noticed thick smoke in the direction of the chapel. On his way back home, he saw people running and crying "The chapel is on fire! The chapel is on fire!" To his grief, he soon realized that it was true. He found a great number of people of all colors carrying water and doing what they could to extinguish the fire. In his house there was a great confusion. His wife—who was pregnant with her twelfth child—was moving things out with the help of others. The militia had turned out to help. So had the captains and crews from vessels in the river. But the chapel burned to the ground in less than two hours. His house also suffered. A large section of the roof collapsed, several windows were broken, and many of his things destroyed. Wray calculated the loss at about two thousand pounds.[112]

The fire had started in a small house a few feet away from his. It belonged to an old black woman who had been out for some days, but had left her own faithful slave to take care of it. Nobody could explain how the fire had started, but everything seemed to indicate that it was no accident. A threat had been made a few weeks before, when the rumor was spread that Smith had been shot. Then there had been a rumor that the governor of

Berbice had shut Wray's school and chapel. After that, the word went around that some people wanted the governor to confiscate Wray's papers because of the correspondence he had maintained with Smith before the rebellion. He had even been called to the fiscal's office to show letters he had received from Smith before his arrest. A few days before the fire, one of the members of his congregation had told him that a "white person" had said that Wray's sermons were being monitored, and that he had a "mark" on the people who came to the chapel and would not give a pin for their lives. People were saying that Wray was going to do in Berbice what Smith had done in Demerara. To make things worse, the newspapers were filled with attacks against missionaries. And all the meetings for catechizing, prayer, and "religious conversations" were being described as meetings for sedition and rebellion.

In spite of his pessimism, Wray admitted in one of his letters to London that several free people had come to help him repair his house and several masters had sent their slaves. He also had received strong support from the governor and the fiscal. The governor, once forced to call Wray to answer some of the accusations made against him, sent a quarter of mutton to Wray's wife, a gesture that could only be interpreted as an apology for the inconvenience. Still, nothing could convince the terrified missionary that his situation was not as bad as he thought. He interpreted every piece of evidence against Smith, every mean-spirited comment, every thoughtless or insolent remark, as part of a sinister plot to involve him and the other missionaries.

During the first week of October, Wray met a man who told him that he had been asked to go to Demerara to appear as a witness in Smith's trial. Wray panicked. He asked the man how a resident in *Berbice* could have any connection with Smith's trial. "Mr. Smith happened to pay us a visit about a year ago," he explained. Smith had stopped to dine and sleep at plantation *Profit* and the conversation had turned to missionaries and the religious instruction of slaves. The manager who had entertained Smith was now called to bear witness against him. The whole story made Wray very depressed. "They are not satisfied with the evidence they can obtain in Demerary," he wrote, "but must even send to Berbice and investigate a private conversation, which took place a year ago. Can anything be so base? All confidence is destroyed. There is no sincerity in man: it is dangerous to speak a word if you do not rail against Missionaries, Wilberforce, and Buxton, and advocate slavery."[113]

Wray had not heard a word from Smith or his wife. But he knew them well enough to doubt that they could have done anything to instigate the slaves. He had visited them several times and had joined Smith to preach in the chapel at *Le Resouvenir.* He thought of Smith as a pious and devoted missionary. But even if Smith had been at fault, why should anyone blame the other missionaries? Hamilton, the manager of *Le Resouvenir,* had also been in jail ever since the insurrection on suspicion of supplying the rebels with firearms. Yet hardly anything was said about him in the newspapers. Was this not a proof that the plot was against missionaries? "What an unjust thing it would be to inflame the minds of all the West Indies against managers of plantations because one has been taken up on suspicion," he wrote. But that was exactly what was being done with the missionaries. The press was castigating them every day, and people looked at them all with suspicion.[114]

Away from the scene of action, condemned to hear nothing but gossip and to read the unreliable colonial press, Wray was profoundly distressed. The *Guiana Chronicle* continued its work of defaming the missionaries from the London Missionary Society. On October 8 and 10, it had long pieces against Davies and Elliot. The former had just returned from a visit to England. But although he had been away during the rebellion, he was not spared the criticism. Elliot had been detained for a few days and then released. Smith was still in jail. His trial was about to start, and no one could guess where it would end. Wray felt lonely and isolated. "A man," he ruefully concluded, "is never safe."[115] A month later, he remarked bitterly that his fellow missionaries were safer among "the savages of Africa" than he was among his own countrymen.[116]

CHAPTER SEVEN

A Crown of Glory That Fadeth Not Away

> And Aaron shall lay both his hands upon the head of the live goat and confess over him all the iniquities of the children of Israel and all their transgressions in all their sins, putting them upon the head of the goat, and shall send him away by the hand of a fit man into the wilderness: And the goat shall bear upon him all their iniquities unto a land not inhabited: and he shall let go the goat in the wilderness. *Leviticus 16:21–22*

Political trials are peculiar trials. Their goal is to reassert power and authority. A political trial is both a ritual of exorcism and a process of excommunication, whose purpose is to expel the one who has threatened the established order and raised doubts about its legitimacy. The defendant's guilt has been decided a priori. The trial is theater, staged to reinforce community bonds, to sacralize rules and beliefs, and to demonstrate the "fairness" of the punishment. In such trials, the accusation, the inquiry, and the sentence expose the ideological foundations of the social "order" and offer important clues to the nature of the conflict rending it. By defining what is criminal, the trial reveals what is the norm. Political "criminals" have few choices. They may conform to the norms, admit their "guilt," and make public penitence, in the hope their judges will be merciful (which is very unlikely considering the purposes of the trial). They may deny the ideas or actions imputed to them and plead innocence (which seldom has a better result). But they may reassert their repudiation of the norm and try to use the trial as a setting for the advocacy of their own ideas, a legitimation of their own norms, and a validation of their rebellion. In such a case, the trial brings to light with unusual clarity the ideological gulf that separates accusers from accused (but also notions they may share). That is precisely what happened in Demerara during the trial of John Smith.

Since Smith's arrest, the colonists had been eagerly waiting for their moment of revenge. His trial offered them the opportunity to condemn indirectly all those who threatened slavery as an institution: the evangelicals, the abolitionists, and those in Parliament and the press who sided with slaves against their masters—those who, in the colonists' view, were sowing discontent and revolt among slaves. By blaming others for the rebellion, the colonists placed themselves above suspicion, exempted themselves from responsibility, and freed themselves from guilt. They advertised to the world that it was not oppression or exploitation but delusion that had caused the slaves to rebel. More important, they reassured themselves that the "bonds" that supposedly united slaves and masters could be restored and the danger that threatened their "community" could be exorcised. Conversely, the trial allowed Smith to reverse the picture, to protest his innocence and to blame not only masters and colonial authorities but the slave system itself, to defend abolitionist ideas and condemn slavery, and, finally, to preach his last sermon. This time, however, he would preach to masters, not to slaves.

By transforming his defense into a sermon in which he exposed the masters' and royal authorities' "sins," Smith signed his own death sentence. What to him seemed to be proof of innocence, to the colonists was proof of guilt. In the end, when accused and accusers confronted each other, they could not but display their contradictory views about the world and reenact the conflict which, from the beginning, had pitted one against the other. Conscious of their colonial situation and their accountability to the mother country, each side played its role with the metropolitan audience in mind. The colonists appealed to British conservatives who, having to contend at home with riots and radicals undermining the foundations of what they perceived as a good society, advocated repression. Smith spoke to British reformers who, to meet such challenges, advocated change at home and abroad. Both audiences would respond enthusiastically.

Smith's trial was handled with considerably more care than the slaves'.[1] It took twenty-seven days, and involved numerous witnesses. But, in the end, his trial was no less a mockery. Instead of being brought before a regular court of justice, Smith was tried by a court-martial. Later, the governor defended this decision by arguing that it would have been otherwise impossible to guarantee Smith a "fair" trial. There was too much hostility in the colony against him, and most colonists were convinced of his guilt. Yet there is plenty of evidence that the trial was staged to convict the missionary. Although he was informally assisted by a local lawyer, he did not have regular counsel inside the courtroom and was left alone to cross-examine prosecution witnesses and conduct his own defense. Some witnesses brought by the prosecution to testify against him had obviously been instructed beforehand. A crucial witness, Michael McTurk, was Smith's personal enemy. Others were slaves or servants in the houses of powerful people and could be easily influenced by their masters. The only prosecution witnesses Smith could expect to be fair were the members of his congregation. But they too were under terrific pressure. During the preliminary inquiries and the slaves' trials, the authorities had been eagerly gathering the evidence they hoped would incriminate Smith. And, as in the case of Jack Gladstone, some people were more than willing to manipulate the slaves' testimony. There was no chance Smith could have a fair trial.

At ten o'clock in the morning, October 13, 1823, in the Colony House in Georgetown, John Smith was brought before the court-martial. He was accused of having promoted "discontent and dissatisfaction in the minds of the negro slaves toward their lawful masters, managers, and overseers," intending to excite them to rebel against authority and "against the peace" of the King and against "his crown and dignity." Smith had "advised, consulted and corresponded with a certain negro named Quamina . . . on matters concerning a slave rebellion." Even after the slaves had broken into open rebellion, "knowing Quamina to be an insurgent, he had made no attempt to detain him, nor had he informed the proper authorities that the slaves intended to rebel." To these charges Smith pleaded not guilty. The court adjourned until the next day.[2]

The judges represented money, power, and standing. Presiding over the court was Lieut. Colonel Stephen Arthur Goodman, a man who collected about £4,500 a year from his job as vendue master of the colony, was commandant of the Georgetown Brigade of Militia, and half-pay 48th Regiment. Victor Heyliger, the colony's first fiscal, acted as judge advocate, and Richard Creser, Esq., Robert Phipps, Esq. and J. L. Smith, Jr., Esq., were assistant judge advocates. Lieut. Colonel Charles Wray, militia staff and president of the colony's court of justice (where he was paid £3,000 a year), was also a member of the court. The others were officers in resplendent uniforms, mostly from the Royal North British Fusileers, the Fourth or King's Own Regiment, and the First West India Regiment, stationed in Demerara. Before these men, the missionary in his dark suit, with his modest origins and his £140 annual stipend would have made a poor figure—were it not for his pride and determination, and the strong sense of mission that gave special eloquence to his words and dignity to his demeanor.[3]

When the court reconvened on October 14, the judge advocate, before introducing evidence in support of the charges, briefly stated the case against Smith. It was his purpose, he said, to demonstrate that, from the moment he arrived in the colony, the prisoner had begun to "interfere" with the complaints of the slaves and their management and with "the acts and deeds of the constituted authorities." His conduct had generated "discontent and dissatisfaction" among slaves, and "his opinion of the oppression under which they labour[ed]" had led him to "expound to them such parts of the Gospel" as bore on "the oppressed state in which he considered them to be." This had finally led "to the tearing asunder the tie which formerly united master and slaves." Revolt was the consequence "of this state of discontent in which they had been taught." Smith had been aware of the intended rebellion, and had advised the slaves on the difficulties they would have to encounter. But he had never attempted to inform the authorities. On the day of the rebellion, he had been in town and had left "without having made that disclosure, which as a faithful and loyal subject, he was bound to do." Later, after the rebellion had started, he had corresponded with one of the insurgents without making any attempt to detain him, or to notify the authorities. He had persisted in such behavior even after a detachment of the militia had arrived at his house.

After these preliminary remarks, the prosecutor introduced as evidence Smith's journal, from which he extracted what he thought to be incriminating passages. The move was sensationalist; it aimed at causing scandal and creating a climate of hostility against the missionary. The unguarded confessions Smith had made in his diary in the privacy of his home, and in moments of frustration, anger, and distress—unaware that one day they would be read by thousands of people in the colony and abroad—were amply publicized by the local press. Catering to the public's prejudices, the *Guiana Chronicle* waged a violent campaign against Smith and the other LMS missionaries, publishing editorials attacking them day after day, raising doubts about their honesty and their decency. Aside from news of the trial and transcripts of Smith's journal, it published insinuations of corruption and immorality, and piquant stories about the "amours" of the Reverend Mr. Elliot and the greed of the Reverend Mr. Davies.

By selecting particular entries from Smith's journal, the prosecution intended to expose Smith's commitment to abolitionist ideas, his repudiation of the slave system, and his condemnation of planters, managers, fellow missionaries, and local authorities. The prosecution also hoped that the entries would demonstrate that Smith had interfered with the management of plantations and had used the Gospel to sow discontent among slaves. The first passage introduced by the prosecution was dated March 30, 1817—a few weeks after Smith had settled at *Le Resouvenir*. It told the story of a conflict the missionary had tried to resolve between husband and wife. The incriminating sentence was Smith's remark that "a missionary must in many instances act the part of a civil magistrate." The implication was that Smith had deliberately trespassed onto terrain belonging to colonial authorities. The next passage referred to Lucinda, an old slave Smith described as so pious that when the manager prohibited her from going to the chapel, she replied that she would go "even if he cuts her throat for it." This episode, which to Smith had seemed moving because it revealed the strength of Lucinda's devotion, to the prosecution was evidence that Smith condoned and perhaps even celebrated the slave's disrespect for her manager. Equally twisted was the prosecution's interpretation of another 1817 passage:

> A great number of people at chapel. From Genesis XV.1. Having passed over the latter part of chapter 13, as containing a promise of deliverance from the land of Canaan, I was apprehensive the negroes, might put such a construction upon it as I would not wish; for I tell them that some of the promises, etc. which are made to Abraham and others, will apply to the Christian state. It is easier to make a wrong impression upon their minds than a right one.

The text was obscure, allowing for contradictory interpretations. Rather than considering Smith's remarks as evidence that the missionary was trying to avoid "dangerous" passages (as Smith later argued), the prosecutor introduced it as a proof of Smith's sinister intentions. Another entry selected by the prosecution with a similar purpose referred to the first Epistle of Peter. But in this case Smith had commented in his journal that he had deliberately chosen this text because it seemed to have been written for the comfort of Christians who were scattered and persecuted, "which is the case with our people." The prosecution was trying to prove that Smith had chosen Biblical texts that were likely to provoke discontent among slaves and lead them to rebel. Another journal entry introduced was one in which Smith, after describing how upset he was by the sound of the whip, commented: "the laws of justice, which relate to the negroes, are only known by name here"—a remark very likely to infuriate the colonists.

Many other such entries were read, and later published in the local newspapers. Some spoke of Smith's wish that slavery be abolished, of his attempts to settle disputes among slaves, of his outrage at seeing slaves persecuted for the sake of religion. Others expressed his astonishment at their capacity to endure so much work and so much punishment. Still others described his emotions when he heard slaves pray aloud "that God would overrule the planters' opposition to religion," his hopes that God would hear the "cry of the oppressed," and his suspicions that certain "whites" were coming to the chapel to spy on him. Finally, some contained bitter remarks about his missionary peers, the fiscal, and other colonial authorities, including the governor—feelings he had confided to his diary in his moments of exasperation.

Only the last four entries presented to the court were directly related to incidents that immediately preceded the rebellion. The entry for June 22, 1823, reported Smith's meeting with Isaac, a slave from *Triumph* who, after Governor Murray's proclamation stipulating that slaves could not go to the chapel without a pass from their managers, had come to ask Smith whether the "new law" forbade the slaves' meeting for the purpose of learning the catechism in the evenings. The manager, Isaac said, "had threatened to punish them if they held any meeting." Smith had written in his journal that he had told Isaac that "the law gave the managers no such power," but that he "advised them [the slaves] to give it up, rather than give offence and be punished, and to take care to ask for their passes only on Sunday mornings, and come to the chapel to be catechized." This was advice that Smith could interpret as an evidence of his innocence, but for the prosecution it was a proof of his guilt. Smith may have recommended compliance but he had implicitly denied legitimacy to the manager's order.

The next two entries read to the court were presented as evidence of Smith's commitment to abolition. Smith, after a visit from Elliot, had recorded that it "appears the same impediments are thrown in the way of instructing the negroes on the west coast as on the east, and it will be so as long as the present system prevails, or rather exists." A few days later, he mentioned a conversation he had with Hamilton in which the manager had said that were he prevented from flogging the women who were not punctual with their work, he would deprive them of food and keep them in solitary confinement. After remarking that the manager seemed to take comfort in the idea that Canning's project would never be carried into effect, Smith commented: "The rigours of negro slavery, I believe, can never be mitigated, the system must be abolished." The final entry the prosecution chose to read to the court was from August 18, 1823, the day the slaves rose up. It said simply: "Early this morning I went to town, to consult Dr. Robson on the state of my health." The prosecution intended to establish that although Smith had been in town that day, he had not informed the authorities of the plot.

The next step for the prosecution was to demonstrate that Smith not only had sown discontent among the slaves and undermined the authority of masters, managers, and public officials, but had also been privy to the conspiracy and sympathetic to the rebels. The prosecution was ready to call as witnesses anyone who at any time had heard anything that could incriminate Smith. Even people like Edmund Bond and William McWatt, who lived miles away in Berbice and could report only a conversation they had had with Smith a year before the rebellion, were brought to testify against him.

Bond was a carpenter by trade, McWatt an overseer. They had met Smith on a visit he had made to Wray in Berbice the year before, when Smith had stopped overnight at plantation *Profit*. Bond testified that they spoke about slavery and that Smith had argued that "the negroes could do as well in the West Indies without white people as with them," and had made some allusions to Haiti. Bond could not remember Smith's exact words, but said that when he asked Smith whether he wanted the same thing to happen in Demerara and Berbice, Smith "appeared confounded." Asked by the court whether he had heard the prisoner say anything about missionaries, Bond said he did not remember. At first, the question put by the court seemed odd. But when the next witness was called, it became clear that the prosecutor already knew the content of the conversation. After repeating more or less the same story, McWatt added that when Bond had asked Smith whether he wished to see scenes like those of Haiti, Smith had answered that he thought "that would be prevented by the missionaries."

During cross-examination, Smith tried to establish the context of his remarks. He asked the witnesses whether they remembered hearing a Mr. Hutchenson (who had also been present during the conversation, but had not been called to testify) say that times were so bad that the whites would have to sell off and go home, and that he wondered what would happen to the "poor negroes"—the remark that had given rise to Smith's observation that they would do as well without the whites. Neither Bond nor McWatt recalled the detail. Smith met with the same difficulty when he asked McWatt whether he remembered that when Smith had said that "such a scene as the one in St. Domingo [Haiti] would be prevented by the missionaries" he also had said that "the effects of the Gospel would prevent such scenes, or words to that effect." Out of context and without the necessary qualifications, Smith's words could be interpreted as the remarks of a hot-headed, incendiary revolutionary.

Hearsay and fragments of conversation were again used as evidence when the prosecution called John Bailey and John Aves, two of the men Smith had welcomed to his house in the middle of the night on August 18, just a few hours after the rebellion started. The prosecutor tried to establish that Smith had known of the plot for about six weeks before it started. Smith attempted to make the witnesses be specific, to bring them to admit that what he had said was that from the moment the word of the new regulations coming from England had spread among the slaves, he could have anticipated a rebellion. But neither Bailey nor Aves seemed to remember the conversation that way.

Two other important sets of evidence were introduced by the prosecution. The first was the testimony of several members of Smith's congregation. The second was the testimony of Michael McTurk. Curiously enough, none of the slaves who during their own trials had seriously implicated Smith was called to testify. Some had already been hanged. Others, like Jack Gladstone, were still alive but were not called by the prosecution, probably because it feared that manipulation and tampering might become obvious during cross-examination. Instead, the prosecution presented only certified copies of the charges and sentences from the trials of Jack, Telemachus, Sandy, and Paul of *Friendship,* and Quamina of *Nooten Zuyl.*

In spite of some contradictions and minor errors, which Smith later pointed out in his defense, the testimony of the members of his congregation seems remarkable for its accuracy, sincerity, and coherence, particularly in light of the extraordinary pressure they were under. After the failure of the rebellion, the brutal repression, the weeks of confinement and distress, the trials and ritual killings, for these people to appear before the court with all its ceremonies and protocols and to testify before a hostile audience must not have been an easy task. That they did it with such skill and dignity testifies to their extraordinary courage and resilience. Most of the slaves called before the court had been deacons or had had important roles in the chapel. They all showed respect and even affection for Smith, although it became clear during the inquiries that the slaves never trusted him entirely—not even Quamina, who was very close to him, had ever forgotten that Smith was a white man.

The first to testify was Azor. Nothing in his testimony could incriminate Smith. The only time Azor said anything that could be used against the missionary was when the court asked him to explain his understanding of Smith's teachings. Azor referred specifically to a Biblical passage involving David and Saul. His interpretation of the passage, like the scriptural interpretations made by other slaves during Smith's trial, opens a window on a reality always very difficult to grasp: the slaves' way of understanding the Bible. Azor's story shows how the Biblical message was filtered through the slaves' experience and, conversely, how they applied the message, as they understood it, to their day-to-day lives. Azor told the court that Saul had driven David into the woods (an obvious analogy for running away to the bush). David went to the woods, Azor explained, "because if he went in a friend's house, he would get trouble; David himself was to get trouble." Another passage of the Bible that seemed to have lodged in his mind—as it did in the minds of countless other slaves, not only in Demerara but all over the New World—was that of Moses crossing the Red Sea (although as Smith later explained to the court, he had read this passage two years before the trial). But, in spite of the obvious metaphorical potential of this passage, Azor—perhaps being careful—made no explicit connection to the slaves' hope of deliverance. He told the court that

> when Moses took the children of Israel, and carried them through the Red Sea, then Pharaoh gathered the soldiers, and went after them to bring them back; and the Lord made darkness and thunder between the King of Israel and Moses; when Moses had gotten over with the children of Israel, Pharaoh was drowned in the sea, and Moses built a temple, and prayed to God. Only that I heard from the prisoner.[4]

Azor further testified that he had heard Smith saying that God had made the Sabbath holy, and "that this country was a very wicked country; in England they were all free, and they all kept the Sabbath holy"; that they should not work on the Sabbath except if there was a fire or a dam breaking. "If a half a row was left in the field it was not fit to be worked on the Sabbath." Realizing that this could be used against him, Smith tried during cross-examination to clarify this issue. But he only managed to incriminate himself even more. Pressed by Smith, Azor admitted that the missionary did not tell them *not* to finish the row, he only had said it was not right to work on a Sunday. Upon further questioning, Azor said that when the slaves justified their absence from the chapel on Sunday by saying that they had been given work to do, Smith told them that they were "fools for working on Sunday, for the sake of a few lashes."

The next witness for the prosecution was Romeo, an old slave from *Le Resouvenir,* a man who had been taught to read and had been a deacon in the chapel since the time of John Wray. His testimony coincided on the whole with Azor's,[5] but he remembered that Smith had also said that if the masters forced them to work on a Sunday they should obey and not grieve or be angry, for the masters would answer for it. Romeo added that Smith said the words of the Bible were all true, and that "he preached very true too." He recollected with surprising accuracy the text Smith had preached from the day before the rebellion, the 19th chapter of Luke beginning at the 41st and 42nd verses. But he could only remember some words of the 41st, which he rendered as "When Jesus came near the city he wept over it."[6] Romeo added that he had seen Smith on Sunday before the rebellion, and then again on Tuesday night, and that on that occasion Smith had expressed a wish to see Quamina or Bristol.[7] Asked by the court if he had ever heard Smith say that the slaves were fools for working on Sunday for a few lashes, Romeo denied it and repeated his original version. Asked whether the deacons held any separate meetings in the chapel, in their houses, or elsewhere, for the purpose of teaching "the negroes," Romeo said that on the estate where he lived they met sometimes, but he did not know what happened on other estates. He confirmed that Smith knew about the meetings and approved of them, although he had never seen Smith at any of them. He was then asked whether Smith had directed him to "explain his sermons to the people," and how often he performed such a task. Romeo replied that he did it every time the missionary preached. Asked whether he had explained the sermon Smith gave on the Sunday before the revolt and what he had said, Romeo told the court that on that day he had not explained the text: "the negroes said that Mr. Smith was making them fools; . . . he would not deny his own colour for the sake of black people. These words grieved me, and I went away straight along, because I was hurt to see them behave so ungratefully." He had explained the text of the Sunday before (Revelation 3:3), but could not recollect very well what he had said. He remembered that it was something about the people on the Mahaica side going to Essequibo (a reference to a sermon Smith had preached when some slaves had been sold away): "What you do know hold fast; God is not so slack in his promises as some men are; I know that you have some children to be instructed, that wherever they go they may not forget God, because when they go to some strange places they will throw away their Christianity. My explanation was that if you deceive God, God will set a curse upon you and your children."[8] Finally, when the court asked him whether the prisoner ever pointed out to him particular chapters in the Bible for him to teach from, Romeo responded, "No, only the catechism."

Two other witnesses were called the same day, Joe from *Success* and Manuel from *Chateau Margo*. Like Azor and Romeo, both remembered very clearly certain passages of the Bible. Both described the sermon the day before the rebellion, but neither Joe nor Manuel established any connection between Smith's words and the uprising. Manuel gave many details about the conversation Quamina had had with Smith about the "new laws." He revealed that, after the rumors about the new laws started circulating, he had suggested to Quamina that he ask Smith for clarification. But Quamina had said he did not believe that Smith would tell him. (This seems to indicate that even Quamina had second thoughts about talking to Smith about such matters.) When Quamina finally did decide to talk to Smith, he was told that "there was no freedom in the papers, and that their masters could not afford to lose so much money as to let them all go free." The missionary told them to "bear patience, if there was any thing good come, it was come for the women, because the drivers were not to carry whips any longer in the field." Manuel spoke of the meeting the slaves had had at *Success* the Sunday before the rebellion, of how Quamina and Bristol had gone to talk to Smith, and how Bristol came back and "gave two bits" to a man from

Vigilance to carry a message to Joseph, "to take care that he did not do any thing in the way of taking away the Buckra's gun." Once again the question of work on Sunday was raised, and Manuel confirmed what others had said before him. He added, however, that the missionary had said that if the master had any work for them on Sunday, it was their duty to tell him Sunday is God's day. He had heard slaves say that if Sunday was to be taken to serve God they ought to have Saturday to work their own ground—or at least Saturday afternoon.

Manuel was asked whether at the morning prayer Smith read only from the Old Testament—the part of scripture that masters everywhere most feared and slaves always preferred—and if so, whether he selected passages or worked his way through the text, chapter by chapter. Until then, slaves had all mentioned the same passages of the Bible, those about Moses, Joshua, and David. The court apparently wanted to learn whether Smith had deliberately chosen such texts. But to their disappointment, although Manuel admitted that for the past two years Smith had read only from the Old Testament during morning prayers, he told the court that Smith did read the passages one after another in regular sequence.

The next witness was Bristol, Jack Gladstone's brother-in-law. Like Romeo, he was a deacon and was familiar with the practices of the chapel. He admitted that slaves made contributions to the chapel and to the London Missionary Society, and that members who could afford it paid for psalmbooks, catechism books, tracts, and Bibles. All contributions were voluntary and sometimes slaves gave Mrs. Smith fowls and yams. "They carry these things," he explained, "not in lieu of money, but as presents to be eaten." Bristol said that Smith had told him to catechize people at home, but did not instruct him to explain his sermons or any other text. Asked by the court who appointed the "teachers" on each estate, Bristol said that the deacons chose the teachers with Smith's approval. He then carefully described the morning service. The prayers, he explained, were spontaneous and aloud, "from our hearts, not learnt out of a book." And when he was asked to give an example he said: "At our prayer meetings we prayed to God to help us and to bless us all, that we may be enabled to seek after him more and more, and that he would bless our masters, and the governor and the fiscal; that we might make good servants unto them, and they might be good masters unto us; and to give us health and strength to do that which it might be our duty to do, and to bless all our brothers and sisters; we pray about our masters' hearts likewise."

Like the slaves who had testified before, Bristol seemed to remember only certain parts of the Bible. He too spoke of Moses and Joshua, but failed to recollect any particular chapter from Exodus. Yet, when pressed by the prosecution, he admitted that the slaves applied the story of the Israelites to themselves: "When they read it then they began to discourse about it; they said that this thing in the Bible applied to us just as well as to the people of Israel. I cannot tell what made the negroes apply it to themselves. What created the discontent in the mind of the negroes was because they had no other time to wash their clothes, or do any thing for themselves, but the Sabbath day."

Bristol also told the court that when slaves complained to Smith that they had been punished for coming to the chapel, he replied that it was not right for the masters to prevent them from going to the chapel but there was nothing he could do about it. In such circumstances Smith often advised them to go to the fiscal or the governor. And when slaves ran away, Smith told them not to let masters catch them again, for they would be punished. Once again the issue of the Sabbath was raised and Bristol made it clear that Smith said that if the masters told them to work on a Sunday they should obey, but that God would punish them if they worked on their own grounds. Interrogated about the conversation Quamina had had with Smith before the rebellion, Bristol did not add much to what had been mentioned before. But what he did say was enough to damage Smith. Although Bristol gave plenty of evidence that Smith had tried to discourage any attempt to rise, he also made it clear that Smith indeed had known something about the slaves' intentions before the rebellion broke out.

Bristol's testimony seemed to disturb Smith, who spent a long time questioning him on minute points. Was the money collected used only for the wine or also to buy candles? Who bought the candles and paid for them? What was the largest and the smallest sum of money ever collected? How much did the slaves pay for different books? He then turned to more relevant questions, hoping to dispel any suspicion that he had held secret meetings or preached things that could have instigated the slaves against their masters. He also tried to convince the judges that when he spoke with Quamina, he could not have known how serious the slaves really were.

Now came Michael McTurk's turn on the stand. He clearly had little to say about the rebellion. Most of his testimony had to do with the smallpox incident of 1819. McTurk described his dealings with the manager of *Le Resouvenir* and the violent altercation he had had with Smith on that occasion. In support of McTurk's testimony, the assistant judge read passages from Smith's diaries relating to the episode. During cross-examination it became clear that McTurk had felt insulted not only by Smith but also by the way the slaves had acted the day he had gone to *Le Resouvenir* to examine them with the alleged purpose of finding whether there were still signs of smallpox. He told the court that although he had warned the manager he was coming, the slaves did not wait for him. When he arrived at *Le Resou-*

venir some had already gone to the field to work and refused to return. Others stood in front of their houses and would not be examined by him. The drivers were sent by the manager to reassemble the slaves but had no success. Only five or six slaves out of a population of almost four hundred had shown up. After waiting for about an hour, McTurk finally had given up. When he returned to *Le Resouvenir* the next day, some slaves pelted him with sticks and hard mud, "and used most abusive language." McTurk attributed the slaves' insubordination to Smith—particularly since Smith during their quarrel had claimed to have power over the slaves.

During cross-examination it became clear that, although McTurk had given orders prohibiting slaves from going to the chapel during the smallpox episode, he had not prohibited them from going to the market. This seems to give credit to Smith's suspicion that McTurk's main purpose had been not to avoid the risk of an epidemic but to teach the missionary a lesson. For the court, however, this detail was irrelevant. McTurk was a man of property and standing in the community and a burgher officer. The authority he derived from his position conferred a semblance of truth on his words.

In the following days, more evidence against Smith was brought in by the prosecution: copies of the letters that had passed among Jack Gladstone, Jacky Reed, and Smith were introduced. New witnesses for the prosecution were called. Smith's doctor testified he had seen him in his office in town on Monday morning, the day of the rebellion. Seaton and Jacky Reed gave details about the plotting of the rebellion. Antje told how Mrs. Smith had asked her to tell Quamina to come to her house. Several other slaves, including Smith's servant Elizabeth, testified that they had seen Quamina going to Smith's house after the rebellion had broken out. Finally, Lieut. Thomas Nurse described his visit to the Smiths a few days after the rebellion, and the circumstances of their arrest. With this the prosecution rested its case.

The court granted Smith five days to prepare his defense. On the fourteenth day of the trial, Saturday, November 1, Smith read his opening statement. While the prosecution had based its arguments on the assumptions that masters had absolute authority over the slaves, and that anyone who raised doubts about the masters' authority or interfered with it committed a crime, Smith based his defense on the assumption that God was the supreme authority and God's law the supreme law. Anyone who disobeyed His law was a sinner and anyone who guided himself by God's teachings could not be a criminal. To justify his behavior, Smith quoted the Bible copiously and tried to establish his professional credentials. Smith said he was a minister of the Gospel, ordained and sanctioned by the Missionary Society, "a most respectable body of men well known to and sanctioned by the Government at home." Their sole purpose was the conversion of the heathen. They had nothing to do with the civil or political state of the countries or the temporal conditions of the people under their missionary care.

He then proceeded to describe the organization of his congregation and religious services, explaining that the "teachers" were chosen by the slaves themselves without his interference, and had no connection with the chapel. Their chief qualification was a knowledge of the catechism. To justify his dealings with slaves, he argued that no minister of the Gospel could "properly discharge his sacred functions without having some other intercourse with his people besides that of public teaching." (Here he cited Ezekiel 33: 7–8: "So thou, O son of man, I have set thee a watchman unto the house of Israel; therefore thou shalt hear the word at my mouth, and warn them from me. When I say unto the wicked, O wicked man, thou shalt surely die; if thou dost not speak to warn the wicked from his way, that wicked man shall die in his iniquity; but his blood will I require at thine hand.")

Smith proudly admitted his aversion to slavery. If this was a crime, he said, then he had as his "associates in guilt the most liberal and best part of mankind." After the recent recognition by the House of Commons and the British government that slavery was repugnant to Christianity, it should not be necessary for him, a minister of the Gospel, to justify his feelings. But he insisted that he had always abstained from making any remarks respecting the masters and had always exhorted slaves to "a dutiful submission"—as had been proven by the testimony of the witnesses for the prosecution, and could be confirmed by many other members of his congregation. In an attempt to minimize the impact of his journal he argued that it was a private document whose contents were unknown even to his wife. Smith denied that he had in any way tried to use his religious teaching to mislead the slaves. His journal showed instead his preoccupation with avoiding misinterpretation. He stressed that even the witnesses "whose memories were so very tenacious on the subject of Moses and Pharaoh and the children of Israel" had stated that they never had heard him applying the history of the Israelites to the condition of the slaves. He then argued that "extempore prayer" was a common practice in many Christian churches, particular among Protestant dissenters. When they prayed the slaves had their eyes closed. The doors were always open during prayer meetings and any "black, coloured, or white person" might have entered the chapel. Surely there could be no improper behavior in such circumstances.

As to the money he collected from slaves, Smith insisted this was not only "according to the usage of all churches, but agreeable to the scripture." Such donations were done spontaneously by the slaves and with the knowledge of their masters. Only rarely did slaves bring a fowl or yam to his wife, and in return he often provided them with a bottle of wine when they were sick.

He admitted he had sold them books, but in this he had also followed common practice. Besides, the slaves purchased them voluntarily. And even if he had forced them, how could this be used as a charge against him? The slaves' discontent would surely have turned against him and not against their masters. Smith then denied that he had interfered in any way with the temporal concerns of the slaves, "save in such cases as were intimately blended with their spiritual concerns," as when he settled their disputes or rebuked them for "immoral" conduct. Quoting from his diary a passage the prosecution had neglected, Smith showed that in fact he would much have preferred not to hear about their troubles.

Except for a statement from Azor, there was no evidence, Smith argued, that he had ever told the slaves to disobey their masters. All evidence pointed to the contrary. Indeed, Smith said, he had taught the slaves that it was sinful to work or traffick on the Sabbath. But "every member of the court will, I am sure, allow, that in doing so I taught one of the first precepts inculcated in that holy book on which they have sworn to do justice." The scripture showed that the violation of the Sabbath by voluntary labor that was not absolutely necessary was a heinous sin. Was he supposed to dispense with the commandments of God? Were masters greater than God?[9]

Smith explained that he had made allusions to England as a free country only to admonish the slaves that "they must not make their condition in life an excuse for breaking the commandments, and neglecting religious duties." He "could never imagine that such an allusion to a free country would be construed into a crime." He then justified the remarks in his journal about the disregard for the law in the colony. These remarks—he explained—had been written soon after he had seen a driver flog a slave in the absence of the manager and overseer. When he spoke of the laws of justice not being respected in the colony, he had in mind "arbitrary punishments inflicted by managers," and "drivers flogging negroes in the absence of masters." That these were common practices on some estates at the time his comments had been written was too well known to be denied.

The tone of Smith's defense could only have enraged the colonists. Instead of apologizing, he was admonishing them. He did not acknowledge any of the charges and seemed to turn them around so as to implicate the colonists. But then he went even further. Ignoring the protocols of class, which required that he treat such a person as McTurk with respect and even subservience, Smith challenged McTurk's testimony, pointing at contradictions, and insisted on treating McTurk as if they were equals. McTurk, Smith said, had mentioned "an attack which I made upon him, but he did not say that he provoked any apparent [*sic*] disrespectful language I may have used on the occasion. He has not told the court, what I shall prove, that he sneered at me, and mocked at the idea of the negroes being instructed in the tenets of our holy religion."

Smith then went on to discuss Lieut. Nurse's testimony. He admitted that he had misunderstood his obligations to enroll in the militia, but refuted all the other charges. He had been accused of having "remained at his house during the whole revolt in safety and without fear." Why had he done that? Because he had no slaves and was not conscious of ever having wronged one. On the first night of the rebellion, when he had gone over to Hamilton's, the slaves had told him to return to his house, "as it was not their intention to hurt any one." He had believed them. Perhaps he had placed more faith in their promise "than it was politic to do, or than others would have done." But that could not be seen as an offense on his part.

Rejecting the charge that he had been acquainted with the slaves' plot for several weeks before the rebellion, Smith gave his own version of his encounter with Quamina. Smith insisted that at the time he had no idea that the slaves intended to revolt. His first inkling had come when he received the note from Jacky Reed. As for his alleged meeting with Quamina after the rebellion had started, Smith argued that there was no proof that Quamina actually was a rebel, or that at the time Quamina had come to his house he had any knowledge of Quamina's being an insurgent. But, more important, Quamina had come to his house at the request of Jane, and although he believed that a husband was "responsible in civil courts for the acts of his wife," he did not think he should be considered responsible for her "crimes." (He did not mean to say that she was guilty, he explained, but that the evidence did not relate to him. Since the court would not allow Jane Smith to testify, there were several points that could never be clarified.)

Probably on the advice of the lawyer William Arrindell, who had been assisting him informally, Smith protested that the charges against him were too vague. The prosecution had not changed him positively with any specific offense, and had not complied with criminal procedures established by British law. After denying every charge, Smith raised doubts about the reliability of the slaves' testimony. The witnesses, he said, were "decidedly under the influence of their owners." Their love of truth and justice could not be stronger than their fear of men they had seen punishing their "fellow-labourers" merely for attending divine worship. Some of the witnesses were extremely "ignorant" and "savage" and did not even understand the nature of an oath—hence the "prevarications, and falsehoods, and contradictions so apparent in their evidence." They had no notions of time or circumstance, and it was all too clear that their evidence had "been made up of shreds and patches, obtained from conversation, from hearsay, and from their own misinterpretations of what had been propounded to them." (Smith was in a difficult bind: by questioning the slaves' credibility as witnesses for the prosecution, he also undermined the value they might have when testifying in his favor.)

Smith concluded his statement in a prophetic and post-millenarian tone. His opinions, not himself, had been tried. But his opinions were "founded upon the Gospel that hath withstood for ages all persecution: its promulgation has increased from opposition, and its truths been made manifest by investigation." The Gospel had prospered, and still would prosper, he said, and would "impart happiness" to all who sought knowledge from it. It already had improved the minds of the slaves. The love of religion was already so deeply "implanted" in them that the power of men would not be able to eradicate it. In the midst of the revolt slaves were heard to say, "We will shed no blood, for it is contrary to the religion we have been taught." And who were the slaves who said such things? "Not the lowest class of Africans—not the heathen, but the Christian negroes." In former revolts in the colony, as in Jamaica, Grenada, and Barbados—there had been bloodshed and massacres. But in this one "a mildness and forbearance, worthy of the faith they professed (however wrong their conduct may have been) were the characteristics." The few attempts at bloodshed had been confined to Africans who had not yet been baptized. (As the prosecution would notice later, when it summed up the charges, Smith's statement could be interpreted as a recognition that members of his congregation had indeed been in control of the rebellion.)

For several days the missionary presented evidence. Most of his witnesses were plantation owners, managers, and missionaries, but there were a few free blacks and slaves. One of the first to be called was Henry Van Cooten, proprietor of *Vryheid's Lust* and attorney for *Le Resouvenir,* a man who had resided for about fifty years in Demerara and had been well acquainted with Wray and Smith. Van Cooten expressed the view of the few planters who had been supportive of the missionaries. He testified that he had given permission to slaves to attend services at Smith's chapel, and that he thought they were "rather more obedient than formerly." He said he did not object to slaves having books because he saw no harm in it. He also admitted he had made contributions to the London Missionary Society, and knew that slaves themselves made such contributions. When Smith asked him whether he thought the slaves were capable of reporting correctly any conversation, Van Cooten said that in general they did it very badly but some were more capable than others. (When asked by the court whether the slaves could "recollect the heads of a short discourse, and accurately take up the meaning of the lectures," Van Cooten answered hesitantly: "Of a short discourse some might, I think.")

Smith's second witness was John Stewart, the manager of *Success.* Although he was particularly evasive and cautious, Stewart did testify that the slaves who attended services at Smith's chapel were for the most part obedient, and that he had recommended several for baptism and he himself had attended services from time to time. He then provided—at Smith's request—information about the conversations he and Cort, the attorney for *Success,* had had with the missionary about rumors circulating among slaves a few weeks before the rebellion. Nothing Stewart said could incriminate Smith. Cross-examined by the court he admitted that several slaves who had been tried as ring-leaders did attend services at the chapel, but he stressed that he had not seen Quamina or Jack do any harm. If anything, they had prevented other slaves from injuring him.

Probably because of his connections with Susanna, Hamilton, the manager of *Le Resouvenir,* was even more cautious and evasive, avoiding answers that might put him in trouble. When Smith asked him to confirm that during a conversation he had with McTurk in Hamilton's presence, McTurk had sneered at him, Hamilton said that he remembered there had been a disagreeable conversation between the two men, but since much time had passed he could not recollect the details. Smith pressed the manager again, but could not get him to say more. The rest of Hamilton's testimony focused mostly on the day of the rebellion, and what he said confirmed in its main lines Smith's version. He admitted that he himself had been in town on August 18 and had been warned of the plot around one or two in the afternoon, many hours before the slaves rose up, but had not warned Smith.

Smith then called John Thomas Leahy, Lieut. Colonel of the Twenty-First Regiment. Leahy gave details of his encounter with the rebels at *Bachelor's Adventure,* and reported among other things that some slaves had complained that when they asked leave to go to chapel on Sundays they were punished for it. Leahy declared that at no time had the slaves mentioned Smith.

The next witness, John Reed, the owner of *Dochfour,* told the court that Smith had come to see him about a piece of land for a chapel, saying that this would "save the negroes from walking so far, which was a subject of complaint among some of the planters"—a fact Smith probably wanted to bring up to show his willingness to accommodate to the planters' demands. Reed's testimony was on the whole favorable to Smith but when the missionary tried to make him give details about the remarks Reed had made on that occasion about the governor's not being favorable to Smith's request, Reed acted as if he could not remember his own words. He obviously did not want to say anything that could upset the highest authority in the colony. Cross-examined by the court he said only: "My permission for the erection of a chapel depended on his Excellency's approval; and his Excellency was pleased to disapprove of it in consequence of complaints made against the prisoner."

Of all the witnesses for the defense, the most effective and the one that most irritated Smith's enemies was Wiltshire Staunton Austin, the minister of the Established Church in Georgetown and chaplain to the garrison. Guided by Smith's questions, he told the court that he considered "familiar intercourse between a minister and his parishioners" one of the most important of ministerial duties. Like Smith, he had frequently been called to settle disputes among slaves or between masters and slaves. During and after the rebellion he had talked to many of the rebels. Having been led initially to believe that Smith had been involved in the rebellion, he had asked them many questions. They had given a variety of reasons, but "in no one instance . . . did it appear, or was it stated, that Mr. Smith had been in any degree instrumental to the insurrection. A hardship of being restricted from attendance on his chapel was, however, very generally, a burthen of complaint." Smith asked him whether he considered verses 41 and 42 of the 19th chapter of Luke—from which Smith had preached the day before the rebellion, and about which there was great controversy—"an improper text for a sermon." Austin answered that he considered it "one of the most beautiful texts in scripture." But when the prosecutor, trying to reduce the impact of Austin's testimony, asked him whether he had ever heard any of the rebels "insinuate that their misfortunes were occasioned by the prisoner's influence over them, or the doctrines he taught them," Austin was cautious: "I have been sitting for some time as a member of the committee of inquiry; the idea occurs to me that circumstances have been detailed there against the Prisoner, but never to myself individually, in my ministerial capacity."

Equally supportive was the testimony of the two missionaries of the London Missionary Society, John Davies and Richard Elliot. Brought together by hardship, both seem to have forgotten whatever differences they once had had with Smith. Everything they said confirmed that Smith's religious practices were not unusual. There was nothing in his behavior that could be considered a crime.[10]

The slaves' testimony pointed to the same conclusion, particularly that of Philip, now a free black man and a cooper who lived in Georgetown but attended services at Smith's chapel. Philip told the court that when he was a slave working on plantation *Kitty,* he had gone to complain to Smith about the way he was treated, and thanks to Smith's advice he had become a faithful servant. "Do you remember any of the doctrines and duties taught you and the people by the Prisoner?" asked Smith. I do, answered Philip. "He told me, if my master sent me any where about his duty, that I must be very particular in seeing it done." Without Smith's advice, Philip concluded. "I should not have been my own man this day."

One after another the slaves confirmed that Smith always told them to obey their masters and perform their duties; that he punished those who ran away by suspending them from the chapel and forbidding them to participate in the Lord's Supper; that he read to them from the Bible, but extracted only lessons of obedience and respect for God's law; and, finally, that far from instigating the slaves to rebel, Smith had tried to dissuade them. The image that emerged from the slaves' testimony was that of a pious and dedicated missionary, concerned only with the slaves' souls and always ready to remind them to obey their masters and do their work well, even though sometimes he disapproved of the masters' behavior. Yet instead of making the best of these statements, Smith—either moved by a natural excessive zeal and obsessive concern with detail, or compelled by fear—proceeded to undermine the credibility of his own witnesses by making a meticulous study of minor and irrelevant contradictions in their testimony, concluding that the evidence was such as "to render it impossible for any one to say, that from it alone, the real truth can be ascertained."

After examining once again the charges against him in the light of the evidence, he concluded that they showed that neither he nor his doctrines were the cause of the revolt. Smith closed his defense by reasserting his innocence: "Gentlemen, I have done; to you my case is now confided; whatever may be your determination, I do, as a minister of the Gospel, in the presence of my God, most solemnly declare my innocence."

Five days later the court gathered again to hear the prosecutor's reply. The strategy the prosecution followed must have been used hundreds if not thousands of times in political trials, before and since, all over the world. The prosecutor employed a version of the classic flawed syllogism against which there was no possible defense: conspiracies are done in secret, so conspirators leave no trace; therefore the very absence of evidence is proof of guilt. If nothing had been found that could prove Smith's guilt, that was in the very nature of his crime.

> The crime presupposes great secresy [*sic*], and great caution; for the criminal is placed in a situation of extreme delicacy, where one false step, one precipitate movement either on his own part or on the part of the negroes, may at once ruin all his projects. He must hold out one character to the world, and another to the negroes; he must endeavour to conceal even from them the end he has in view, else their rashness may betray him, and he must thus strive to poison the minds of his victims without their being themselves aware of the hand which administers the potion.

The prosecutor went on to describe the way Smith's congregation was organized, stressing its democratic procedures, the leadership role deacons and teachers played, and most of all the missionary's power over his congregation. As evidence, he mentioned the contributions the slaves made to the chapel and to the London Missionary Society. The prisoner had said that such contributions were voluntary and "were given in consequence solely of his addresses from the pulpit." But that only established the more clearly his influence. "It proves that it was so great as to make the negroes, of all people on the face of the earth, part with their money freely, and not on any principle of force. Vast indeed, must have been his ascendancy over the negro mind, when he could induce them to contribute their money to a society for spreading the gospel through distant regions, the very names of which were unknown to them."

One of the ways Smith had managed to acquire such "influence," the prosecutor argued, was by hearing the slaves' complaints and settling disputes among them—a role that always had belonged to masters and colonial authorities. The implication was that by doing this, Smith not only had acquired an undue influence over the slaves, but had also stripped masters and local authorities of their own power. After referring again to evidence from Smith's journal to prove his involvement with the slaves, the prosecutor remarked: "[A] man, who really meant to support the authority of the master, would never do any thing to lessen this confidence in the mind of the slave; he never would teach him to look to any one but his master for the settlement of the disputes between him and his fellow slave." (The statement, which indeed expressed the point of view of most slaveowners, contradicted the very spirit of the laws of the colony, which had given the slave the right to appeal to a third party, the fiscal.)

The prosecutor missed no chance to extract from the slaves' testimony whatever could be used to show that Smith, instead of teaching the slaves obedience, had taught them subversion. While he may have told the slaves to obey their masters, in fact he undermined the slaves' respect for their masters, depicting them as "a thing to be dreaded, or despised." Smith considered voluntary work on a Sunday such a crime "as to render the negro unworthy of partaking of the Sacrament. In what light must the masters have been held!" The very punishment Smith inflicted on the slaves for Sunday work made them look on their masters "as being under the curse of Heaven." The prosecutor pointed out that Smith had told Romeo: "Work, if the masters force you, for they will have to answer for it. Could this lowering of the master in the eyes of the slaves be intended to make them more obedient? Were they more likely to be submissive to men whom they believed exposed to the wrath of God?" Moreover, Smith, by prohibiting the slaves from working on Sundays, had sown discontent among them and led them to rebel.

The prosecution invoked example after example to demonstrate that Smith had instigated the slaves to disobey their masters. He had received them in the chapel, "though at the time he knew they came in direct contradiction to their masters' orders." He taught the slaves to consider any attempt on the part of their masters to restrain them from coming to his chapel—whatever the masters' motives might be—as an act of gross injustice and oppression. He aimed, in fact, to make them believe they were "an oppressed and persecuted race." To prove this, the prosecutor referred to a remark Smith made in his journal about having chosen a passage of scripture "which he conceived addressed to persecuted Christians, as being best suited to their condition." To reinforce his point, he cited another passage from Smith's journal that showed that he had allowed the slaves to pray in his presence that God would overrule the planters' opposition to religion.

The prosecution then blamed Smith for the slaves' interpretation of Biblical passages, particularly those on David, Moses, and Joshua. He questioned Smith's preference for the Old Testament. Referring specifically to the slaves' interpretation of the story of Moses, he noticed that they all had talked about "slave" and "slavery" and the Pharaoh's "soldiers," but that there was no reference in the actual passage to such words. This he attributed to Smith's having deliberately used words that "brought the tale most completely home to the negroes."

Moving to the incident between Smith and McTurk, the prosecution argued that Smith had challenged McTurk's orders in the smallpox episode, and had shown "the same spirit of rank disobedience to the orders of those in authority" on several occasions. He had refused to obey orders to join the militia; he had criticized the governor's proclamation establishing rules governing slaves' attendance at religious services; and he had told the slaves that their managers had no right to prohibit them from going to the chapel. Was this not a way of telling slaves that their masters broke the laws and oppressed them in violation of all justice?

To give more weight to its accusations against Smith, the prosecution portrayed the slaves on the East Coast as relatively privileged, compared with those in other parts of the colony. "Of all the negro population of this extensive colony, there are, perhaps, none who have fewer difficulties to contend with, than the negroes of the east coast; there are but few sugar-estates there, comparatively speaking, the greater part being in cotton." Despite their advantages, the slaves had rebelled. The principal leaders of the rebellion were deacons, teachers, members, and attendants of Bethel Chapel. They were "the principal tradesmen on their estates, men in the confidence and favour of their masters, who knew the hardships of slavery only by name." What did these people have in common? They all belonged to Smith's chapel. The day before the rebellion, they had met after services to lay down the final plans. What could explain their rebellious behavior if not their connection with the chapel?

The prosecutor argued that Smith's statement was full of inconsistencies.

The missionary described the slaves as the most oppressed and persecuted of human beings. But it had been proven that even field slaves could afford to make monetary contributions. He had accused the planters of preventing the slaves from attending religious services, yet had exhibited "a host of passes from these planters to their negroes to have them baptized." And the chapel, although enlarged, could not contain all the members of the congregation. The prisoner had asserted "that he made it a rule to admit no negroes to his chapel or baptism unless recommended by their masters as good and obedient servants." But "if these negroes were obedient when they first went to listen to his doctrines, and these same men afterwards rose in rebellion against their masters, what must we think of the doctrines which have been preached to them?" How did they know about the "instructions" that had come from home? Who had first told them? "At present," said the prosecutor, "all credit of doing so rests with him; all efforts to trace it farther back are unavailing."

Through a curious reversal of the normal practice, the slaves' testimony—usually considered unreliable by the whites—was amply used by the prosecutor to demonstrate Smith's guilt. He argued that past decisions of the Court of Justice of the colony would show that both white and free criminals had been tried and convicted on "negro evidence." And Smith's arguments against the credibility of their testimony were "rebutted by the tales" they had told from the Bible. The "correctness as to the substance of the tale," in men who could not read, totally disproved Smith's assertion. Selecting from the slaves' testimony what could incriminate Smith and dismissing what could prove his innocence, the prosecutor tried to demonstrate that Smith had had prior knowledge of the revolt and had made no attempt to warn the authorities or to detain Quamina when he came to his house during the rebellion. That Smith knew the slaves on *Success* were "in a state of rebellion" was amply proved by several witnesses, including Smith's own maid, who had been threatened with punishment if she told anyone of Quamina's having been there. Smith had argued that all these were the acts of his wife. But "if he had not sent for Quamina, it was at all events in unison with his wish," as had been proved by Romeo's testimony.

Regarding Smith's aversion to slavery, the prosecutor said that no man had a right to "publish sentiments which can only tend to the subversion of the society in which he lives." The prosecutor concluded his summary in an anti-climactic way. He rejected Smith's interpretation of the Mutiny Act (which Smith had argued prohibited the use of evidence referring to events that occurred more than three years before a trial), and once again endorsed McTurk's testimony. The court was "cleared for deliberation, and subsequently adjourned." Five days later, on November 24, the judges found Smith guilty of all charges (with some qualifications) and sentenced him to be hanged by the neck until dead. But, under the circumstances of the case, the court begged "humbly to recommend the prisoner, John Smith, to mercy."

Writing from Berbice to the London Missionary Society a few days after Smith was sentenced to death, Wray commented that people were dreadfully enraged and hostile to Smith. "I am told they say if the Governor should send him home they will murder him before he gets on board. I am told they have hung him in Effigy and that a cap has been fixed to the gallows which they say is for him. What the end of these things will be God only knows. . . . I have been thinking of going down but I am advised not as it is not safe and it is uncertain whether I should be able to see him or not. Our hearts are deeply afflicted on account of those things." At this point, Wray was not certain of Smith's innocence, although he obviously wanted to believe in it. He thought he should give the Smiths some kind of support, but kept postponing his trip to Demerara. By mid-January, after receiving more information about the trial and a detailed letter from Jane Smith, Wray wrote that he felt "greatly confirmed in my opinion of Mr. Smith's innocence."

John Smith lingered in jail, waiting for the decision of the King-in-Council. But when the news that mercy had been granted finally arrived, he was dead. The tension of the trial and the "pulmonary consumption" that was undermining his body had finally defeated him. He died quietly on February 6, 1824. After several doctors and colonial authorities, including his old nemesis McTurk, paraded to his corpse to certify his death, John Smith was buried secretively, in the middle of the night.[11] Jane Smith was left alone to make arrangements for her return to England. Since she and her husband had been arrested, the only person she had had on her side was Mrs. Elliot—the woman she once had criticized for her "free" manners and "unbecoming" behavior.

After Smith's death, Wray finally overcame his fear and offered to help. He traveled to Demerara, spoke to a number of people, and reported to the London Missionary Society everything he heard that confirmed Smith's innocence. He managed to get hold of Smith's church books and frantically searched through the baptism registers to find out whether the slaves who had been condemned really had been members of Smith's congregation. After laborious research, he finally wrote to the directors that, except for five or six, the slaves executed as ring-leaders either had not been baptized or had belonged to plantations where most slaves were not Christians. Only Telemachus was a communicant. Bristol, Jason, and Romeo, who were deacons in the chapel, had not been tried as ring-leaders. At *Success, Chateau*

Margo, and *Beter Verwagting* (where many slaves had been baptized) the men executed as ring-leaders had not received that ordinance—which Wray thought was "a good proof that they had not attended to religion." Great bloodshed had occurred at *Golden Grove*—a plantation run by John Pollard, whose cruelty was well known. But Wray chose to attribute the violence to the fact that no one there had been baptized. Similarly, at *Plaisance, Triumph, Coldingen, Porter's Hope, Non Pareil, Enterprise*, and *Nabaclis*, where twelve ring-leaders had been executed and one had received a thousand lashes, Wray noted that no slave had been baptized. By contrast, on plantations where Smith had baptized several slaves, they had sided with their masters. What better proof did anyone need of the good effects of religion? Wray listed as examples plantations *Brother, Vryheid's Lust, Industry, Mon Repos, Endraght, Vigilance, Montrose*, and *Dochfour*.

What Wray either did not know or did not take into account was that on several of these plantations Quamina had warned the slaves not to rebel. Wray also dismissed the fact that punishment had been so random and whimsical that an innocent man might have been executed, while a ring-leader might have escaped punishment. (Bristol, for example, who had been privy to the conspiracy from the beginning, was never convicted.) All this raises some questions about Wray's interpretation of his evidence. Besides, if slaves seemed more inclined to commit violence on plantations where the managers or masters had opposed religious instruction, was the slaves' aggressive behavior a consequence of their lack of religious instruction—as Wray concluded—or a response to the restrictions imposed by managers? Still, whatever his biases, Wray's efforts to demonstrate Smith's innocence paid off. Much of the information he gave the directors was amply publicized by the British press. His findings would be used by the London Missionary Society and by all those who favored missionary work and slave emancipation to demonstrate Smith's innocence and to condemn once again the evils of the slave system.

John Smith's trial had powerful repercussions both in the colony and abroad.[12] In Demerara, a large number of whites attended the trial. The evidence adduced both for and against the missionary was reviewed, day after day, in private conversations.[13] Not even Smith's death put the subject to rest, for the colonial press continued its campaign against missionaries and abolitionists.

Even though he was an Anglican, the Reverend Mr. Austin, who had testified in Smith's favor, became the target of open public hostility. His house was vandalized, and a petition was signed by more than 250 people asking that he be suspended from clerical functions. Even members of his own family broke relations with him.[14] So severe was the pressure that Austin finally was forced to quit his post.[15] There was also a meeting to expel William Arrindell, the lawyer who had advised—though not actually represented—Smith. Davies and Elliot were constantly under attack. And in Berbice, after Wray's chapel was set on fire, he continued to be harassed by people who suspected him of sowing subversion among slaves.

Jane Smith left the colony a few weeks after her husband's death. Unable to recover the money taken from her house by the authorities and seized by the colonial government (on the pretext that it would be used to pay for her husband's lodging and food in jail), she had to rely on the support of her fellow missionaries for her trip back to England. The London Missionary Society organized a collection for her, but after three months only four hundred pounds had been gathered. The *Guiana Chronicle* used the opportunity to attack the missionaries once again. It ridiculed the small amount of money collected and, betraying its class bias, commented sarcastically:

> This is no doubt enough for the only rational purpose to which it can be applied, and that is to place the woman in some grocer or grocery haberdasher's shop, where she may earn an honest livelihood as we presume (although one writes in ignorance of the fact) that such a solution would be the most fitted for her from the station in life in which she has been brought up as we cannot for a moment imagine that any women, but of the lowest class would have been the wife of such an illiterate, low bred man as Demerara's Smith was known to be.[16]

Demerara's white colonists held several public meetings to honor their "heroes." Colonel Leahy, who had distinguished himself during the repression by liberating managers and masters from the stocks and killing many slaves, was decorated for his "brave" and "loyal" services to the colony and rewarded with 200 guineas "for the purchase of a sword." He also received another 350 guineas from "the inhabitants of the West Bank and Coast of the River Demerara." And the Court of Policy voted to offer him another 500 guineas for the purchase of "Plate for the use of their regimental mess" as "a mark of their esteem and approbation of the 21st Fusileers." Captain Stewart, commandant of the First Indian Regiment, was given 200 guineas "to be laid out in Plate." Lieutenant Brady, who had commanded a detachment of the Twenty-First Fusileers in Mahaica, received 50 guineas. Militia colonels Goodman and Wray, who had played important roles in the trials, were also rewarded. Goodman received, for his "arduous duties" as commandant of the Georgetown militia, £400, "to be laid out in plate of his own selection," and £100 "for the purchase of a sword."[17]

For days, members of the local elites toasted their victory and reasserted their solidarity. Dutch, Scots, English, and Irish, Presbyterians, Anglicans, and Catholics, planters and merchants, all momentarily forgot their conflicts, congratulated each other, and expressed their gratitude to the militia and soldiers who had defended their property and their lives. With their consciousness of shared interests rekindled by events, a group of more than a hundred merchants, planters, attorneys, and managers gathered to petition the British government for compensation for the losses they had suffered and to protest against "undue" interference in the life of the colony.

Three weeks after Smith's death, the *Guiana Chronicle* published a report of a general meeting of the "Inhabitants of the United Colony of Demerara and Essequebo." The colonists stated, among other things, that it had been established "by the most unquestionable proof" that the immediate cause of the insurrection had been rumors circulated among the slaves of discussions in Britain contemplating changes in the internal regulations of the colony. The effect of such rumors had been greatly aggravated by the "pernicious" predisposition occasioned by missionary instruction and influence, particularly by "one individual of that class," whose "discourses and studied perversions of portions of the sacred writing," as well as examples which he incessantly exhibited of opposition to "constituent authorities," created "in the minds of negroes" feelings of discontent. That same "individual"—John Smith—had impressed on the "negroes" a belief that "rights and privileges incompatible with the existence of the colonial system, were unjustly and unlawfully withheld from them." The colonists complained about the "clamours" in England of "a faction hostile to the existence of the colonies," and boasted that in the year ending in January 1823, there had been cleared at the Custom House 74,317 tons of shipping, employing 3,910 seamen, and the colony had yielded in excise and custom duties upward of a million pounds sterling. They stressed that the colony had not been incorporated into the British dominion by conquest or by force of arms. It had been "consigned by the deliberate and voluntary act of its Inhabitants, represented by their colonial Legislature, to the Protection of Our most Gracious Sovereign," under a formal treaty whose first article guaranteed that the law and the usages of the colony would remain in force. They claimed that during the trials no slave had complained of bad treatment and that any amelioration of their conditions of living could only come from their masters. Finally, they concluded that the colony had a just claim to indemnification from the British government for the severe losses it had sustained. The Demerara colonists' forceful statement found echo throughout the Caribbean and became the subject of comment and criticism in the British press.[18]

News of the Demerara rebellion spread fear throughout the Caribbean. Everywhere, colonists took the opportunity to accuse the British government and the missionaries of disturbing the "peace" of the colonies, and of threatening their properties and their lives. The worst incident occurred in Barbados, where for days local newspapers teemed with invective against "certain hypocritical characters who, under the pretence of giving religious instructions to the slaves, were introducing principles entirely subversive of those foundations on which the comfort and happiness of society rested." Local newspapers reproduced excerpts from a letter that William Shrewsbury, a Methodist missionary who had a chapel in Bridgetown, had published in the *Missionary Notices* three years earlier. There were insinuations that the published letter was not the real one, and that the real letter contained "calumnies" against the Barbadians. This became a subject of passionate debate. Shrewsbury's sermons sparked dispute and criticism every time he preached. People in the streets were shouting "That fellow ought to have a rope tied round his neck! Hang him!" During an evening service, some people threw bottles containing "some offensive chemical mixture" into the chapel. The following Sunday—when Shrewsbury preached in spite of renewed threats—two men wearing masks and armed with swords and pistols came riding swiftly down the street. As they passed the chapel they fired pistols. In spite of constant harassment, Shrewsbury continued to preach, but evening after evening people gathered around his chapel, throwing stones, riding to and fro on horseback, and "saluting the congregation with catcalis, whistles" and other offensive noises. To make things worse, Shrewsbury received a summons for failing to enroll in the colonial militia. The summons said that the Toleration Act (which had exempted nonconformist ministers from militia service) did not apply to the West Indies. Finally, on October 20, a mob of about a thousand "headstrong fellows" began to demolish the chapel, and by midnight nothing was left. Shrewsbury and his family were forced to flee for their lives.[19]

The West Indian colonists' reaction was as much a response to events in Demerara as to the British government's policies. In fact, as soon as the news of Buxton's motion advocating the gradual abolition of slavery in the colonies had reached the Caribbean, whites had started imagining slave plots everywhere. The news of the Demerara rebellion only heightened their paranoia. In Saint Lucia, an alleged plot was uncovered and three "ring-leaders" arrested. In Jamaica, two or three people were detained for carrying "inflammatory documents," supposedly brought in from Haiti. In Trinidad—where things had initially been peaceful—there were rumors of an intended rising, and twenty-three slaves were arrested. (Later it was discovered that the whole thing had been nothing but a hoax.) Another case of an alleged insurrection occurred in Dominica in December. Finally, there were again rumors of rebellion in Jamaica, and several slaves were tried and sentenced to be hanged.[20]

In Britain the Demerara rebellion and Smith's trial had an even greater impact. Abolitionists and anti-abolitionists alike used the opportunity to promote their own causes. Predictably they gave opposite versions of the events.

At a London meeting of "gentlemen connected with Demerara," held in October 1823 at the counting house of Hall and Co., delegates decided to trigger a public campaign to inform the public of the "real character" of the "insurrection." A week later, another meeting praised the conduct of the governor and of the military forces. It attributed the rebellion to Parliament's discussions of slavery in the West Indies, and to "evil designing persons" who inculcated in the minds of the slaves the belief that the King had granted them freedom. Alleging that there were strong reasons to believe that one missionary had been deeply implicated in the insurrection, the meeting decided to "implore his Majesty's Government" to restrict the transit of missionaries from Great Britain to "British Guiana." It also claimed that serious losses of property had occurred as a consequence of the rebellion and that the "sufferers" were entitled to a full compensation from Parliament.[21]

While proprietors and merchants assumed a bold stance in defense of their interests, using all the resources they had to promote their cause, the London Missionary Society reacted cautiously.[22] As late as December 1823, the *Evangelical Magazine* was still informing its readers that "The Directors are concerned that they are still unable to relieve the anxiety of their Friends, on the subject of the events in Demerara." To reassure its readers it had published in November a passage from the instructions given to Smith when he left for Demerara.[23] In January, the magazine reproduced parts of a letter from Elliot saying that the only crime the missionaries had committed was "their zeal for the conversion of the negroes." They had "neither been so weak nor so wicked as to excite the negroes to rebellion."[24]

News from Demerara was slow to arrive, and when the February 1824 number of the *Evangelical Magazine* was published, the London Missionary Society was still unaware that Smith had been sentenced to death.[25] The directors had tried without success to have him removed from Demerara.[26] When they finally got hold of the trial proceedings and gathered enough evidence, they launched a campaign to clear Smith's reputation—and preserve the prestige of their missions. They published letters they had received from John and Jane Smith, and several documents relating to his arrest and trial. They publicly criticized the way the trial had been conducted, and privately protested in letters to their supporters in government against the restriction imposed in the crown's grant of mercy (which had required that Smith never return to the West Indies). To the directors this seemed to imply an acknowledgment of culpability. So they attempted—again without success—to have the restriction removed. When the news of Smith's death arrived, the *Evangelical Magazine* published a long editorial calling him a martyr to the cause of spreading the Gospel,[27] and started the publication of a serial biography of Smith that stressed his piety and devotion to missionary work. Confident of Smith's innocence, the directors of the LMS decided to petition the government for "redress."

While the LMS prepared its petition to the House of Commons, letters of support came from the "Associated Members of the Friends of the Society" and from a variety of denominations throughout the country, revealing the important and efficient networks the evangelicals had managed to build. Wesleyans, Baptists, Independents, and even some ministers of the Church of England organized petitions and collected signatures from members of their congregations. During 1824 several evangelical magazines included editorials on the slave rebellion in Demerara and the trial of John Smith, tying them to "the cause of humanity," missionary work, and emancipation. The editorials expressed outrage at the way the trial had been conducted and demanded in the name of British law and British principles that Parliament open an inquiry. "The cause of Humanity, as it respects the projected amelioration of the condition of the Negro-slaves and the cause of Missions among that long neglected class of our fellow men have both been, in some sense, so implicated in this transaction, as to render an accurate investigation of it highly necessary," said a report on Smith's trial published in *the Wesleyan Methodist Magazine.* The editor expressed his hopes that when the facts were known they would show that "the notion of *Interminable Slavery,* is as incompatible with the *security* of the West Indians themselves, as with the righteous claims of our Negro Bondsmen on the justice and liberality of this professedly Christian empire." Another article reported that a meeting of the West India "interest" had stressed the legal right of slaveholders to their slaves, and had claimed indemnification from the government for any injury "that kind of property" might sustain as a result of the plans for amelioration. Although the editor granted legitimacy to such claims, he also insisted that it was "due to justice, to humanity, to their own honour, and to the public feeling of this country," that the West Indian proprietors express "their readiness to concur, with the unquestionable righteous measures of amelioration contemplated by his Majesty's Government."[28]

The Methodists were particularly troubled because the incidents in Demerara seemed to have endangered their own work everywhere in the West Indies. So it was very important for them to insist on Smith's innocence. At an anniversary meeting, the Wesleyan Methodist Missionary Society acknowledged that the events in Demerara had caused great damage to their mission. Chapels were nearly deserted, and though the missionaries had escaped "the hand of legal violence," they had been exposed "to obloquies and insults," and one of them had narrowly escaped a personal attack from certain white people "who waylaid him on his return by night from his duty in the country." At the society's annual meeting of 1824, the chairman in his opening speech proclaimed that it had been generally admitted that Smith was "entirely innocent."[29] The Baptists, who perhaps even more than other denominations were engaged in permanent struggles in England to further measures to protect dissenters and to defend itinerant preachers from persecution, were equally moved to support Smith.[30]

Even stronger support came from some evangelicals in the Established Church, who were particularly outraged by the Demerara colonists' attacks against Austin for his role in Smith's defense. Like the Wesleyans and the Baptists, they openly advocated emancipation and did not spare criticism of the West Indian planters. The *Christian Observer,* a monthly publication by evangelical members of the Established Church, said that the "most extravagant and incredible charges" had been made against Smith. It also criticized a proclamation issued in Jamaica against proceedings in Parliament in favor of amelioration, and condemned the behavior of the colonists in Trinidad and Barbados. It called upon every man in the kingdom "who has the fear of God before his eyes, and who has any regard for the obligations of humanity and justice," to support Parliament in "its righteous purpose" of admitting the slaves to full participation in "those rights and privileges which are enjoyed by other classes of his Majesty's subjects, at the earliest period which is compatible with the well being of the slaves and the safety of the colonies, whatever may be the sacrifice the country may be required to make in order to afford a fair and equitable compensation to the parties immediately concerned." The editor expressed his hope that the nation would not be willing to continue "at a large expense in bounties and protecting duties, and in other ways, to support the present system which has been clearly proved to be as unprofitable and impolitic as it is unconstitutional and unchristian."[31]

No document could have displayed better all the clichés that were being woven together in antislavery rhetoric: the conviction that slaves were entitled to the same rights as those enjoyed by other subjects; that to support such a "humanitarian" and "just" cause was the obligation of every Christian; that the slave system had to be abolished; and that it was possible to find a solution which would not only satisfy both masters and slaves but also reconcile profit and Christian morality. It was precisely because a growing number of people espoused such notions that the Demerara rebellion and Smith's trial could have such an impact. On the other hand, debate over the events helped to popularize a rhetoric that redefined the concepts of "humanity" and "citizenship" and enhanced national pride in "British wisdom" and "liberties."

The *Christian Observer* was ready to blame events in Demerara, at least in part, on the persistent influence of Dutch customs and laws. In this regard it did nothing but rehearse a long tradition that had always opposed the enlightened British to the cruel Dutch. The *Observer* argued that although the numbers of English proprietors had increased, they were for the most part non-residents, while a very large proportion of the overseers and "petit-blancs" were still Dutch. But even if this were not so, there were certain habits of feeling, thinking, and acting which became the inheritance of a community and were not easy to eradicate. This helped to explain both the particular ruthlessness of the slaveowners in Demerara and the rebelliousness of their slaves—which was incorrectly being attributed to abolitionists' speeches and pamphlets and to "incendiary" discourses of missionaries.[32] On another occasion, the *Christian Observer* rejoiced at the reference in the King's Speech to the issue of slavery, saying that the concurrence of the government with the general feeling of the public could not fail to bring about the adoption of wise and prudent measures that would safely end "that monstrous system of oppression, in spite of the furious clamours of the colonial taskmasters, and the mendacious statements of their hired advocates in this country." It concluded by saying that the Demerara rebellion had been the result of cruelty and oppression, of immoderate labor, of severity of treatment, of religious persecution, and of a most wanton disregard of the feelings of the slaves.[33] (Interpretations of this sort, which found parallels in other evangelical and abolitionist tracts, helped to consolidate one view of the rebellion that future historians would follow.)

But perhaps the most effective argument used by the *Christian Observer*—because it was addressed to the hearts and minds of British laborers—was to compare their lot with that of slaves. This rhetorical strategy had a double edge. It condemned slavery, and it portrayed the British laborer as privileged. A long 1824 article about the Demerara rebellion made the connection:

> Let us suppose that the miners of Cornwall or the iron-workers of Wales, or the keelmen of the Tyne, or the weavers of Lancashire had conceived themselves (whether justly or not) to have been aggrieved by their masters, whom they suspected, on what appeared to them good grounds, of withholding from them the advantages which the law allowed them; that in consequences of this apprehension they had struck work, and refused to resume it until they had obtained the requisite explanations, and that they had even gone the length of threatening violence to their masters, and of maltreating such of their body as continued to work in the usual way. . . . Would it be tolerated that these men should be forthwith attacked by a military force, killed in cold blood by hundreds, hunted down like wild beasts, tried and executed by scores as traitors?

What if large bodies of Spitalfields weavers had crowded at Westminster (as they did the year before) "imploring the members of the Legislature to protect them from the unjust purposes, as they deemed them, of their masters," and Parliament, "instead of lending a patient ear to their complaints and suspending even the intended course of legislation, in deference to their perhaps unreasonable fears and misapprehensions (for such was the line of its policy), had called out the military to saber and hunt them down by hundreds, and had them tried and executed the survivors by scores; what would have been the general feeling amongst us? Should we not have raised our voices as one man against such insufferable tyranny and oppression?" What if the agricultural laborers who were destroying threshing machines, and the Luddites, and the Blanketeers, many of them "most criminal individuals should have been dealt with as the poor, ignorant, oppressed, cart-whipped slaves of Demerara?"[34]

The article spoke of British workers and laborers—many of whom had risen in different parts of the country against poverty and oppression—as a privileged group of people, protected by the courts and by the laws. But it was silent about the arbitrariness they had suffered. And the persecution. And the gag laws. And the Home Office prohibition of workers' combinations and secret meetings. And the harassments endured by radical leaders. And the many workers who had been arrested, tried, and sentenced to be transported or hanged. Instead the *Christian Observer* spoke of how British "labourers" and workers were privileged to live in a nation where they could find justice and freedom under a magnanimous Parliament and the rule of law. This rhetoric had a contradictory effect. It might legitimize the status quo, but it could also give workers and laborers a powerful argument to claim full citizenship. This explains its appeal.[35]

Not every worker, of course, would respond positively to such rhetoric. Among the more radical labor leadership there were many who suspected that the abolitionist campaign was designed to divert attention from class struggles in Britain.[36] By this time, however, antislavery had gained such a widespread support among laborers of all sorts that, recognizing the efficacy of such rhetoric, the *Christian Observer* would resort to it again. After condemning the use of the whip and denying validity to the argument often invoked in its defense (that flogging was used in the army, navy, and courts of justice at home—"one bad practice is ill defended by another"), it contrasted the two cases. This time it chose the example of Cornwall:

> Let us suppose that in that country, every proprietor of land, or of mines, or of manufactures; every bailiff, or overseer, or head of an establishment having servants under him; every attorney, guardian, executor or administrator; every supervisor of a work-house, and every keeper of a gaol, might at their own discretion for any offence, real or imaginary, cause to be stripped naked, any man, woman or child employed under them, and either publicly or privately inflict upon the bared body 39 lacerations of the cart-whip, and might then subject the sufferer with his bleeding wounds to confinement and hard labor

> at pleasure. . . . Let us further suppose that the whole of the labouring classes was debarred by law from giving evidence in the case of any abuse of power committed by their superior. . . . And yet if some benevolent individuals deeply affected with the cruel and brutalising effect of such a system, were to propose to ameliorate the condition of their Cornish brethren, and to raise them "to a participation in those civil rights and privileges which are enjoyed by other classes of his Majesty's subjects," our ears would probably be dinned with representations of the humanity of the owners, and bailiffs, and supervisors and gaolers of Cornwall, and of the superlative happiness of its laboring population. See how fat and sleek they are, how well fed, how well lodged, how much better off than the wretched labourers in other parts of England, who have no kind masters to look after them!! Should we listen to such representations for a moment?

The *Christian Observer* concluded that in such a case, the voice of the people of England, to which "some cold hearts in high places" were disposed to pay so little regard, certainly would and should prevail.[37]

The success of this strategy can be measured by the use that some of the most famous rhetoricians of the time made of it. In 1824, the *Christian Observer* transcribed an article by Thomas Clarkson refuting the argument constantly invoked by slaveowners that colonial slaves were better treated than British laborers.[38] The well-known abolitionist's powerful words filled page after page of the magazine. Slaves were sold as cattle, he said; but could any man, woman, or child be sold in Britain? Slaves were sold for their masters' debts. Could British laborers or servants be sold for the same cause, or "on account of the imprudence or wickedness of their employers"? Slave families endured the pain of being sold and separated. Could such afflicting scenes occur among the "peasantry" of Britain? Who could interrupt their "domestic enjoyments" with impunity? It was not in the power of the King himself to separate the husband from the wife, the mother from the child, or the parents from their children. Slaves were branded. Could this be done in England? Yet, all those acts, all "these enormities were perpetrated by persons who considered themselves to be Britons" and Christians. Slaveowners always said that slaves were better clothed, lodged, and fed than the British laborers. But such things did not constitute the most important aspect of "a man's happiness." What did constitute "the best part of man's happiness" was liberty, personal protection, the unmolested enjoyment of family and home, the recognition of one's citizenship, of one's humanity, the sympathy of one's fellow creatures, the freedom and enjoyment of religious exercises, and "hope, blessed hope, that balm and solace of the mind." Such were the principal components of a "rational" human being.

Clarkson's words epitomized some of the most important tenets of the liberal ideology that he, among many others, was helping to consolidate by weaving it together with another, equally powerful notion, that of the nation. Clarkson claimed that the British nation was a land of freedom and law—that to be a Briton was to be a free man,[39] and to be a British citizen was to be protected by the law. Such ideas were being espoused by individuals belonging to different social groups. They were used for different and sometimes contradictory purposes by both powerful and powerless. The powerful used them to legitimize the social "order," the powerless to challenge it. Proclaimed in parliamentary speeches, spread in evangelical magazines, reproduced in the British press, repeated over and over throughout Great Britain, and hailed in popular songs, these notions of freedom and law merged with the imperial ideology of men like Macaulay, who proclaimed that British rights should be extended to all subjects of the empire. They found echo in the four quarters of the world, including Demerara, where colonists, unaware of the contradiction, sang in a land of slaves: "Rule, Britannia, Britannia, o'er the waves. Britons never, never, never shall be slaves."[40]

This combination of militant humanitarianism, evangelicalism, liberalism, nationalism, and imperialism was highly corrosive of the ideological assumptions that had validated slavery. Once the humanity of the slaves was acknowledged and they were perceived by the government as subjects of the British empire, and once the empire was defined as the realm of freedom and law, it became difficult for slaveowners to defend their "right" to own slaves. But even more decisive were the concrete struggles waged in the name of such ideas on both sides of the Atlantic. When slaves rose in the name of their "right" to be free, and their struggles in the colonies found parallels in Britain, where rural and urban people were struggling for the recognition of their own rights of citizenship, both struggles gave new impetus to those who were arguing in Parliament and the press that slaves should be emancipated. What was left was only for Parliament to decide how and when emancipation should come. (The limits of the slaves' freedom would be tested after emancipation, but in the aftermath of the Demerara rebellion, the campaign against slavery momentarily obscured other questions.)

If, by exposing the oppressive nature of colonial society and contrasting it with the mother country, antislavery rhetoric appeared to ignore some forms of oppression at home, and told British laborers that their hardships were nothing compared with the horrors of slavery, by stressing the privileges they supposedly enjoyed, it also provided them with a rhetoric they could use to claim their own full rights of citizenship. It is against the background of the British workers' struggles (as well as evangelicals' and women's struggles) for an ampler concept of citizenship that reactions toward events in Demerara can best be understood. To them the slaves' struggle for freedom in Demerara and Smith's trial were evidence of the evils of an oppressive system that was still defended by some members of Parliament who not only were representing West Indian interests, but were also deciding the fate of the British people at home. They saw Smith as a martyr for the cause of freedom and justice, and the slaves as victims of the arbitrariness of masters and royal authorities. And by rallying to support emancipation, working men and women gained a new impetus to fight against their own oppression.[41]

Smith's trial and the Demerara rebellion became matter of public debate not only among evangelicals, but in Parliament and the press. Predictably, newspapers and magazines were divided. Some, like *John Bull, Blackwood's,* and the *Quarterly Review,* expressed the point of view of the planters and their supporters. They attributed the rebellion to debates over emancipation in Parliament, to the "revolutionary" rhetoric of the press, and to reformists, abolitionists, evangelicals, and fanatic missionaries, particularly John Smith.[42] Others, like the *New Times* and the *Edinburgh Review,* expressed the point of view of those who supported the cause of missions, emancipation, and reform. They condemned the slave system, attributed the rebellion to abuses committed by managers and masters, and called for a review of Smith's sentence by Parliament. Even local newspapers like the *Derby,* the *Leeds Mercury,* the *Preston Chronicle,* the *Lancashire Advertiser,* the *Dorset County Chronicle,* and the *Norwich Mercury* carried articles on Smith and the Demerara rebellion.[43] Later, when the issue was finally debated in the House of Commons, both sides publicized lengthy editorials quoting from speeches and commenting on the final vote—each side interpreting the outcome as evidence supporting its own position. Both sides transcribed parts of the proceedings of the trials in Demerara, and quoted from Smith's journal and from other sources. Smith's case became a *cause célèbre.* It seemed to served everyone's cause. He had finally won his crown of glory.

Smith's trial and the Demerara slave rebellion became subjects of debate among rich and poor, commoners and gentry. They became an important topic in 1824 in the annual meeting of the Society for Mitigating and Gradually Abolishing the State of Slavery Throughout the British Dominions, which was attended by the Duke of Gloucester, several MPs, including Wilberforce, and a large public of gentlemen and ladies "elegantly dressed, among whom there was a great number from *The Religious Society of Friends.*" Before this select audience, the indefatigable Macaulay with his usual eloquence made a long speech condemning the incidents in Barbados and Demerara and repudiating the colonists' accusations against missionaries and abolitionists. He questioned their idea that the rebellion had been instigated by debates over emancipation or the preachings of missionaries. When did speeches, when did pamphlets, when did meetings inflame people to extensive insurrection, if they enjoyed plenty, comfort, and security? Macaulay asked. For many years, hundreds had been employed in telling the people of England that they had been deprived of their just rights, that they were degraded, that they were enslaved. Every day this was heard and read—perhaps even believed—by thousands. More appeals were made to their passions in a week than to those of the West Indian slaves in a year. Yet who in England lived in fear of a rebellion? It required no very skillful interpreter to translate the West Indian clamor into confessions of tyranny. Macaulay then went on to condemn Smith's trial as characteristic of the injustice of a slave society, and concluded his speech celebrating England and the empire. Several other speakers absolved Smith of any guilt and tied his case to the cause of emancipation. They too argued that the cause of religion, humanity, and justice had been brought into question by his trial. And they boasted that the number of petitions to Parliament for the abolition of slavery had increased from 225 in the previous session to 600.[44]

Threatened by the new emancipationist wave that Smith's case had triggered, proslavery groups mobilized all their traditional arguments. They invoked their right to property and stressed that slavery was a legal institution. They exposed the oppression that lay behind the rhetoric of free labor. They claimed that slaves in the West Indies were better treated than anywhere else in the world—even better than workers in Great Britain—and were content with their lot. They argued that the movement in favor of emancipation was promoted either by the "enemies of the West Indies," or by revolutionaries, visionaries, and fanatics who knew nothing about the situation in the colonies and were always ready to propagate false notions to achieve their own purposes. Their main culprits were Wilberforce, Macaulay, Brougham, Buxton, and James Stephen.

Typical of the anti-emancipationist rhetoric was an article in *Blackwood's*. It identified three groups of people as "dangerous" to the West Indies. First, "a body" of persons who acted or supposed themselves to be acting under the influence of no motives whatever but those of general philanthropy and religious zeal. "Extreme imprudence," was their main characteristic, and Wilberforce was their "facile princeps." The second was a "more cool-headed body" of people who agitated the public mind in the hope of seriously injuring the West Indian colonies, with the sole purpose of gaining commercial benefits. This group included many "characters" within the East India Company and a still larger number of well-known individuals deeply connected with free trade to India and the coast of Africa (as well as many eminent leaders of the African Institution). The third group was "neither a religious or a commercial one." It consisted of politicians—men like Brougham—who seemed but too willing "to disturb existing establishments of every kind," provided they saw any chance of thereby gaining popularity "to prop up the ruinous reputation of their own sorely degraded faction, the Whigs."[45]

John Bull was equally violent. The paper promised to unmask Wilberforce,

Macaulay, and the African Institution. It accused the "Saints" of being responsible for the bloodshed in the colonies. It raised the specter of Cromwellian revolution and insinuated that the petitions in favor of abolition of slavery and for the reversal of Smith's sentence were orchestrated by "a well adjusted, orderly, and regulated machine, which as a sort of spiritual steam engine possesses a power equal to that of government itself in enforcing obedience to its edicts," and which when the day of conflict arrived would supersede and absorb, as it had done once before, every other power of the country. *John Bull* condemned the "perverted principles" of Christian philanthropy and considered the rebellion as the first fruit of the "philanthropic" efforts of Wilberforce's faction in the House of Commons. It described the reformers as the offspring of the Puritans who "overthrew the Government and Constitution" in the reign of Charles I, bringing the King to the block in order to place an "independent demagogue tyrant on his vacant throne." Like their adversaries in the *New Times,* the editors of *Blackwood's* and *John Bull* were at the same time making history and rewriting it.

The signs that "Wilberforce's faction," although still a minority, was making significant progress in Parliament became clear on June 1, 1824, when, after numerous petitions had been presented to the House of Commons for an inquiry into Smith's trial, Brougham moved that an address be presented to the crown, saying that the House of Commons contemplated "with serious alarm and deep sorrow the violation of law and justice which is manifest in those unexampled proceedings." Brougham's motion also asked that the King adopt measures "for securing such a just and humane administration of law in that colony as may protect the voluntary Instructors of the negroes, as well as the negroes themselves, and the rest of his Majesty's subjects, from oppression."[46]

The motion triggered an intense debate. Brougham's main opponent was Wilmot Horton, the under-secretary for the colonies. Several speakers followed. For hours both sides rehearsed the arguments that already had been raised numerous times for and against Smith and his trial. Once again the story of the rebellion was told. And once again both sides presented a picture in which the slaves appeared not as historical agents in their own right, but as passive victims either of manipulation by a misguided missionary or exploitation by cruel masters. The speeches went on for so long that when Lushington tried to speak cries for adjournment silenced him. Further discussion was scheduled for the following day, but it took ten days for the motion to regain the floor.[47]

Lushington was the first to speak. He analyzed the trial proceedings and the body of evidence laid out before the court, and condemned the behavior of the colonists. He told his peers that if Commons did not express in the strongest and most decided terms its disapproval of the Demerara proceedings, it "would let it go abroad that you do not mean to govern your colonies upon principles of law and Justice." In an apocalyptic statement, he predicted that if the House of Commons failed, severity against the "negroes" would increase a hundredfold, the cause of religion would fall to the ground, government would lose its authority, and "all the hateful and degrading passions of man would be brought into full and unrestrained action." Several members of the House of Commons then spoke for and against, including Wilberforce in favor and Canning against. The debate focused less on whether Smith was innocent than on whether adequate procedures had been followed in his trial. It was the Demerara colonists, the governor, the members of the court-martial who were now on trial, not Smith.

Implicit in the debate was the question of how much power the British government should have over the colony. Wilberforce made this clear when he argued that Parliament must act, because the issue being discussed concerned "the rights and happiness" of a British subject, the administration of justice in the West Indian colonies, and the amelioration of the conditions of the slaves there. Wilberforce's eloquent speech was neutralized by the conciliatory strategy of Canning, who spoke immediately after. He suggested that since it was impossible to reach a "satisfactory judgement," the House should simply drop the question. He was clearly unwilling to embarrass the colonial authorities and alienate even further the colonists and the West India lobby. But he reassured the Commons of the government's commitment to the amelioration policies. "I am satisfied," he said, "that the discussion itself will have answered every now-attainable purpose of public justice; and that we cannot be misinterpreted, as intending by our vote to shew any lukewarmness in the cause of the improvement of our fellow creature, or in our belief that religion is the instrument by which that improvement is to be effected." Canning's speech was received with loud cheers.[18] He had found a compromise that could satisfy a majority. Nothing could save Brougham's motion.

Brougham did make one more effort to turn things around. The session had been long and difficult. It was late at night, but he continued to lay down his reasons, trying to refute his opponents and to sound compromising. It behooved the House of Commons, he said, to teach a memorable lesson: "that the mother country will at length make her authority respected—that the rights of property are sacred, but the rule of justice paramount and inviolable—that the claims of the slave owners are admitted but the dominion of Parliament indisputable—that we are sovereign alike over the White and the Black." He tried to present the case as an act of defiance by the Demerara colonists to British authority, but in spite of all his rhetorical skills his strategy failed. When the question was finally put to the House of Commons, the result was 146 in favor, 193 opposed. Brougham's motion had lost by 47 votes.

Smith's trial and the Demerara rebellion had become part of a complex political game. In spite of their defeat, the cause of Brougham, Lushington, Buxton, and Wilberforce came out strengthened. The publicity given to the parliamentary debates increased their popularity. And even though Parliament had refused in the end to condemn the Demerara authorities, Governor Murray was recalled and the new governor, Benjamin D'Urban, proceeded to implement the instructions that originally had provoked so much debate in the Demerara Court of Policy. This was an assertion of the power of the British government over the colonists, but it was also a validation of the ideas Smith had struggled for. The repercussions of his trial and death gave a new boost to the abolitionist movement.[49]

Smith was absolved in Britain for the ideals that had condemned him in Demerara, where he had come to represent everything the colonists most feared. Demerara's slaveowners and masters had watched with growing apprehension and irritation the movement toward emancipation. They had resented the abolitionists' critiques and had been hostile to British policy toward the slaves. They had looked with suspicion on the evangelical missionaries, whose notions and practices they disapproved of and whom they saw as spies ready to send home tales that reinforced the worst prejudices against slavery and slaveowners. They had followed with deep concern the debates on free trade taking place in England, which threatened the monopoly they had enjoyed in the British market. They had worried about the decline of the prices of commodities in the international market, and feared that they would not be able to compete with Cuban and Brazilian producers who continued to import slaves. They had dreaded these new trends that menaced both their slave property and their profits. In 1823, when the slaves rose up, the colonists were already prepared to demonstrate to the world that the rebellion was a consequence of the abolitionists' debates and the subversive activities of evangelical missionaries.

Smith was their scapegoat. They condemned him not only because of his supposed involvement with the rebels, or his sense of self-righteousness and missionary zeal, but because he had through the years challenged the foundations of the slaveowners' authority. The misunderstanding and conflicts that had pitted Smith against the colonists were born of different ways of defining the social order, different notions about justice, about law and citizenship, about social control, about crime and punishment, about the role of education and religious instruction, and different notions about what was proper and improper, right and wrong, fair and unfair. Time and again, in spite of his conscious efforts to comply with established norms, Smith had trespassed boundaries and violated the code of propriety, the rules, protocols, and rituals that were meant symbolically to reassert in everyday life the superiority of masters over slaves and of whites over blacks. He had undermined the system of sanctions and assertions that maintained slavery. He had dared speaking the unspeakable, raised questions about the legitimacy of the masters, questioned their fairness, and exposed their sins. He had constantly reminded them that in their dealings with slaves they were subordinated to the authority of God, King, and Law, and that slaves were human like themselves. In his daily relations with the slaves, Smith had redrawn the line between permissible and forbidden. He had disregarded the master-slave etiquette that aimed at consolidating power and authority and at breaking down resistance. In his chapel, he had created an alternative space, where the social distinctions that kept blacks and whites, free and slave, apart were subverted.

The system of authority characteristic of the slave system required the humiliation of the slaves; Smith spoke of their dignity. It postulated the slaves' dependence; he encouraged their autonomy. It aimed at destroying their leaders; he gave them power. It worked to destroy group solidarity, to prevent the formation of networks of social cooperation; Smith gave them a community of brethren. Instead of fear, he gave them hope. By behaving the way he did, Smith challenged the myth of the benevolent master and contented slaves, validated slaves' dreams of freedom, and legitimized their rebellion. In the eyes of the colonists he was guilty. He had broken the rules of propriety and had to be punished. But it was precisely what led to his punishment in the colony that secured him support in the mother country. There, Smith had to be absolved for the sake of the order he somehow represented: "the rights of the British citizen," the "enlightenment" of the state, the "superiority" of its law and legal procedures, the "sanctity" of the missions abroad, the supremacy of the imperial government, the "civilizing" mission of the empire. He had to be found innocent to reassert the superiority of one political faction in England over another, one way of life over another. But since there was no unanimity on the issues the trial had raised, and the British were still divided on emancipation, the final solution adopted by Parliament was compromise. How long the compromise could be maintained would be determined by future struggles on both sides of the Atlantic.

In Demerara the slaves returned to their day-to-day forms of resistance. Sometime after the rebellion, Joe Simpson's wife died of poison. Joe, also known as Packwood, was the slave who had betrayed the rebellion to his master on the morning of August 18—the day the slaves rose up. Suspicions of poison were also raised when a number of soldiers and members of the militia fell seriously sick after one of the banquets given in their honor. The slaves had been defeated, but had not surrendered. They continued to fight their battles in their usual ways. When, in compliance with the new instructions from the British government, the first records of slaves' "offences" and punishments were publicized, Governor D'Urban reported with astonishment that 20,000 punishments had been inflicted in a year, in a total slave population of 62,000. He calculated that at that rate more than six million physical punishments would have been inflicted on working people in England. A few years later, commenting on conditions in Demerara and Essequibo, a certain Captain Elliot, protector of slaves for British Guiana, testified before a House of Commons committee that the number of punishments was increasing. "This state of things cannot continue to subsist," he said. "The slave has advanced beyond such a system of government." Trying to find a more efficient way of controlling labor, managers had already adopted a task-system in the early 1820s. This system had become generalized. But Elliot thought this arrangement did not always work. It all depended on how fair the managers' demands were. And to judge by the number of punishments recorded, the task-system was failing rather badly. To increase productivity and discipline, Elliot suggested a profit-sharing system![50] The Demerara slaves' long struggle for freedom was approaching its end. Ten years after the rebellion in Demerara, the British government abolished slavery in its colonies and created a system of apprenticeship.[51] Ex-slaves and ex-masters entered a new contest over the meaning of freedom. The struggles of the past would lead them into the future.

A NOTE ON SOURCES

Because this book focuses on the history of a British colony in the New World, it has led me not only to a different time but also to different places. For a Brazilian who all her life had been a specialist on Brazilian and Latin American history, this journey was full of surprises. The wealth of documents and the extraordinary organization of the Public Record Office in England contrast markedly to the precarious state of most Latin American archives. The abundance of primary and secondary sources was at the same time a blessing and a curse. This explains, in part, why this trip to the past that I started in 1983–84, expecting that it would take me just a few years, lasted a decade.

For the history of England from the turn of the eighteenth century to the 1820s I relied to a considerable extent on secondary sources. I deliberately tried to avoid involvement in the interminable scholarly controversies that inevitably characterize a literature that counts among its writers some of the best historians the world has known. I limited myself to selecting from historians what seemed necessary to make my point, and to describing briefly in the footnotes other points of view. It was also extremely interesting to see the history of Britain as it was perceived by the colonists, through their writings and comments and the selection of things to be discussed in their newspapers. This inversion of the usual approach to British history was quite illuminating and refreshing, and sometimes seemed to point in directions I had not expected from reading the historiography.

The rich literature on evangelicals, dissenters, and abolitionists prepared me to understand the London Missionary Society, its missionary project, and the work of missionaries in Demerara. But it was the London Missionary Society's Archives and the *Evangelical Magazine* which were my most important sources. I found a complete collection of this and other evangelical magazines such as the *Wesleyan-Methodist Evangelical Magazine*, the *Chris-*

tian Observer, and the *Baptist Magazine* at the Mudd Library at Yale University. I read the *Evangelical Magazine* from 1795 to 1824, and the others for the years 1823–24. This allowed me to sense the atmosphere that surrounded missionary work, and to understand missionaries' concerns and struggles both in England and abroad. But much more important for my work were the Society's archives, which I found in microfiche at the Yale Divinity School Library. (This archive is now held by the Council for World Mission Archives.) Yale Divinity School Library has a vast collection of microfiches from the London Missionary Society Archives. I consulted the Board of Directors' Minutes, 1810–21; the Demerara and Berbice Incoming Letters, 1808–25; the journals of John Wray, John Smith, and John Cheveley; Minutes of the Committee of Examination, 1812–16; Candidates' Papers (John Smith's case only); and several boxes referring to John Smith and the Guyana rebellion under the label "Odds, West Indies." Particularly useful were materials from Boxes 5 and 6 that contained excerpts from colonial and British newspapers for 1823–24.

I have adopted the following conventions for the purpose of citation of the material in the Council for World Mission Archives, London Missionary Society Archives:

Incoming Letters, West Indies and British Guiana, Demerara 1807–94 and Berbice 1813–99, are cited as LMS IC, with the place, author, and date of the letter. (I have used microfiches 188–204 containing letters from Berbice, and 346–74 containing letters from Demerara.)

Board of Directors' Minutes are cited as LMS Board Minutes. (I have used microfiches 20–52.)

Journals, West Indies and British Guiana, 1807–25, are referenced with the last name of their authors, as in "Wray, Journal" or "Smith, Journal," or "Cheveley, Journal" (microfiches 784–93).

The other citations of the LMS materials are self-explanatory: "Candidates' Papers" (microfiche 574); "Minutes of the Committee of Examination" (microfiches 3–5); "Odds, West Indies," cited as "Odds" (microfiches 794, 798–800, 835–59, and 866).

For the study of Demerara and slave life there were two important sources, aside from the invaluable Colonial Office records in the Public Record Office. The first were the Demerara newspapers. I read in microfilm the *Essequebo and Demerary Royal Gazette* for the years 1806, 1807, 1810, 1813, 1815, 1816, 1819, 1820, 1821, and 1822, although for some of these years the collections I had access to had gaps. This newspaper changed hands and names during this period; to avoid confusion, I have chosen to refer to it throughout simply as the *Royal Gazette.* For the *Guiana Chronicle* I was able to locate only the numbers for 1823 and 1824, in the London Missionary

Society Archives (among the materials labeled Odds, West Indies). Other quotations from this newspaper are from excerpts in missionaries' letters or the *Royal Gazette,* or from numbers found in the Public Record Office. For the *Colonist* I was able to consult in microfiche the few numbers for 1823–24, also in the LMS Archives material, Odds, West Indies. The second repository of information about Demerara is, of course, the extraordinary collection of Colonial Office records at the Public Record Office in Kew, cited throughout as P.R.O.C.O. The most important series of documents I have consulted in that archive for this book were: C.O. 111/11, 28, 36–39, 41–46, 53, and 56; C.O. 114/5 and 7–9; C.O. 116/138, 139, 155, 156, 191–93; C.O. 112/5; C.O. 323/40 and 41; and ZHCI/1039.

For the trials of John Smith and the slaves, I used several publications ordered by the House of Commons, including transcripts of the proceedings and documentary evidence, and also a version of the proceedings published by the London Missionary Society, which includes some documents not contained in the House of Commons publications. Because the titles of all these publications were extremely long, they are cited in abbreviated form, as indicated in notes to the relevant chapters. This is, of course, also true for other publications of the period, including some travelers' accounts and local guides and almanacks I found in the library of the University of Guyana and in Georgetown's public library, and in libraries in the United States.

The Guyana National Archive was in pretty bad condition when I visited it, and I was relieved to discover that there were copies of most of the documents I needed in the Public Record Office. But one should not be surprised to discover that empires not only try to write the living history of their colonies, but end up by controlling their documents and records as well.

A word should be said about the lack of uniformity in the spelling of words in early nineteenth-century documents. The names of people and places were written in shifting ways, and sometimes it becomes next to impossible to maintain consistency through the text. To start with, the name of Guyana itself has changed over time. It was "Guiana," "British Guiana," and, today, "Guyana." For many Guyanese, to refer to their country as Guiana smacks of "imperialism." Today's Georgetown was once called George Town, and before that Stabroek. Demerara was Demerary, and Essequibo, Essequebo. People's names very often appear in the documents under different forms. Van der Haas, for example, also appears in the documents as Van Der Haas, van der Hass, van den Haas, and Van Den Haas. Archibald Brown was sometimes Browne. Cumings appears also as Cummings and Walrond as Walrand. Slaves' names were even more confusing, because the same person could have one African name and one English, and still another nickname. And the colonists often spoke of Haiti as St. Domingo, or Saint Domingue. In all such circumstances, I tried to bring some uniformity to the text by adopting one form and keeping to it throughout, but even this was difficult because in quotations I felt it necessary to be faithful to the documents.

Finally, for a Brazilian, trying to cope not only with the ordinary difficulties of the English language, but also with the differences between early nineteenth- and twentieth-century orthography, as well as with differences between contemporary British and American English, the task was very nearly overwhelming. But I found some pleasure in thinking that changes taking place today announce a world in which more and more people will be crossing boundaries and facing the same sorts of challenges, redressing the frontiers of past and present. In such a world, much tolerance is needed—and not just of "mis"-spellings.

NOTES

Introduction

1. These two contending discourses defined the parameters within which the early histories of Smith and the rebellion were written. According to their own political biases, historians either blamed the missionary and the abolitionists or blamed the planters and the slave system. Although Smith's life was inextricably related to the rebellion, the focus of the earlier historiography was on Smith, not on slaves. And for a long time after the events, the missionary's story continued to be disturbing to Guyana's ruling classes and in other corners of the Empire where analogous structures of power prevailed. In 1848, when Edwin Wallbridge, an evangelical missionary stationed in Demerara, wrote a biography of John Smith under the suggestive title *The Demerara Martyr, Memoirs of Rev. John Smith,* he provoked an angry reaction from colonial authorities who accused him of sowing subversion and "instigating racial and class hatred." In 1924, when David Chamberlin, the secretary of the London Missionary Society, published *Smith of Demerara, Martyr-Teacher of the Slaves* (London, 1923), a quite innocent biography, the book was received with the same suspicion. In a letter written in 1925 (and kept in the Archives of the London Missionary Society) a certain John Kendall from Northdene, Natal, excused himself for not distributing Chamberlin's book because there were many "bolsheviks" around, and he feared that the book could be used against the constituted authorities. The life of John Smith and the revolt in Demerara had been converted from memory into pure metaphor. But the historiography continued to center on Smith and to ignore the slaves. A primary concern with missionary activities drove Stiv Jakobson to devote a chapter to Smith in his book, *Am I Not a Man and a Brother? British Missions and the Abolition of the Slave Trade and Slavery in West Africa and the West Indies, 1786–1838* (Gleerup, Uppsala, 1972). And it was still Smith, not the slaves, who was at the center of a book by Cecil Northcott, *Slavery's Martyr: John Smith of Demerara and the Emancipation Movement* (London, 1976).

A renewed interest in slavery triggered by the black movement in the United States, and the process of decolonization in Africa, as well as the increasing number of Guyanese scholars committed to recovering their past, brought the slaves to the forefront. Four different kinds of interpretations emerged: the first, although recognizing the importance of a variety of factors, stressed the impact of the "bourgeois revolution" on slaves (Eugene D. Genovese, *From Rebellion to Revolution: Afro-American Slave Revolts in the Making of the New World* [Baton Rouge, 1979]); the second attributed the rebellion to increased labor exploitation resulting from the introduction of sugar in the East Coast of Demerara (Michael Craton, *Testing the Chains: Resistance to Slavery in the British West Indies* [Ithaca, 1982]); the third emphasized the African roots of the rebellion (Monica Schuler, "Ethnic Slave Rebellions in the Caribbean and the Guianas," *Journal of Social History* 3 [Summer 1970]:374–85); the fourth, by contrast, attributed more importance to the process of creolization. It credited the missionaries with communicating a sense of moral worth and personal dignity, and it stressed the importance of slaves' acculturation, literacy, and growing knowledge of the outside world. It also emphasized the negative impact the transition from cotton to sugar had on slaves (Robert Moore, "Slave Rebellions in Guyana" [Mimeo., University of Guyana, 1971]).

Chapter 1.
Contradictory Worlds: Planters and Missionaries

1. Albert Goodwin, *The Friends of Liberty: The English Democratic Movement in the Age of the French Revolution* (Cambridge, Mass., 1979); J. D. Cookson, *The Friends of Peace: Anti-War Liberalism in England, 1793–1815* (Cambridge, Eng., 1982); Derek Jarrett, *England in the Age of Hogarth* (1st ed., 1974; rev. ed., New Haven, 1986); George Rudé, *Hanoverian London, 1714–1808* (London, 1971) and *Wilkes and Liberty: A Social Study of 1763 to 1774* (Oxford, 1962); S. MacCoby, *The English Radical Tradition, 1763–1914* (London, 1952) and *English Radicalism, 1786–1812: From Paine to Cobbett* (London, 1955); Malcolm I. Thomis and Peter Holt, *Threats of Revolution in Britain 1789–1848* (London, 1977); Patricia Hollis, ed., *Pressure from Without in Early Victorian England* (London, 1974); Carl B. Cone, *The English Jacobins: Reformers in Late-18th-Century England* (New York, 1968).

2. Philip Corrigan and Derek Sayer's *The Great Arch: English State Formation as Cultural Revolution* (Oxford, 1985) offers a brilliant synthesis of the changes taking place in British society.

3. For an extremely interesting literary analysis of the power of Paine's rhetoric, see James T. Boulton, *The Language of Politics in the Age of Wilkes and Burke* (London, 1963). For the impact of Paine in the United States, see Eric Foner, *Tom Paine and Revolutionary America* (New York, 1976).

4. It has been estimated that two-thirds to three-fourths of the working people in Britain had at least a minimal reading ability. See J. M. Golby and A. W. Purdue, *The Civilisation of the Crowd: Popular Culture in England 1750–1900* (London, 1984), 127.

5. Protestant dissenters were prohibited from holding public office. They were excluded from the direction of chartered companies such as the Bank of England or the East India Company. They were also excluded from office in some local

hospitals, almshouses, workhouses, etc. Albert Goodwin says that the two fundamental issues raised by the campaign against the Test and Corporation Acts were the "participation" of lay Protestant dissenters in the so-called offices of "trust" or "profit," and their condemnation of the continued use of the sacramental test as the means of this exclusion. The first issue led dissenters to claim political equality with Anglicans and in doing so to have recourse to neo-Lockean theories of natural rights. Goodwin, *The Friends of Liberty,* 66–97.

6. Goodwin, *The Friends of Liberty,* 264–65.

7. David Brion Davis, *The Problem of Slavery in the Age of Revolution, 1770–1823* (Ithaca, 1975); idem, *The Problem of Slavery in Western Culture* (Ithaca, 1966); C. Duncan Rice, *The Rise and Fall of Black Slavery* (New York, 1975), 221. See also Lowell Joseph Ragatz, *The Fall of the Planter Class in the British Caribbean, 1763–1833: A Study in Social and Economic History* (New York, 1928); Betty Fladeland, *Men and Brothers: Anglo-American Antislavery Cooperation* (Champaign-Urbana, Ill., 1972); Jack Gratus, *The Great White Lie: Slavery, Emancipation and Changing Racial Attitudes* (New York, 1973); David Eltis, *Economic Growth and the Ending of the Transatlantic Slave Trade* (New York, 1987); David Eltis and James Walvin, eds., *The Abolition of the Atlantic Slave Trade: Origins and Effects in Europe, Africa and the Americas* (Madison, Wisc., 1981); James Walvin, ed., *Slavery and British Society, 1776–1846* (Baton Rouge, 1982). For a review of the recent historiography on abolition, see Seymour Drescher, "The Historical Context of British Abolition," in David Richardson, ed., *Abolition and Its Aftermath: The Historical Context* (London, 1985), 4–24. See also Barbara L. Solow and Stanley L. Engerman, eds., *British Capitalism and Caribbean Slavery* (New York, 1987); David Turley, *The Culture of English Anti-Slavery, 1780–1860* (New York, 1991).

8. Ragatz, *The Fall of the Planter Class,* 260.

9. Robin Blackburn, *The Overthrow of Colonial Slavery, 1776–1848* (London, 1988), 137–41. See also D. H. Porter, *The Abolition of the Slave Trade in England, 1784–1807* (Hamden, Conn., 1970).

10. The appeal the antislavery crusade had for women is stressed by Edith F. Hurwitz in *Politics and Public Conscience: Slave Emancipation and the Abolitionist Movement in Britain* (London, 1973), 89; Seymour Drescher, *Capitalism and Antislavery: British Mobilization in Comparative Perspective* (London, 1986; New York, 1987), 78–79; Walvin, ed., *Slavery and British Society,* 61. The subject was very much on the minds of the colonists. In 1819, Georgetown's *Royal Gazette* was particularly attentive to what it called "Female Reformers." A series of references was climaxed on September 7, 1819, in an article headlined "Seditious Meeting." Citing a "London paper," and quoting the "Reformers" at length, the *Gazette* said that the "Female Reformers . . . have already begun to take rank with the Poissardes of the French Revolution. The first public exhibition of their shamelessness, and unwomanly depravity is thus recorded in a seditious publication called *The Manchester Observer,* very interesting document of the Blackburn Female Reform Society, stressing their determination of 'instilling into the minds of their offspring a deep-rooted abhorrence of tyranny, come in what shape it may, whether under the mark of Civil or Religious Government, and particularly of the present Borough

mongering and Jesuitical system which has brought the best artisans, manufacturers and labourers of this vast community in a state of wretchedness and misery, and driven them to the very verge of beggary and ruin, for by the gripping hand of the relentless tax gatherer, our aged parents who once enjoyed a comfortable subsistence, some of them are reduced to a state of pauperism, whilst others have been sent to an untimely grave.' " The women also "askt 'that the people may obtain Annual Parliaments, Universal Suffrage and Election by Ballot, which alone can save us from lingering misery and premature death. . . . Who will believe that to this wretched state we are reduced, while it is a notorious fact, that two thousand three hundred and forty four persons receive yearly 2,474,805 pounds for doing little or nothing.' "

11. For a review of the literature on the abolition of the slave trade, see Drescher, *Capitalism and Antislavery;* Roger Anstey and P. E. H. Hair, eds., *The Historical Debate on Abolition of the British Slave Trade in Liverpool—the African Slave Trade and Abolition: Essays to Illustrate Current Knowledge and Research* ([Liverpool], Historic Society of Lancashire and Cheshire, 1976); Eltis and Walvin, eds., *The Abolition of the Atlantic Slave Trade;* Eltis, *Economic Growth and the Ending of the Transatlantic Slave Trade.* For the support the Manchester working class gave to emancipation, see Seymour Drescher, "Cart Whip and Billy Roller, or Anti-Slavery and Reform: Symbolism in Industrializing Britain," *Journal of Social History* 15:1 (Fall 1981):7.

12. James Walvin, "British Radical Politics, 1787–1838," in Vera Rubin and Arthur Tuden, eds., *Comparative Perspectives on Slavery in New World Plantation Societies* (New York, 1977), 343–53. See also Walvin's "The Rise of British Popular Sentiment for Abolition, 1787–1832," in Roger Anstey, Christine Bolt, and Seymour Drescher, eds., *Anti-Slavery, Religion, and Reform* (Hamden, Conn., 1980), 152–53; Seymour Drescher, "Public Opinion and the Destruction of British Colonial Slavery," in Walvin, ed., *Slavery and British Society,* 22–48; Drescher, *Capitalism and Antislavery;* Eric J. Hobsbawm, *Labouring Men: Studies in the History of Labour* (New York, 1964); Howard Temperley, "Anti-Slavery as a Form of Cultural Imperialism," in Anstey, Bolt, and Drescher, eds., *Anti-Slavery, Religion, and Reform,* 335–50.

13. Charles Dickens, *Bleak House* (Cambridge, Mass., 1956), 299.

14. James Walvin, "The Rise of British Popular Sentiment for Abolition, 1787–1832," in Anstey, Bolt, and Drescher, eds., *Anti-Slavery, Religion, and Reform,* 152.

15. Ragatz, *The Fall of the Planter Class,* 384–87.

16. Patricia Hollis remarks: "The movement of parliamentary reform stretched back to Wilkes, Wyvill, and Major Cartwright . . . in the 1770s and to the artisans of the London and provincial corresponding societies who studied Paine in the 1790s." But the tightening of the government's grip on public and political "order" sent the movement underground. Not until the founding of the Hampden Clubs in 1812 did a reform movement of working- and middle-class radicals come out into the open. Working-class unrest culminated at Peterloo. Patricia Hollis, *Class and Conflict in Nineteenth-Century England, 1815–1850* (London, 1973), 89.

17. Rhys Isaac characterizes evangelicalism in Virginia "as a popular response to a mounting sense of social disorder." *The Transformation of Virginia, 1740–1790* (Chapel Hill, 1982), 168.

18. E. P. Thompson, *The Making of the English Working Class* (London, 1976), 386. See also Alan D. Gilbert, *Religion and Society in Industrial England: Church, Chapel, and Social Change, 1740–1914* (London, 1976).

19. While there were among the Methodists those "who wanted to make it an adjunct to the state and subservient to the Established Church, there were others who desired to abide by its original purpose to reform the Church and the nation. They wanted to transform the framework of society." The New Connexion asserted the equality of all in Christ and instead of duties emphasized people's rights. The Primitive Methodists were also dissenters. In rural areas, the formation of Methodist societies often "meant the organization of a body of people independent of the squire, parson and landlord." These distinctions have sometimes been lost in the debate over the impact Methodists had on the working classes. Robert F. Wearmouth, *Methodism and the Working-Class Movements of England, 1800–1850* (London, 1937), 206, 263.

20. Deborah Valenze, "Pilgrims and Progress in Nineteenth-Century England," in Raphael Samuel and Gareth Stedman Jones, eds., *Culture, Ideology and Politics: Essays for Eric Hobsbawm* (London, 1982), 114.

21. Eric Hobsbawm says that the "effectiveness of official Wesleyan conservativism has often been exaggerated." This he attributes to a misunderstanding of the reasons that turned workers toward various sects. Hobsbawm stresses that the various seceding Methodist groups did not sympathize politically with the Wesleyans, and that even among the Wesleyans the rank and file were less conservative than their leaders. There was no sign of conformity in districts they controlled. *Labouring Men,* 23–33.

22. Vittorio Lanternari, *The Religions of the Oppressed: A Study of Modern Messianic Cults* (Lisa Sergio, trans., New York, 1963).

23. For a discussion of the impact of Methodism and other sects, see E. J. Hobsbawm, "Methodism and the Threat of Revolution in Britain," *History Today* 7 (February 1957):115–24.

24. The convergence of evangelicals and radicals in America in the 1770s has been pointed out by Eric Foner. "Both evangelicals and rationalists," he says, "spoke the language of millennialism and of the primacy of the individual conscience, and both came to envision an internal transformation in American society as a desirable, even necessary counterpart of separation from Britain." *Tom Paine and Revolutionary America,* 117.

25. Edith Hurwitz says: "There were 7,116 non-Anglican places of worship licensed between 1688 and 1770, while between 1771 and 1830 there were over 32,000. . . . Methodists of all connections experienced the greatest growth. Baptists and Independents were next." *Politics and the Public Conscience,* 81. For a later period see Donald M. Lewis, *Lighten Their Darkness: The Evangelical Mission to Working-Class London, 1828–1860* (New York, 1986).

26. Deborah Valenze sees popular religion mostly as a cry against modernity,

a result of the product of the encounter of village culture with modern industrial capitalism. We can also see it as a weapon of the oppressed against their oppressors in times of repression. This would explain the success of liberation theology in Latin America since the 1960s. For more details on popular religion in England at the beginning of the nineteenth century, see Valenze, "Pilgrims and Progress in Nineteenth-Century England," 113–25; John Walsh, "Methodism and the Mob in the Eighteenth Century," in G. J. Cumming and Derek Baker, eds., *Popular Belief and Practice* (Cambridge, Eng., 1972), 213–27; James Obelkevich, *Religion and Rural Society* (Oxford, 1976); Eric J. Hobsbawm, *Primitive Rebels* (New York, 1959); Hobsbawm, "Methodism and the Threat of Revolution in Britain," 115–24, and *Labouring Men*.

27. Thompson, *The Making of the English Working Class*, 437.

28. This newspaper began publication in Georgetown, Demerara, in 1803 as the *Essequebo and Demerary Gazette,* under the joint ownership of two men—one most likely Dutch and the other English—Nicholas Volkertz and E. J. Henery. In 1806, Volkertz sold out to Henery, who adopted the somewhat grander and certainly more British title, *Essequebo and Demerary Royal Gazette.* In 1814, the paper was bought by William Baker, who renamed it the *Royal Gazette, Demerary and Essequibo.* Over the years, the paper spelled colony names differently: "Demerary" and "Demerara"; "Essequibo" and "Essequebo." For the sake of clarity and simplicity, and since no confusion is possible, the newspaper will be referred to throughout this book as the *Royal Gazette.* See note 94 below, and Mona Telesforo, "The Historical Development of Newspapers in Guyana, 1793–1975" (B.A. thesis, University of Guyana, 1976).

29. LMS IC, Berbice, Wray's letter of July 31, 1824, cites the article published in the newspaper in 1808.

30. The Toleration Act licensed many Protestant sects while still denying individual non-conformists many civil rights. Dissenters were prevented from acting as legal trustees, guardians, executors, and they were banned from office-holding. It is thus not surprising that dissenters were involved in struggles for civic equality and human rights, and that they defended toleration for all modes of thinking. See Blackburn, *Overthrow of Colonial Slavery*, 73, 136.

31. Wearmouth, *Methodism and the Working-Class Movement of England.*

32. In 1811, Lord Sidmouth introduced a bill amending the Toleration Act, which had granted freedom to dissenters. His purpose was to eliminate what he perceived as "abuses" of the Toleration Act, such as black-smiths, chimney-sweepers, pig-drovers, peddlers, cobblers, and the like, becoming itinerant preachers. He expressed his concern that soon England would have a nominal Established Church, but a sectarian people. His bill triggered a tremendous opposition on the part of dissenters. A shower of petitions fell upon Parliament and the bill was defeated. See *Evangelical Magazine* 19 (June 1811), 237–47.

33. Ragatz, *The Fall of the Planter Class*, 281–85.

34. The African Association (or Institution) was founded in 1806 under the auspices of evangelical groups with the purpose of watching the African coast to prevent slave trade. Asa Briggs, *The Age of Improvement, 1783–1867*

(London, 1959), 174; Fladeland, *Men and Brothers,* 86; Ragatz, *The Fall of the Planter Class.*

35. Mary Turner says that in Jamaica planters were "divided on the issue of mission work." Mission patrons gave priority to meeting the imperial government's requirements on religious toleration; their opponents were more concerned with the missionaries' connections with the antislavery movement. In 1802, the Jamaican assembly passed a law to curtail mission work. It prevented preaching by persons not "duly qualified by law." The act was disallowed by the imperial government. In 1807, the assembly passed a new consolidated slave code which made missionary work illegal and gave to the Anglicans the exclusive right to preach to slaves. Methodists and all other sectarians were prohibited from preaching. Once again the British government intervened in favor of the missionaries. It was only in the second decade of the nineteenth century that conditions for mission work became more favorable in Jamaica. Under growing pressure from the British government some planters came to recognize that it was better to cooperate with imperial policies. But even then missionaries had to walk a tightrope. Mary Turner, *Slaves and Missionaries: The Disintegration of Jamaican Slave Society, 1787–1834* (Urbana, Ill., 1982), 15–18, 26.

36. Eric Williams, *Documents on British West Indian History, 1807–1833* (Trinidad, 1952), 226.

37. A few years later the Wesleyan Mission House wrote to Bathurst an appeal on the subject of restrictions laid upon the labors of missionaries in Trinidad by the local government. The letter stressed that the missionary had been forced to withdraw and the chapel had been shut. Williams, *Documents,* 244.

38. Rhys Isaac, commenting on the activities of Baptists in Virginia in the eighteenth century, stresses these "democratic" tendencies, born of shared emotions. He remarks that "these people, who called one another brothers and sisters, believed that the only authority in their church was the meeting together of those in fellowship. They conducted their affairs on a footing of equality so [*sic*] different from the explicit preoccupation with rank and precedence that characterized the world from which they had been called." Isaac, *The Transformation of Virginia,* 165.

39. Haiti was sometimes referred to as Saint Domingue, sometimes as St. Domingo, and rarely as Haiti. To avoid confusion I have adopted the term Haiti. For the impact of the Haitian rebellion in England, see David Geggus, "British Opinion and the Emergence of Haiti, 1791–1805," in Walvin, ed., *Slavery and British Society,* 123–50. See also Geggus, *Slavery, War and Revolution;* and Blackburn, *The Overthrow of Colonial Slavery.*

40. Joseph Marryat, *Thoughts on the Abolition of the Slave Trade and Civilization in Africa . . .* (London, 1816).

41. An order in council of August 1805 and an act of Parliament a year later prohibited the importation of slaves into Demerara for the purpose of opening new estates and limited annual entries to 3 percent of the existing number of slaves. Ragatz, *The Fall of the Planter Class,* 278.

42. C. Silvester Horne, *The Story of the LMS* (London, 1904), 12–13.

43. *Royal Gazette,* November 20, 1813.

44. Thompson, *The Making of the English Working Class,* 393. Barbara Hammond, in *Town Labourer,* suggests that evangelicals had "chloroformed the people" against any revolutionary tendency. James Hammond and Barbara Hammond, *Town and Labourer, 1760–1860: The New Civilization* (London, 1920). This idea had long ago been espoused by Elie Halévy (1905). For an appraisal of this debate, see Elissa S. Itzkin, "The Halévy Thesis—a Working Hypothesis? English Revivalism: Antidote for Revolution and Radicalism, 1789–1815," *Church History* 44 (1975):47–56. In *Labouring Men,* Hobsbawm argued that popular evangelicalism and popular radicalism grew up at the same time and in the same places, as responses to rapid social change.

45. Mary Turner in her study of Jamaica remarks that the missionaries appealed "to the intellectual and moral qualities the slave system in principle denied its chattels. They provided a new philosophical and organization framework. . . . They also unwittingly encouraged the slaves to extend the customary rights they enjoyed as independent producers and traders." Turner, *Slaves and Missionaries,* 199.

46. *Demerara. Further Papers, Copy of Documentary Evidence Produced Before a General Court Martial. . . .* (House of Commons, 1824), 27–29. See also Public Record Office, Colonial Office, 111/42, cited hereafter as P.R.O. C.O.

47. In the instructions to Richard Elliot in February 1808, the LMS warned him that although the slaves should be the object of his commiseration, it was out of his power to relieve them from their "servile" condition, "nor would it be proper, but extremely wrong to insinuate anything which might render them discontented with a state of servitude, or lead them to any measures injurious to the interest of the masters." He was not to excite the sort of opposition that might damage any future missions. "These poor creatures are slaves in a much worse sense—they are the slaves of ignorance, of sin, and of Satan. It is to rescue them from their miserable condition by the gospel of Christ, that you are now going." P.R.O. C.O. 111/43.

48. *Royal Gazette,* June 6, 1822.

49. In the nineteenth century, when the new ideas of the rights of man were propagated in different parts of the world, an increasing number of people, even in Catholic countries like Cuba or Brazil, started questioning the compatibility between Christianity and slavery. See, for example, Gwendolyn Midlo Hall, *Social Control in Slave Plantation Societies: A Comparison of St. Domingue and Cuba* (Baltimore, 1971).

50. There was nothing compelling, however, about these notions. As Mary Turner has shown for Jamaica, Wesleyan missionaries continued to defend the compatibility between slavery and Christianity, sometimes going as far as to oppose the most progressive guidelines coming from England. For more detail, see Turner, *Slaves and Missionaries.*

51. Several scholars (Roger Anstey in particular) have noticed a correlation between evangelicalism and abolitionism in England. They attributed it to theological trends. But as Donald Mathews and others have shown, most evangelicals in the American South did not become abolitionists. In fact, the more evangelicals became slaveowners, or slaveowners became evangelicals, the more evangelicalism lost the antislavery drive that had been characteristic of the early years. Simultaneously what

had once been a shared religion that brought together whites and blacks soon became, in the eyes of the whites, a mission to catechize blacks. Meanwhile, blacks created their own churches, independent from whites. This seems to suggest that the correlation between evangelicalism and abolitionism in England requires a different explanation. Most likely, as Donald Mathews has argued, it is the combination of artisanal radicalism with evangelicalism that gives impetus to abolitionism in England. For a discussion of these issues see my comments on Roger Anstey's "Slavery and the Protestant Ethic," in Michael Craton, ed., *Roots and Branches: Current Directions in Slave Studies* (Toronto, 1979), 173–77. See also Donald G. Mathews, "Religion and Slavery: The Case of the American South," in Anstey, Bolt, and Drescher, eds., *Anti-Slavery, Religion and Reform,* 207–32. For more details, see note 52.

52. For the connection between evangelicalism and abolitionism, see Roger Anstey, *The Atlantic Slave Trade and British Abolition, 1760–1810* (London, 1975) and "Slavery and the Protestant Ethic," in Craton, ed., *Roots and Branches.* See also my comments, in the same volume (pp. 173–79), on Anstey's "Slavery and the Protestant Ethic." For the debate on abolitionism, see David Brion Davis, "Reflection on Abolitionism and Ideological Hegemony," *American Historical Review* 92 (October 1987): 797–812; John Ashworth, "The Relationship Between Capitalism and Humanitarianism," ibid., 813–28; Thomas Haskell, "Convention and Hegemonic Interest in the Debate over Antislavery: A Reply to Davis and Ashworth," ibid., 829–78. For further detail, see Seymour Drescher, *Econocide: British Slavery in the Era of Abolition* (Pittsburgh, 1977) and *Capitalism and Anti-Slavery: British Mobilization in Comparative Perspective* (New York, 1987); Solow and Engerman, eds., *British Capitalism and Caribbean Slavery;* David Eltis, *Economic Growth and the Ending of the Transatlantic Slave Trade* (New York, 1987); David Brion Davis, *Slavery and Human Progress* (New York, 1982); Eltis and Walvin, eds., *The Abolition of the Atlantic Slave Trade;* Stanley Engerman, "The Slave Trade and British Capital Formation in the Eighteenth Century: A Comment on the Williams Thesis," *Business History Review* 46 (Winter 1972):430–43; Howard Temperley, "Capitalism, Slavery, and Ideology," *Past and Present* 75 (1977):94–118; Drescher, "Cart Whip and Billy Roller," 3–24; Anstey, Bolt, and Drescher, eds., *Anti-Slavery, Religion, and Reform;* and Betty Fladeland, *Abolitionists and Working-Class Problems in the Age of Industrialization* (Baton Rouge, 1984).

53. Roger Anstey stresses that antislavery owed much to some theological trends within Protestantism in the eighteenth and nineteenth centuries: Arminianism, redemption, sanctification, post-millenarianism, and denominationalism. The first because it required that the Gospel be preached to all men, since God's saving grace was available to all, the second because redemption was redefined to apply to the present, "post-millennialism" because of their prophetic belief in human perfectibility. The call was linked to benevolence, compassion, perfectionism, and "millenarian expectations." Anstey believes that together such trends explain the evangelicals' participation in the antislavery campaign. Anstey, "Slavery and the Protestant Ethic," in Craton, ed., *Roots and Branches.* For a development of Anstey's thesis, see also his *The Atlantic Slave Trade and British Abolition;* Hurvitz,

Politics and the Public Conscience, also examines the connections between evangelicalism and abolitionism.

54. Francis Cox, *A History of the Baptist Missionary Society 1792–1842* (2 vols., London, 1842); Horne, *The Story of the LMS;* William Ellis, *The History of the London Missionary Society* (London, 1844); George G. Findlay and William W. Holdsworth, *The History of the Wesleyan Missionary Society* (5 vols., London, 1921–24); James Hutton, *History of the Moravian Missions* (London, 1922); Stiv Jakobsen, *Am I Not a Man and a Brother?* (Gleerup, Uppsala, 1972); John Owen, *The History of the Origin and First Ten Years of the British and Foreign Bible Society* (2 vols., London, 1816); Eugene Stock, *The History of the Church Missionary Society* (4 vols., London, 1899–1916); Augustus Thompson, *Moravian Missions* (London, 1883).

55. Horne, *The Story of the LMS,* 2.

56. Full political rights for all denominations were obtained only in 1828. Turner, *Slaves and Missionaries,* 69. For a view of the dissenters' struggle for the abolition of the Test and Corporation Acts, 1787–1790, see Albert Goodwin, *The Friends of Liberty: The English Democratic Movement in the Age of the French Revolution* (Cambridge, Mass., 1979), 65–97.

57. *Evangelical Magazine* 3 (October 1795):425.

58. Ibid. 4 (January 1796):35–39.

59. Ibid. 4 (January 1796):35–39.

60. Ibid. 12 (April 1804): 181.

61. Ibid. 1 ([July?] 1793):1.

62. Ibid. 3 (December 1795):509.

63. The LMS was not the only organization to recruit missionaries from the artisan ranks. Similar policy was adopted by the Baptists. See Duncan Rice, "The Missionary Context of the British Anti-Slavery Movement," in Walvin, *Slavery and British Society,* 155.

64. In August 1796 the vessel *Duff* carried thirty missionaries with the destination the South Seas. Only four were trained and ordained ministers, twenty-five were artisans, and one was a surgeon. Among the artisans were bricklayers, carpenters, tailors, weavers, a blacksmith, and a gunner of the Royal Artillery. Horne, *The Story of the LMS,* 23.

65. *Evangelical Magazine* 22 (January 1814):11.

66. In a letter to the editor of the *Evangelical Magazine* the author wrote: "It should likewise never be lost sight of by those who are interested in the subject, that there is a great diversity in the fields of Missionary labour, and while China, Hindostan, and other civilized nations, may require persons of superior talents and endowments, there are many other countries where men of humble capacities and acquirements may most usefully enjoy their zeal. Among Hottentotes, Negroes and a multitude of other rude tribes of mankind, they will find ample scope for their exertions." *Evangelical Magazine* 21 (November 1813):415.

67. For a correlation between nonconformity and antislavery petitions, see Gilbert, *Religion and Society in Industrial England.* For the support workers gave to abolitionism, see Betty Fladeland, "Our Cause Being One and the Same: Abolitionism and Chartism," in Walvin, ed., *Slavery and British Society,* 69–99. For a

different view see Patricia Hollis, "Anti-Slavery and British Working-Class Radicalism in the Years of Reform," in Anstey, Bolt, and Drescher, eds., *Anti-Slavery, Religion, and Reform,* 294–318. While Fladeland stresses the support given by workers to abolitionism, Hollis stresses workers' antipathy to the abolitionist movement. This contradiction may be solved if one considers the links between evangelicalism and abolitionism. Support for the abolitionist campaign may have come predominantly from artisans and workers who joined evangelical movements, although there might have been working-class radical groups that also denounced slavery. This interpretation is put forward by James Walvin in "The Propaganda of Anti-Slavery," in Walvin, ed., *Slavery and British Society,* 64–65. See also his "The Impact of Slavery on British Radical Politics: 1787–1838," in Rubin and Tuden, eds., *Comparative Perspectives on Slavery in New World Plantation Societies,* 343–55. A similar point has been made by Seymour Drescher in *Capitalism and Antislavery.*

68. *Evangelical Magazine* 13 (January 1805):25, (March 1805):123, (May 1805): 212, (June 1805):276.

69. For the connection between antislavery and evangelicalism, see, in addition to notes 51 and 52 above, David Brion Davis, *The Problem of Slavery in the Age of Revolution 1770–1823* (Ithaca, 1975); Drescher, *Capitalism and Antislavery,* chap. 6.

70. P.R.O. C.O. 111/28 has an 1819 "List of Dutch Proprietors of Plantations in Demerara, Essequebo and Berbice, Whose Properties are Mortgaged to British Citizens."

71. A list of plantations for the year 1810 shows that approximately one-third carry Dutch names. *Royal Gazette,* November 3, 1810.

72. Henry Bolingbroke, *A Voyage to Demerary, 1799–1806* (Vincent Roth, ed., Georgetown, 1941). Cited hereafter as Bolingbroke, *Voyage.*

73. In 1815 the colonists faced again the possibility of Demerara's changing hands. The *Royal Gazette* of April 15, 1815, commented with irony, "What a delightful life of uncertainty we again enjoy. At first, with commendable resignation we made up our minds to become Dutch in consequence of the restoration of these realms and all that there is—then again to continue in our present state of 'betweenity,' the scale somewhat preponderating in favour of being British!" Now the colonists were asking themselves what would come from discussions taking place in Europe after Napoleon's defeat. According to the new treaties, Demerara was to remain British. Yet there was a bitter tone when the editor of the *Gazette* commented: "At last they have condescended to inform us to whom we belong." *Royal Gazette,* April 15, 1815.

74. James Rodway, *History of British Guiana, from the Year 1668 to the Present Time* (3 vols., Georgetown, 1891–94), 2:164. See also Cecil Clementi, *A Constitutional History of British Guyana* (London, 1937); Henry G. Dalton, *The History of the British Guyana* (London, 1855); V. T. Daily, *A Short History of the Guyanese People* (New York, 1975); D. A. G. Waddel, *The West Indies and the Guianas* (Englewood Cliffs, N.J., 1967); Roy Arthur Glasgow, *Guyana: Race and Politics Among Africans and East Indians* (The Hague, 1970); E. S. Stoby, *British Guiana Centenary Year Book, 1831–1931* (Georgetown, 1931). A.

R. F. Webber, *Centenary History and Book of British Guiana* (Georgetown, 1931).

75. *Royal Gazette,* April 5, 1807.

76. Orders in Council issued in 1805 prevented any further imports of slaves from Africa into Guyana. Blackburn, *The Overthrow of Colonial Slavery,* 306.

77. Rubin and Tuden, eds., *Comparative Perspectives on Slavery in New World Plantation Societies,* 184.

78. *Royal Gazette,* March 22, 1806.

79. This process was selective. It reproduced the ambiguities and contradictions of the metropolitan ideological representations. The colonial elites found support in British conservative thought, while people like Smith found comfort in the new liberal ideology. But they all seemed to have pride in stressing their "Europeanness." They all in fact invented a "mother country" by selecting some of its forms of representation and rejecting others. Colonial elites, for example, rejected ideological trends that threatened the colonial social order. From their perspective the new "bourgeois" ideology acquired a particular transparence. A good example is the way colonists denounced the ideology of "free labor." At the same time the identification with the "mother country" was essential to the maintenance of colonial forms of domination.

80. *Royal Gazette,* September 7, 1819.

81. The *Royal Gazette,* July 21, 1821, Transcribes Marryat's remarks stressing that much of the misery the slaves endure in Demerara should be attributed to the Dutch laws still in use. See analogous opinion in P.R.O. C.O. 111/23; ibid., 112/12; also Bolingbroke, *Voyage.*

82. On May 8, 1821, the *Royal Gazette* published an article (typical of many such) "The Political Influence of England." Its point of departure was that there was no other nation in Europe in which the "principles of liberty" were so "well understood." And, the *Gazette* boasted, no other nation had such a large proportion of its people "qualified to speak and act with authority . . . at all times ready to take a reasonable, liberal, and practical view."

83. *Report of the Committee of the Society for the Mitigation and Gradual Abolition of Slavery* (London, 1824), 76.

84. William Blackstone in his commentaries on the laws of England, first published in 1765, wrote: "The idea and the practice of this civil and political liberty flourish in the highest degree in these Kingdoms, where it falls little short of perfection, and can only be lost or destroyed through the elements of its owner; the legislature, and of course the laws of England, being peculiarly adapted to the preservation of this inestimable blessing even in the meanest subject. . . . And this spirit of liberty is so implanted in our constitution, and rooted even in our very soil, that a slave or a negro, the moment he lands in England falls under the protection of the laws and becomes in one instant a freeman." Quoted in Blackburn, *The Overthrow of Colonial Slavery,* 81.

85. See Christopher Hill, *The Century of Revolution, 1603–1714* (London, 1980); Edward P. Thompson, *Whigs and Hunters* (London, 1975); Philip Corrigan and

Derek Sayer, *The Great Arch* (Oxford, 1985); Roy Porter, *English Society in the Eighteenth Century* (London, 1982); and Blackburn, *The Overthrow of Colonial Slavery;* Douglas Hay et al., *Albion's Fatal Tree: Crime and Society in Eighteenth Century England* (New York, 1975).

86. See, for example, Rough's and Johnstone's appeals in P.R.O. C.O. 111/43. See also London Missionary Society Archives, Odds: The Case of Sergeant Rough.

87. *Royal Gazette,* March 27, 1821. Transports and Mortgages on Plantation Good Hope, "without prejudice to a certain mortgage on said estate in favor of Sarah Barnwell, free coloured woman."

88. According to Higman, all districts of Georgetown had similar proportions of slaves, whites, and freemen. Barry W. Higman, *Slave Populations of the British Caribbean, 1807–1834* (Baltimore, 1984), 99.

89. LMS IC, Berbice, Wray's letter, November 4, 1824.

90. *Royal Gazette,* August 7, 1810.

91. Ibid., January 17, 1807.

92. Ibid., August 24, 1822.

93. Ibid., September 11, 1819.

94. The debate between the *Guyana Chronicle* and the *Royal Gazette* exhibited two different and contending notions about the press. On the one hand, the press depends on the patronage of parties or government and is supposed to serve "public interest." Censorship is accepted as necessary to maintain the political commentary within "tolerable" limits, as defined by ruling groups. On the other hand, it allegedly depends on "public" patronage (the market), and serves "individual" not party interests. It repudiates censorship and claims to be a "free press." In practice, of course, things were different. The *Guiana Chronicle,* which catered to the market, became the voice of wealthy planters and merchants, and a staunch enemy of the missionaries. See *Royal Gazette,* August 8 and 10, 1822.

95. In May 1799, the States General, at the instance of the planters of Guyana, resolved to adopt vigorous measures in support of the slave trade. Accordingly, they voted 250,000 guilders to the West India Company and enacted several regulations for encouraging the importation of slaves into their colonies. *Royal Gazette,* June 15, 1820.

96. R. E. G. Farley, "Aspects of Economic History of British Guyana 1781–1852," quoted by Alan H. Adamson, *Sugar Without Slaves: The Political Economy of British Guyana, 1838–1904* (New Haven, 1972).

97. In 1796, Liverpool imported 6,000 bales of cotton from Essequibo and Demerara. In 1804 it imported four times as much. Similar increases were also registered in London, Glasgow, and Bristol. Bolingbroke, *Voyage,* 139.

98. For more detail see Chapter 2 below. For a later period see two extraordinary books: Adamson, *Sugar Without Slaves,* and Walter Rodney, *A History of the Guyanese Working People, 1881–1905* (2nd printing, Baltimore, 1982).

99. LMS IC, Demerara, Wray's letter, October 9, 1812.

100. Years later, when he moved to Berbice, Wray complained that to build a chimney in Berbice would cost as much as to build a whole church in India. He had to pay five shillings a day for a mason who did only a third of the work a mason would do in England. He also complained that he could not find a pair of shoes for less than 18 to 20 shillings, and a maid cost him about two pounds a month. LMS IC, Berbice, Wray's letter, October 7, 1824.

101. Williams, *Documents*, 319–20.

102. Ibid.

103. Ibid., 335.

104. Similar complaints were heard throughout the Caribbean. Ragatz, *The Fall of the Planter Class*, 327.

105. Rodway, *History of British Guyana*, II, 196.

106. Estimates of the number of West Indians in the British Parliament in the 1820s vary from 39 to 56, while there were about thirty peers in the upper house between 1821 and 1833. Roger Anstey, "The Pattern of British Abolitionism in the Eighteenth and Nineteenth Centuries," in Anstey, Bolt, and Drescher, eds., *Anti-Slavery, Religion, and Reform*, 24. See also C. Duncan Rice, *The Rise and Fall of Black Slavery* (New York, 1975), 133. The best essay is Barry Higman, "The West India 'Interest' in Parliament, 1807–1833," *Historical Studies* 13 (October 1967): 1–19. See also G. P. Judd, *Members of Parliament, 1734–1832* (New Haven, 1955); Douglas Hall, *A Brief History of the West India Committee* (Caribbean History Pamphlets, 1971); Ragatz, *The Fall of the Planter Class*.

107. *Royal Gazette*, June 13, 1820.

108. Ibid., June 29, 1820.

109. Ibid., April 24, 1821.

110. Ibid., June 19, 1821.

111. Ragatz, *The Fall of the Planter Class*, 390–95.

112. *Royal Gazette*, March 2, 7, April 16, July 27, 1816. The bill presented by Wilberforce in 1815 was defeated, but colonial legislatures were "invited" to introduce registration mechanisms of their own. Rice, *The Rise and Fall of Black Slavery*, 249.

113. The West Indian group, led by Charles Ellis, Keith Douglas, and Joseph Marryat, confronted Ricardo and Wilberforce. And although the West Indian spokesmen won by a large margin, the debate irritated the colonists who found themselves once again the target of severe criticism. Ragatz, *The Fall of the Planter Class*, 364.

114. Alexander McDonnell, *Considerations on Negro Slavery, with Authentic Reports, Illustrative of the Actual Condition of the Negroes in Demerara* (2nd ed., London, 1825).

115. No one exemplifies this class better than John Gladstone. S. G. Checkland, *The Gladstones: A Family Biography, 1754–1851* (Cambridge, Eng., 1971). There was a profound difference between resident planters and planters like Gladstone who lived in Britain, and that difference only grew with time.

116. Among those who represented the Caribbean in the House of Commons between 1785 and 1830 were William Beckford, Bryan Edwards, Charles Ellis,

George Hibbert, Joseph Marryat, John Gladstone, Alexander Grant, William Young, and many others. Ragatz, *The Fall of the Planter Class,* 52–53. Barry Higman calls attention to internal tensions within the West Indian lobby itself, first, between planters and merchants, and then between representatives of the "old" and of the "new" colonies. At the time of the abolition of the slave trade, representatives of the old colonies were more inclined to support the motion to reduce the competitiveness of the new colonies. And later, when the amelioration laws were discussed, only James Blair, who had sizable investments in Demerara, opposed it. Besides, sixteen West Indian MPs also had interest in East India. See Barry W. Higman, "The West India 'Interest' in Parliament, 1807–1833." See also Eric Williams, *Capitalism and Slavery* (New York, 1966).

117. John Gladstone was a typical representative of this group. A Lowland Scot born in 1764, he centered his commercial activities in Liverpool. During the war with France he did very well. By 1797 he was doing considerable business as an insurer of ships bound for the Baltic, America, Africa, and the West Indies. By June 1799 he was worth no less than £40,000. He then moved into real estate. In 1800, after his first wife died, he married into a Highland family. During the war years he expanded and diversified his business. His ships were sent to the Baltic and to Russia to buy wheat, but he also traded in tropical products such as cotton, sugar, and coffee. In 1803 he and an associate advanced a £1,500 mortgage on the estate *Belmont* in Demerara. He also became a supplier of timber, salt herring, and other necessities, and became the agent for other plantations. Although he was a slaveowner, he supported the abolition of the slave trade. In 1807 he became a partner of the Liverpool *Courier.* Soon he was sending ships to Argentina, Brazil, and India. In 1809 he was elected chairman of the Liverpool West Indian Association. Gladstone expanded his business in Demerara and acquired a half-interest in the plantation *Success.* His wealth and family ties brought him important political connections. He became the friend of ministers and the confidant of prime ministers, and his political ties made him even wealthier. By 1815 he was worth about £200,000. In 1818 he was elected to the House of Commons and established a London base. In partnership with John Wilson and Charles Simson and others, he expanded his business in Demerara, and bought the other half of *Success,* converting it to sugar and doubling the number of slaves. He also acquired *Vredenhoop,* an estate with 430 slaves. By that time he had already earned in Britain a reputation as a philanthropist. A pious evangelical, he managed to make money from his many charities, constructing churches and schools for the poor. He then turned to the construction of canals and railroads. From 1821 to 1828, his fortune grew from £350,000 pounds to £500,000 and his annual income from £30,000 to £40,000 a year. Checkland, *The Gladstones.*

118. McDonnell, *Considerations on Negro Slavery,* 17, 26.

119. Ibid., 36.

120. Ibid., 60.

121. Ibid., 76.

122. Ibid., 235–46.

123. *Evangelical Magazine* 17 (February 1809):83–84.

Chapter 2.
Contradictory Worlds: Masters and Slaves

1. Adaptation of a remark attributed to a slave, quoted in *Great News from the Barbadoes, or A True and Faithful Account of the Grand Conspiracy of the Negroes Against the English and the Happy Discovery of the Same. With the Number of Those That Were Burned Alive, Beheaded, and Otherwise Executed for Their Horrid Crime. With a Short Description of That Plantation,* cited by Michael Craton, in *Testing the Chains: Resistance to Slavery in the British West Indies* (Ithaca, New York, and London, 1982), 109.

2. The first (?) edition of *A Voyage to Demarary* appeared in 1807. A second edition was published in 1809, a third in 1813. In 1941, an edition prepared by Vincent Roth with a Foreword by J. Graham Cruickshank was published in Georgetown. I have used both the 1807 and the 1941 editions; the notes refer to the latter.

3. Eugene D. Genovese and Elizabeth Fox-Genovese, *Fruits of Merchant Capital: Slavery and Bourgeois Property in the Rise and Expansion of Capitalism* (New York, 1983).

4. Bolingbroke, *Voyage,* 207.

5. Ibid., 23, 207.

6. It should be noticed that although British historians today say that by this time there was no longer an English "peasantry," Bolingbroke uses the expression. *Voyage,* 31. For a critique of the notion of an English peasantry, see E. J. Hobsbawm and George Rudé, *Captain Swing* (New York, 1968).

7. Bolingbroke, *Voyage,* 31.

8. This contrasting picture was as old as English colonization in America. See Edmund S. Morgan, *American Freedom, American Slavery* (New York, 1975). It appeared again, much later, in the conflicts between the British and the Dutch in South Africa. Leonard Thompson, *The Political Mythology of Apartheid* (New Haven, 1985). It became part of the history of Guyana. Henry G. Dalton, *The History of British Guiana* (2 vols., London, 1855), 2:325. The *Royal Gazette,* July 21, 1821, transcribes an article by Marryat attributing the misery of slaves in Demerara to Dutch laws. In a letter to the London Missionary Society, John Wray contrasted the ruthless behavior of the Dutch with the enlightened behavior of the British commissioner in Berbice. Out of convenience or naiveté, the myth was created and re-created whenever the British and the Dutch competed with each other for the control of colonial territories.

9. Bolingbroke, *Voyage,* 146. Among the planters who will play salient roles in this book, Bolingbroke's distinctions between Dutch and British habits and attitudes did not hold. The man who brought the first missionaries to Demerara, Hermanus Post, had made a fortune working side by side with his slaves in his early years. As for brutality, if anything, planters with Dutch names like Hermanus Post and Henry Van Cooten would prove less brutal than men like Alexander Simpson or Michael McTurk, who played key roles in the suppression of the rebellion.

10. Walter E. Roth, ed., *The Story of the Slave Rebellion in Berbice, 1762.* Translated from J. J. Hartsinck's *Beschryving van Guiana* . . . (Amsterdam, 1770), in *Journal of the British Guiana Museum and Zoo,* Nos. 21–27 (December 1958–September 1960); Robert Moore, "Slave Rebellions in Guyana" (Mimeos, University of Guyana, 1971).

11. Analogous contrasting pictures are also found in books by travelers who visited other plantation societies in the nineteenth century. In Brazil the same phenomenon was noticed. Historians stressed differences between plantation owners who lived in the first half of the century and those who became plantation owners in a later period. This difference was sometimes characterized as an opposition between different regions: coffee planters of the Paraiba Valley were contrasted to coffee planters of the West. In recent years, however, historians have repudiated such distinctions. They have stressed that planters had always been interested in profit—which is true—and seem to infer from this that planters were all alike and related to slaves in the same way, which of course is not true. Such historians minimize the complexity of master-slave relations. They seem to forget that the impact the system of production has on slave life is necessarily mediated by different institutions and ideologies, and that the system of production also changes, depending on the degree of technological development.

12. Of course there had always been rebellions. But it seems that the intensity of day-to-day resistance, sabotage, and insubordination was greater with the passage of time. Increasing technological complexity and growing rates of capital investment made the plantations much more vulnerable to the slaves' rebelliousness, and this vulnerability might have heightened the whites' awareness of the threat.

13. When W. S. Austin, who had been a minister of the Anglican Church in Demerara at the time of the rebellion, testified before the House of Commons in 1832, he said that the Dutch were more severe than the English, but that the workload among the Dutch was not as heavy. Select Committee on the Extinction of Slavery Throughout the British Dominion, with the Minutes of Evidence, Appendix and Index. House of Commons, August 1832, P.R.O. C.O. ZHCI/1039.

14. LMS IC, Demerara, Van Gravesande's letter, 1811.

15. One of the weakness of the dependency theory has been to neglect the fact that the impact the center has on the periphery depends on political, economic, and social structures as well as on the intensity of class struggle that takes place both on the periphery and at the center. Similar criticism can be made of "world systems" approaches. For one such critique, see Steve Stern, "Feudalism, Capitalism, and the World-System in the Perspective of Latin America and the Caribbean," *American Historical Review* 4 (October 1988):829–72; reply by Immanuel Wallerstein, "Comments on Stern's Critical Tests," ibid., 873–86; and response by Stern, "Ever More Solitary," ibid., 886–97. For an example of a successful synthesis that manages to bring together world and local trends, and human agency as well, in the study of a slave society, see Dale W. Tomich, *Slavery in the Circuit of Sugar: Martinique and the World Economy, 1830–1848* (Baltimore, 1990). For a similar attempt to achieve a creative synthesis of these different approaches in a study of a post-

emancipation society, see Michel-Rolph Trouillot, *Peasants and Capital: Dominica in the World Economy* (Baltimore, 1988).

16. In fact, when emancipation came it did not fulfill the slaves' hopes, and although it did bring bankruptcy for some masters, others like John Gladstone used the money received as compensation to expand their plantations. Two excellent books portray what happened after emancipation in Guyana: Alan H. Adamson, *Sugar Without Slaves: The Political Economy of British Guiana, 1838–1904* (New Haven, 1972), and Walter Rodney, *A History of the Guyanese Working People, 1881–1905* (Baltimore, 1981).

17. A fully developed irrigation and drainage system requires some 55 miles of waterway for each square mile of cultivation. See Clive Y. Thomas, "Plantations, Peasants and State: A Story of the Mode of Sugar Production in Guyana" (Center for Afro-American Studies, University of California at Los Angeles, 1984). A detailed description is given in *The Overseer's Manual, or, A Guide to the Cane Field and the Sugar Factory for the Use of the Young Planters, Revised and Enlarged* (1st ed., Demerara, 1882; 3rd ed., 1887).

18. Walter E. Roth, ed. and trans., *Richard Schomburgk's Travels in British Guiana, 1840–1844* (2 vols., Georgetown, 1922), 1:55.

19. Although the density of the slave population for the whole of Guyana was low because the constituent colonies of Berbice, Demerara, and Essequibo covered a large area, most of which was unoccupied, the density on the East Coast was extremely high. In the area between Georgetown and Mahaica Creek, the rebels of 1823 were able to rally more than 12,000 slaves. The density was probably as high as that of Barbados, about 500 slaves per square mile. See Barry W. Higman, *Slave Populations of the British Caribbean, 1807–1834* (Baltimore, 1984), 85.

20. Schomburgk wrote that in a rich soil and with good attention and care, one planting of cane could supply eighteen crops before new cuttings had to planted. He calculated that in the 1840s the annual yield of an acre was 2.5 tons of sugar, 250 gallons of syrup, and 100 gallons of high-proof rum. On the larger estates, one-sixth of the whole area under cultivation was newly planted every year, and the main crop gathered in January, February, March, or during October, November, December, and January. *Schomburgk's Travels,* 1:63.

21. The Rule enacted in 1772 was renewed and amplified in 1776 and again in 1784. It is a copy of the 1784 version that is in P.R.O. C.O. 111/43.

22. Managers and masters knew that slaves used satirical and even revolutionary songs to challenge or terrorize them. Barbara Bush reports that "following the example of Saint Domingue, for instance, women on a Trinidadian plantation intimidated their master by singing an old revolutionary song. As they walked along a path balancing plantation baskets on their heads they rattled chac-chac pods and danced in rhythm to this chorus: *Vin c'est sang beque* (Wine is white blood), *San Domingo, Nous va boire sang beque* (We shall drink white blood), *San Domingo.*" Barbara Bush, "Towards Emancipation: Slave Women and Resistance to Coercive Labour Regimes in the British West Indian Colonies, 1790–1838," in David Richardson, ed., *Abolition and Its Aftermaths: The Historical Context, 1790–1816* (London, 1985), 42.

23. When Bolingbroke was in Demerara, the law permitted only thirty-nine lashes at one time. This seems to indicate that the punishment had become more severe since the Rule had been enacted. Bolingbroke, *Voyage,* 39.

24. There was no explicit reference to slaves, but we know from other sources that it had been the practice in the earlier period not to give work to slaves on Sundays.

25. For a brief history of this office, see P.R.O. C.O. 116/155, Appendix 1. For proceedings of the fiscal's office see "Copies of the Record of the Proceedings of the Fiscals of Demerara and Berbice in Their Capacity of Guardians and Protectors of Slaves, and Their Decision in All Cases of Complaints of Masters and Slaves, with Explanation and Documents, Presented to Parliament by His Majesty's Command" (London, n.d.), P.R.O. C.O. 116/156 and 116/138.

26. Apparently the Spanish had adapted it from the Romans. The Demerara fiscal had functions similar to those of the Protector of Indians in the Spanish colonies. An analogous institution existed in the Dutch African colony of the Cape. See Robert Ross, *Cape of Torments: Slavery and Resistance in South Africa* (Boston and London, 1983).

27. Bolingbroke characterized the fiscal as "the chief magistrate, public accuser, and attorney general, to prosecute in all cases for the sovereign." Besides a stipulated salary, the fiscal received a portion of all fines. According to Bolingbroke, "this appointment, exclusive of perquisites, is estimated at three thousand pounds yearly." Bolingbroke, *Voyage,* 52.

28. After 1824 the fiscal's functions in Demerara merged with those of the protector of slaves created by the British government as part of its scheme to ameliorate the slaves' conditions of living and "prepare" them for emancipation.

29. Sometimes, however, even in the earlier period, slaves who appealed to the fiscal found redress. Bolingbroke reported a case of the conviction of a planter for ill-treating his slaves. Apparently slaves had been left without provision for a week or ten days. The slaves sent a deputation to the fiscal and an extraordinary court was called. Charges were made by the slaves, and supported by witnesses. The court declared the proprietor an "improper person" to manage his affairs and appointed curators for his estates, imposing a "severe penalty." A similar case was rreported in Essequibo, where a member of the Court of Justice was fined 15,000 guilders. Bolingbroke, *Voyage,* 230.

30. In 1824 there were only 3,500 whites (half living in Georgetown) and 4,000 free blacks in a total population of 82,000. Between 1811 and 1824, the population of free blacks almost doubled (although most of this increase was of children), while the slave population had declined and the white population had grown by only about a third. But, with few exceptions, the acquisitive power of the free blacks was low.

31. On absentee owners, see the classic essay by Douglas Hall, "Absentee-Proprietorship in the British West Indies, to About 1850," originally published in *Jamaican Historical Review* 4 (1964): 15–34, and reprinted in Lombros Comitas and David Lowenthal, eds., *Slaves, Free Men, Citizens: West Indian Perspectives* (New York, 1973), 106–35. The article includes a lengthy bibliography on the subject.

Hall argues that absentee ownership began with the first English colonization. Another source of absenteeism emerged with the increasing profitability of sugar production. A third source was the inheritance of West Indian property by people resident in Britain. A fourth source was bankruptcy, with property confiscated by creditors in Britain.

32. Roger Anstey, *The Atlantic Slave Trade and British Abolition, 1760–1810* (London, 1975), 375–76; Seymour Drescher, *Econocide: British Slavery in the Era of Abolition* (Pittsburgh, 1979), 78, 95. Higman gives even more striking figures. From 1797 to 1805, some 40,607 slaves were imported into Demerara. *Slave Populations,* 428–29. It is interesting to notice that the tables provided by Higman show that the two Caribbean colonies that imported more slaves during this period were Jamaica and Demerara and Essequibo, precisely those colonies where the two largest slave rebellions occurred. However, in Barbados, where there was a rebellion in 1816, the influx of slaves from Africa was not very significant during the same period. This means that a simple correlation between the presence of Africans and rebellions cannot be established.

33. African slaves also came from other Caribbean colonies. The *Royal Gazette,* October 3, 1807, for example, had an interesting advertisement: "The subscribers inform their Friends who Commissioned them to Purchase Negroes in Barbadoes that they have received by the ship . . . 200 very Prime Gold-Coast Slaves." There was a notice in the issue of July 18, 1807, about "Ebbos." As late as February 27, 1808, the newspaper was advertising "New Negroes from Barbadoes, Windward and Gold Coast," and on June 11, "100 seasoned Angola (having been in the colony for three months)" were advertised for sale.

34. The records are for the year 1819. They also show that while 12,867 slaves had been born in Africa, another 10,000 had been born in other Caribbean colonies and transported to Berbice. Higman, *Slave Populations,* 454–56.

35. George Pinckard landed in April 1796 and left Demerara in May 1797. In 1806 his book was published in London in three volumes. A second edition appeared in 1816, with additional chapters. Pinckard died in 1835. I have used a more recent edition: Vincent Roth, ed., *Letters from Guiana, Extracted from Notes on the West Indies and the Coast of Guiana by Dr. George Pinckard, 1796–97* (Georgetown, 1942), 331.

36. Plantation *Cuming's Lodge,* for example, had 209 slaves in 1813 and produced sugar, rum, and cotton.

37. See "List of Estates in Demerary and Essequibo with the Number of Slaves in Each and the Quantity of Produce Made During the Year 1813," *Royal Gazette,* April 8, 1815. Plantations *Vrees en Hoop* and *Unvlugt,* for example, had respectively 313 and 447 slaves and produced sugar, rum, coffee, and cotton. So did *Hague,* which at the time had 641 slaves, and *Good Hope, St. Christopher, Vergenoegen,* and *Blakenburg,* with, respectively, 210, 251, 299, and 402 slaves. Others like *Vive La Force* (216) and *Hermitage* (172) had only sugar, rum, and coffee. In the East Coast at this time most plantations had only cotton, though a few like *Le Resouvenir* (396 slaves), *Goed Verwagting* (276), *Plaisance* (179), *Beeter Hoop* (199), *Vryheid's Lust* (217), *Industry* (223), *Wittenburg* (114), and *Le Reduit* (144) all had coffee and

cotton. Clearly, arrangements varied, but gradually plantations did tend to shift to sugar. On those that did not, the number of slaves tended to diminish, while on those that shifted to sugar it tended to increase, despite the overall decline in the slave population.

38. This was calculated on the basis of fifty plantations, from a list published in the *Royal Gazette,* April 1, 1815.

39. Higman shows that from 1810 to 1820 a growing number of slaves was occupied in sugar in the United Colonies of Demerara and Essequibo. In 1810, 58 percent worked on sugar estates, 10 percent on coffee, and 20 percent on cotton, another 8 percent on urban activities, and the remaining slaves were involved in other types of agriculture, livestock, timber extraction, fishing, and the like. By 1820 the number of slaves working on sugar plantations had increased to 72 percent, while those working on coffee and cotton had declined to, respectively, 6 and 10 percent. Higman, *Slave Populations,* Table 3.8.

40. Higman finds that at the end of the eighteenth century Demerara, Essequibo, and Berbice were the leading British colonial producers of cotton and coffee. These crops reached their peak about 1810; thereafter they were increasingly eclipsed by sugar. Between 1810 and 1834, sugar output increased more rapidly than in any other colony, and production per slave more than tripled. Between 1810 and 1831, coffee production dropped from 19.2 to 1.4 million pounds in Demerara and from 2.3 million to 27,000 pounds in Essequibo. Cotton showed a similar pattern, falling from 5.8 million to 400,000 pounds in Demerara and from 1.3 million to 41,000 pounds in Essequibo. The change to sugar was more noticeable in Essequibo. On the Demerara coast, sugar, cotton, and coffee plantations all remained significant until emancipation. In the United Colonies of Demerara and Essequibo in 1813, about 33 percent of the slaves lived on sugar estates, 31 percent on cotton, and 22 percent on coffee plantations. Higman, *Slave Populations,* 63.

41. It is difficult to assess the change taking place on the East Coast because most figures are aggregated for the United Colony of Demerara and Essequibo or for Demerara as a whole but do not break down into regions. I managed to identify some plantations by using a variety of sources, including travelers, advertisements for plantation sales in the newspapers, missionaries' diaries, and tax lists published in the newspapers. The tax lists are particularly interesting because they list the plantations, the number of slaves, and production per plantation. See as an example those published in the *Royal Gazette,* April 8, 1815. Several lists in the reports of the protectors of slaves give figures for the numbers of slaves on particular plantations. P.R.O. C.O. 116/156. Various other sources include information about the size of plantations and their production. Tax lists (*Royal Gazette,* November 3, 1810, and April 8, 1815) and almanacs are the most important. See, for example, *Almanack and Local Guide of British Guiana Containing the Laws, Ordinances and Regulations of the Colony, the Civil and Military Lists, with a List of Estates from Corentyne to Pomeroon Rivers* (Demerara, 1832), 441. For a list of plantations on the East Coast, with number of slaves and an indication of crops produced in 1823, see Joshua Bryant, *Account of an Insurrection of the Negro Slaves in the Colony of Demerara* (Georgetown, 1824).

42. Copies of Reports from Protectors of Slaves: Particular Returns, P.R.O. C.O. 116/156.

43. S. G. Checkland, *The Gladstones: A Family Biography, 1764–1851* (Cambridge, Eng., 1971).

44. The signs of this process were already noticeable at the time of the rebellion. Plantations *John* and *Cove* were merged, as were *Mon Repos* and *Endraght,* and *Enterprise* and *Bachelor's Adventure.* Further Papers, 15. The returns of January 1830 showed that *Bachelor's Adventure, Elizabeth Hall,* and *Enterprise* had a total of 694 slaves. Copies of Reports from Protectors of Slaves. P.R.O. C.O. 116/156.

45. Compare data provided in *The Local Guide, Conducting to Whatever Is Worthy of Notice in the Colonies of Demerary and Essequebo for 1821* (Georgetown, 1821) with the information supplied by the *Almanack and Local Guide of British Guiana* for 1832. Both list estates and numbers of slaves. The *Local Guide* also indicates the crops produced by each plantation, plus an account of sugar, rum, cotton, coffee, and molasses shipped from Demerara and Essequibo every year since 1808. It indicates that the number of vessels doubled during the period. See also Noel Deer, *History of Sugar* (2 vols., London, 1949–50), 1:193–201.

46. At that time there were 750,000 slaves in British colonies producing 4,600,000 cwt., an average of 6 cwt. each. In Saint Vincent, production per slave reached 11 cwt. and in Trinidad, 13. In Jamaica the production was 6 cwt. per slave. Minutes of Evidence Before Select Committee on the State of the West India Colonies, P.R.O. C.O. ZMCI/1039.

47. About Peter Rose, see Cecilia McAlmont, "Peter Rose: The Years Before 1835," *History Gazette* 19 (April 1990):2–9.

48. The Select Committee on the Extinction of Slavery Throughout the British Dominions with the Minutes of Evidence. . . . House of Commons, August 1832, P.R.O. C.O. ZHCI/1039.

49. In 1808–09, at the time John Wray had arrived in the colony, the works and machinery of a sugar estate cost about £10,000. Twenty years later the cost had doubled.

50. Select Committee on the Extinction of Slavery, P.R.O. C.O. ZHCI/1039.

51. Most of the time there were only two gangs, and women belonged to the second gang.

52. Peter Rose's testimony is confirmed by William Henery, who was a proprietor in Berbice, the owner of three estates, two in sugar and one in coffee, with a total of 950 slaves. He had resided twenty years in the colony before he moved to Liverpool. He testified that only a third of the slave population was "effective for labour"; the others were either too young or too old to be useful. Henery also claimed that where the slave trade had continued, as in Suriname, the slave population was on the whole more productive, and slaves were cheaper. In 1830, he said, he had paid an average of £110 for slaves he could have bought in Suriname for £40. Select Committee on the Extinction of Slavery, P.R.O. C.O. ZHCI/1039, p. 94.

53. It is possible that in the long run, with increasing creolization and, conse-

quently, a more even balance of males and females, not only would the number of women grow but also the number of children.

54. There are several discrepancies in the figures given in different documents because some include Essequibo and others do not. The protectors of slaves' records estimated a total slave population of 62,092 in 1828, and 59,492 in 1830. It is important to notice, however, that the number of slaves living on estates did not change much from 1817 to 1823, even though production increased.

55. By the end of this decade, however, slave prices started going down. This seems to indicate that, facing growing rebelliousness and fearing that emancipation would soon come, planters had become less willing to invest in slaves.

56. Several dispatches from the Colonial Office indicate that the British government tried to restrain this trade, but without much success. See P.R.O. C.O. 112/5. A letter from the Colonial Office dated June 1823, for example, noted that "His Majesty" had authorized John Henry and James (surname illegible) to remove 389 slaves from the Bahamas to Demerara. Another letter, written March 18, 1823, reports on a dispatch from the governor of Demerara dated the first of January respecting the proposed importation of slaves into the colony by Henry Curtis Pollard, who wanted to bring slaves from Barbados. P.R.O. C.O. 111/43.

57. Return of Slaves Imported Under License Between 14 January 1808 and 15 September 1821. P.R.O. C.O. 111/37.

58. Correspondence of the Controller of the Customs of Demerary, relating to an illicit importation of "negroes" from Martinique, many of whom were "free negroes." P.R.O. C.O. 111/43.

59. Mortality among whites was even higher than among blacks. If the death returns are reliable, 1,098 whites died from 1817 to 1821, while only 1,306 deaths occurred in the much larger slave population. It is possible that the figures for slave deaths were much higher than the returns indicated, but were not accurately reported. Still, even allowing for a very large margin for error, the difference is still astonishing.

60. Curiously enough, when disease is correlated to sex and occupation it becomes obvious that people working in the fields were less likely to have tuberculosis and respiratory diseases than those listed as "domestics." Those working in the fields were more likely to be affected by diarrhea. Higman, *Slave Populations,* 678.

61. Richard B. Sheridan, *Doctors and Slaves: A Medical and Demographic History of Slavery in the British West Indies, 1680–1834* (Cambridge, Eng., 1985) and "The Crisis of Slave Subsistence in the British West Indies During and After the American Revolution," *William and Mary Quarterly,* 3d series, 33 (1976):615–41; Kenneth F. Kiple and Virginia H. Kiple, *Another Dimension to the Black Diaspora: Diet, Disease and Racism* (Cambridge, Mass., 1982), "Deficiency Diseases in the Caribbean," *Journal of Interdisciplinary History* 11:2 (Fall 1980):197–215, and "Slave Child Mortality: Some Nutritional Answers to a Perennial Puzzle," *Journal of Social History* 10 (1977):284–309; Richard Sheridan, "Mortality and the Medical Treatment of Slaves in the British West Indies," in Stanley L. Engerman and Eugene D. Genovese, eds., *Race and Slavery in the Western Hemisphere* (Princeton, 1974), 285–310; Higman, *Slave Populations,* 260–378; Robert Dirks "Resource

Fluctuations and Competitive Transformations in West Indian Slave Societies," in Charles E. Laughlin, Jr., and Ivan Brady, eds., *Extinction and Survival in Human Populations* (New York, 1978), 122–80.

62. This practice was also common in the United States. See George Rawick, *From Sundown to Sunup: The Making of the Black Community* (Westport, Conn., 1972), 69–70, and *The American Slave: A Composite Autobiography* (Westport, Conn., 1978); Eugene D. Genovese, *Roll, Jordan, Roll: The World the Slaves Made* (New York, 1974), 535–40; and Eugene D. Genovese and Elinor Miller, eds., *Plantation, Town, and Country: Essays on the Local History of American Slave Society* (Champaign-Urbana, Ill., 1974). But it was in the Caribbean plantation societies that gardens and provision grounds were a common feature, particularly in Jamaica. A. J. G. Knox calculated that in 1832 in Jamaica, 71 percent of the total agricultural output was provided by plantation exports, with the bulk of the remainder (27 percent) "coming from the slaves' provision grounds." "Opportunities and Opposition: The Rise of Jamaica's Black Peasantry and the Nature of Planter Resistance," *Caribbean Review of Sociology and Anthropology* 14:4 (1977):386, cited by Sidney W. Mintz in "Slavery and the Rise of Peasantries," in Michael Craton, ed., *Roots and Branches: Current Directions in Slave Studies* (Toronto, 1979), 231. See also Sidney W. Mintz, *Caribbean Transformations* (Chicago, 1974), 146–56; Tomich, *Slavery in the Circuit of Sugar*. The best historiographical survey of this issue is Ciro Flamarion S. Cardoso, *Escravo ou Campones? O Protocampesinato Negro nas Américas* (São Paulo, 1987).

63. For a very useful description of the situation in different British colonies in the Caribbean, see Higman, *Slave Populations*, 204–12. For Martinique, see Tomich, *Slavery in the Circuit of Sugar*, 261–90.

64. See, for example, Pinckard, *Letters from Guiana*, 25; and Bolingbroke, *Voyage*, 76. Similar practices were found in other slave societies, generating what one author has called an informal economy. See Loren Schweninger, "The Underside of Slavery: The Internal Economy, Self-Hire, and Quasi-Freedom in Virginia," *Slavery and Abolition* 12:2 (September 1991):1–22; Betty Wood, "White 'Society' and the 'Informal' Slave Economies of Lowcountry Georgia, circa 1763–1830," *Slavery and Abolition* 11:3 (December 1990):313–31. A special issue of *Slavery and Abolition* 12:1 (May 1991), edited by Ira Berlin and Philip D. Morgan, was devoted entirely to this question.

65. Some historians have seen in this practice a "peasant breach." Some have seen the slaves as part-time peasants, and petty traders, and commodity producers. For a discussion of this debate see Cardoso, *Escravo ou Campones?* and "The Peasant Breach in the Slave System: New Developments in Brazil," *Luso-Brazilian Review* 25 (1988):49–57; Mintz, *Caribbean Transformations;* Genovese, *Roll, Jordan, Roll;* Sidney Mintz and Douglas Hall, *The Origins of the Jamaican Internal Marketing System* (Occasional Papers, New Haven, 1960); and the following essays in the special issue of *Slavery and Abolition* 12:1 (May 1991), edited by Ira Berlin and Philip Morgan: Hilary McD. Beckles, "An Economic Life of Their Own: Slaves as Commodity Producers and Distributors in Barbados," 31–48; Woodville K. Marshall, "Provision Ground and Plantation Labour in Four Windward Islands:

Competition for Resources During Slavery," 48–67; Dale Tomich, "Une Petite Guinée: Provision Ground and Plantation in Martinique, 1830–1848," 68–92; Mary Turner, "Slave Workers Subsistence and Labour Bargaining: Amity Hall, Jamaica, 1805–1832," 92–106; John Campbell, "As 'A Kind of Freeman'? Slaves' Market-Related Activities in the South Carolina Upcountry, 1800–1860," 131–69; John T. Schlotterbeck, "The Internal Economy of Slavery in Rural Piedmont Virginia," 170–81; Roderick A. McDonald, "Independent Economic Production by Slaves on Antebellum Louisiana Sugar Plantations" 182–208; Tomich, *Slavery in the Circuit of Sugar,* 179.

66. Bolingbroke, *Voyage,* 76. This was certainly an exceptional case, but there is other evidence that slaves managed to accumulate some money by working on Sundays, and by raising chickens and pigs and growing vegetables in their gardens and provision grounds and selling the surplus in the markets or to their owners. The contributions that slaves collectively made to the chapel through the years Wray and Smith were in Demerara ran between £100 and £200 pounds annually. The reduction of their free time, however, would have increasingly limited their income. So it is not surprising that slaves at the time of the rebellion demanded more time for themselves.

67. *Royal Gazette,* October 18, 1821. Such orders were periodically reactivated, which seems to indicate that many people were neglecting them.

68. There is also some contradictory evidence which suggests that on some plantations there were still provision grounds, but on the whole the practice was being discontinued.

69. The physician Alexander McDonnell used the expression "locked-jaw" for tetanus. *Considerations on Negro Slavery* (London, 1825), 177. The doctor also said that many children died because their mothers breast-fed them for too long and "stuffed" them, causing their bellies to grow, a remark that revealed both his prejudices and his ignorance. On child mortality, see Kenneth F. Kiple and Virginia H. Kiple, "Slave Child Mortality," 284–309.

70. Table of "Registered Births and Deaths by Colony, Demerara and Essequibo," Higman, *Slave Populations,* 611.

71. Sheridan, *Doctors and Slaves,* 244.

72. There were, however, some promising signs in the period ending in 1829, probably as a result of the British government's increasing pressure and the planters' growing concern with the decline in the number of slaves.

73. This was suggested in 1817. P.R.O. C.O. 112/5. See also advertisements in the *Royal Gazette,* March 8, 1819, and March 3, 1821.

74. Population figures for 1824 showed that there were 3,153 whites and 4,227 free blacks. P.R.O. C.O. 116/193. Higman notes that between 1807 and 1834 the total number of slaves in the British Caribbean declined at an average rate of 0.5 percent a year. But the decline was greater in the new sugar colonies (Dominica, Saint Lucia, Saint Vincent, Grenada, Tobago, Trinidad, Demerara-Essequibo, Berbice) than in the older or marginal colonies. Demerara and Essequibo during the same period showed a total decline of 20.6 percent, while in Berbice the drop was 32 percent. Higman, *Slave Populations,* 72. One of the reasons for the disparity

between Berbice and Demerara-Essequibo is the large number of slaves who were transferred from Berbice to Demerara and Essequibo.

75. The new ordinance abrogated previous acts of 1793 and 1804. See *Royal Gazette,* March 11 and 14, April 10, and July 8, 1815. A list of slaves manumitted from 1809 to 1821 indicates that only 335 slaves had been emancipated during this period. P.R.O. C.O. 111/37. On May 20, 1815, a petition of "Colored People" concerning manumission was submitted to the Court of Policy. P.R.O. C.O. 114/8. For an indication that manumission was still a great concern in 1826, see the discussion in the Court of Policy in response to the pressures of the British government in favor of abolition. Minutes of the Court of Policy, July 3, 1826. P.R.O. C.O. 114/9.

76. This provoked a response from free blacks who had been enjoying "nominal freedom" for several years. Petition of March 20, 1815. P.R.O. C.O. 114/8. The subject was discussed in the Court of Policy, and after some deliberation it was agreed that a scale of classification should be established, by which all those who had been living free from ten to twelve years and whose "character" and "good conduct" the court was satisfied with, should be allowed to take out letters of manumission, after a petition to the Court of Policy. Minutes of the Court of Policy, March 20, 1815. P.R.O. C.O. 114/8.

77. The number of manumissions was negligible, an average of thirty to forty a year, and women and children were over-represented. In a total of 131 manumissions, 66 were below 14. Fifty-eight percent were female, 42 percent male; 62 percent were judged to be "colored," 42 percent black. Sixty-seven percent of the manumissions were given by masters; 16 percent were bought by the slaves themselves or by their relatives; 17 percent were people who were reputed to be free at birth. From 1809 to 1821, some 372 slaves were manumitted—99 of them male and 273 female. P.R.O. C.O. 111/37.

78. Bolingbroke, *Voyage,* 76.

79. Higman, calculating the rates of manumission per 1,000 slaves per annum, gives 0.1 for Demerara-Essequibo in 1808, in 1820 0.2, and 2.3 in 1834. *Slave Populations,* 381. The increase after 1820 was almost certainly due to the new "amelioration" policies adopted by the British government.

80. P.R.O. C.O. 111/37 and 116/156.

81. By 1830 Trinidad freedmen represented 18.9 percent of the population of Dominica, 38 percent in Trinidad, 21.7 percent in Saint Lucia, and 10.6 percent in Jamaica. Higman, *Slave Populations,* Table 4.2, p. 77.

82. Bolingbroke mentions that free people of color came to Georgetown from Barbados and Antigua. *Voyage,* 84.

83. The 1832 *Almanack and Local Guide,* p. 456, indicated that from January 1826 to June 1830 the Court of Policy granted 1,582 manumissions: 595 to males and 987 to females. Of those, 1,243 were by bequest or deed of gift, and 339 by purchase.

84. On July 3, 1826, the Court of Policy discussed a dispatch from Lord Bathurst saying that the slave should be able to purchase freedom "by the fruit of his honest earnings." P.R.O. C.O. 114/9. These sorts of intrusions by the British

government apparently had an effect, for the Reports from the Protectors of Slaves for Demerara and Essequebo for the Year Ending in December 1829 showed a dramatic increase in manumissions. From January 1, 1826, to October 31, 1829, some 1,402 people were freed, of whom 523 were males and 879 were females. From May 1 to October 31, 1829, there were 131 slaves manumitted. They had received their freedom, allegedly, for a variety of reasons: "natural affection," deed of gift, last will, being born in a state of reputed freedom, and "faithful" service. Fourteen slaves purchased their own freedom, at an average price of £94. It should be noted that by this time the price of slaves was going down. "Statement Exhibiting the Number of Slaves Manumitted in the Colony of Demerara and Essequebo, from the 1st of May to the 31st of October 1829, Inclusive, for Each of the Reasons or Considerations Specified in the Record (No. 5) of Manumissions for that Period; the Total Amount of Sums Paid by Them for the Purchase of their Freedom, and the Average Price of Each Freedom Purchased," and Copies of Reports from Protectors of Slaves, P.R.O. C.O. 116/156.

85. A "List of Free Coloured Persons Who Have Paid Their Colonial Tax Levied on Slaves for the Year 1808" registered 271 individuals. Most had fewer than 5 slaves; twenty-six had from 10 to 20; six from 20 to 30; and one had 38. *Royal Gazette,* September 25, 1810.

86. See, for example, dispatch of November 22, 1821, P.R.O. C.O. 112/5.

87. Free blacks also served in the militia. The Militia Regulations of June 1817 showed that the First Demerary Battalion had ten companies including four colored companies. The Second Demerary battalion had five companies, one "coloured." All white and free "coloured" male inhabitants ages 16 to 50 were to serve in the militia, except the members of the Court of Policy and Justice, fiscals, kiezers, persons in holy orders, and other persons of high standing in the colony. See *The Local Guide Conducting to Whatever Is Worthy of Notice in the Colonies of Demerary and Essequebo for 1821* (Georgetown, 1821), 11–14; Hugh W. Payne, "From Burgher Militia to People's Militia," *History Gazette* 17 (February 1990):2–11. For the participation of free blacks in the British West India regiments, see Roger Norman Buckley, *Slaves in Red Coats: The British West India Regiments, 1795–1815* (New Haven, 1979).

88. This explains why the price of slaves increased so dramatically during this period. It is important to notice that the expansion of production generated a serious problem of labor supply. Trying to save labor and increase productivity, planters introduced technological improvements, replacing wind-mills with steam engines, and managers put more and more pressure on slaves.

89. From polder, an area of low-lying land that has been reclaimed from a body of water and protected by dikes.

90. Bolingbroke gave a detailed description of the various steps involved in the production of sugar. *Voyage,* 66–67. See also Thomas Staunton St. Clair, *A Soldier's Sojourn in British Guiana,* Vincent Roth, ed. (Georgetown, 1947), 29. After the abolition of the slave trade many other improvements were introduced in the mills. When Richard Schomburgk visited the colony in the 1840s he noticed the generalized use of steam power, and of the vacuum pan. The "megass," which in the

past was transported manually, was mechanically transferred to a "megass logie," where it was dried to be later used for firing the boiling vats. Once the cane was brought to the mill, it was "squeezed between three iron rollers turned on their axis by steam power." But the most important innovation was the introduction of vacuum pans in the early '30s. The liquid, after passing through the succession of copper vats, was put into a steam-powered vacuum pan—which allowed the crystal sugar to be quickly and completely separated from the molasses, so that it was no longer necessary to cure the sugar. Now it could be immediately packed in large casks and the separated molasses fermented and distilled. These changes saved time and labor. A process that formerly required eight days "in addition to undivided attention and labour," could be done within fifteen hours. Obviously, however, the introduction of such sophisticated technologies required even greater investments. *Schomburgk's Travels,* 1:62–64. Demerara newspapers were constantly advertising steam mills and coffee pulping machines.

91. Peter Wood has shown how the slaves' previous experience in growing rice helped to shape the economy of South Carolina. *Black Majority: Negroes in Colonial South Carolina, from 1670 Through the Stono Rebellion* (New York, 1974); see also David Littlefield, *Rice and Slaves: Ethnicity and the Slave Trade in Colonial South Carolina* (Baton Rouge, 1981); Hilary McD. Beckles, *Natural Rebels: A Social History of Enslaved Black Women in Barbados* (London, 1989).

92. There is a growing literature on slave women in the Caribbean. Among the recent works see Barbara Bush, *Slave Women in Caribbean Society, 1650–1838* (Bloomington, 1990), and Marietta Morrissey, *Slave Women in the New World: Gender Stratification in the Caribbean* (Lawrence, Kan., 1979). See Chapter 5, note 54, below.

93. In 1832, of the slaves in Demerara and Essequibo 78.5 percent worked on sugar; 5.9 percent on cotton; 4.4 percent on coffee; 0.7 on cattle ranches; 0.2 on timber extraction; and 10.3 percent in urban activities. But if we consider only the Demerara parishes, the results are somewhat different. Only 68.5 percent worked on sugar; 14.4 percent still worked on cotton and coffee, 1.0 percent on cattle, 0.1 on plantain, 0.2 on timber, and 15.8 percent were engaged in urban activities. This seems to indicate that as late as 1832 there were still many slaves on cotton and coffee plantations. "Distribution of Slaves by Crop and Parish: Demerara, Essequibo, 1832," Higman, *Slave Populations,* 702.

94. Higman, *Slave Populations,* 48. In 1834, according to Higman, 84.2 percent were employed, 11.8 percent were children, and 4.5 percent disabled (Table 3.3). Higman's estimate of slaves who were working is much higher than those supplied by planters. They insisted that only a third of their slaves were really productive. Dale W. Tomich stresses the privileged position enjoyed by tradesmen, drivers, and domestic servants. He notices that this "may have created an ambivalent but not necessarily conservative response to the slave system . . . indeed the contradiction between individual dignity and self-worth, on the one hand, and slave status, on the other, may have been experienced more palpably by these slaves than others." *Slavery in the Circuit of Sugar,* 227. A similar point has been made by historians such as Eugene Genovese for the United States and Orlando Patterson for

Jamaica. They all stress that the contradictions inherent in the position of craftsmen, drivers, and domestics may explain their participation in rebellions.

95. I have used the record for Berbice in 1819 provided by Higman (*Slave Populations,* 570–89), to which I have added data from newspapers and missionaries' diaries.

96. At the time Bolingbroke visited Demerara, many estates were hiring slaves at three, four, and five shillings per day during picking time.

97. The *Royal Gazette* in the earlier years published lists of slaves per estate and separate lists of slaves owned by "individuals." The first showed the "List of the Estates That Have Paid Their Colonial Tax Levied on Slaves"; the second, the "List of Persons Who Have Paid Their Colonial Tax Levied on Slaves." This makes it possible to separate the two groups. See, for example, the *Royal Gazette,* September 22, November 19, 1810; April 8, 1815.

98. Higman notices that the case of Georgetown is unusual. "Between 1812 and 1824 its population increased from about 6,000 to 10,500." *Slave Populations,* 97.

99. The list published in the *Royal Gazette,* September 22, 1810, shows that most "individuals" owned from 2 to 10 slaves; a few from 10 to 20; and a only a very few owned more than 20.

100. Many advertisements in the *Royal Gazette* offered slave gangs on short notice, such as the following, published November 7, 1807, "A person with a gang of 30 to 40 negroes is wishful of undertaking any job or task work in the River or the East or the West Coast, if applied for any time in the course of the month." Others were from people wanting to hire gangs: "A task gang to prepare some land for canes on an Estate on Wakenhaam [Island]. Any person wishful of undertaking the same, may learn further particularly applying to William King." On August 24, 1810, Hugh Mackenzie offered to "hire a gang of 40 to 45 strong healthy negro men." On October 30, 1810, Stephen Cramer gave notice that he wanted to hire "one hundred able working negroes" to work in November and December. On November 24, 1810, another notice offered to hire fifty coffee pickers and promised to give "each negro" a weekly allowance of two bunches of plantain and one-and-a-half pounds of fish, plus two drams of rum. See also May 26, August 24, October 20, November 24, December 22, 1810; January 3, March 28, 1815; April 21, July 5, 1821; February 26, 1822—and many more.

101. P.R.O. C.O. 116/156.

102. For a detailed description of work on sugar plantations see *The Overseer's Manual, or A Guide to the Cane Field and the Sugar Factory for the Use of the Young Planters* (1st ed., 1882; 3rd ed., revised and enlarged, Demerara, 1887).

103. McDonnell, *Considerations on Negro Slavery,* 147–67.

104. Ibid.

105. Ibid., 156. Similar stories were told in the United States. Raymond A. Bauer and Alice H. Bauer, "Day-to-Day Resistance to Slavery," *Journal of Negro History* 27 (October 1942):388–419.

106. In a letter of May 21, 1812, to the LMS, John Wray reported a case of slaves from *Success* who refused to receive their weekly allowances of saltfish because it was no larger than a normal allowance and it was common to give more on

holidays. They also refused to perform their duties and were punished. LMS IC, Demerara, Wray's letter, May 21, 1812. As we shall see, when slaves discussed strategy before the rebellion of 1823, some suggested that they should lay down their tools to force the governor to satisfy their demands. See Chapter 5 below.

107. The use of strike by slaves in Jamaica has been pointed out by Mary Turner in *Slaves and Missionaries: The Disintegration of Jamaican Slave Society, 1787–1834* (Champaign-Urbana, Ill., 1982), 153–59; "Chattel Slaves into Wage Slaves: A Jamaican Case Study," in Malcolm Cross and Gad Heuman, eds., *Labour in the Caribbean: From Emancipation to Independence* (London, 1988), 14–31; and "Slave Workers, Subsistence and Labour Bargaining: Amity Hall, Jamaica, 1805–1832," in Ira Berlin and Philip D. Morgan, eds., The Slaves' Economy: Independent Production by Slaves in the Americas, special number, *Slavery and Abolition* 12:1 (May 1991):92–106.

108. McDonnell, *Considerations on Negro Slavery,* 153. This opinion is confirmed by other sources.

109. That the British government's interest in the welfare of the slaves had grown after the rebellion is visible in the colonial correspondence but also in the character and extent of the fiscals' and protectors' records. There are very few surviving fiscal books for the period before the rebellion, so it is probably true that until then fiscals were not careful in keeping records. The few that still exist show that most often slaves who complained got punishment rather than redress. After the rebellion, the governor was constantly harassing fiscals for not performing their duties adequately, and fiscals appear to have become more inclined to punish managers when they overstepped the limits established by law. The British government continued to put pressure on governors. See several dispatches from Downing Street, particularly one dated September 2, 1829, from Sir George Murray to Governor Sir Benjamin D'Urban, P.R.O. C.O. 112/5. See also ibid., 116/156.

110. Although the interest of the British government increased after the rebellion, there is evidence that even before that there had been complaints regarding the appointments of fiscals by the Court of Policy. When Heyliger was appointed fiscal, Bathurst criticized the appointment, arguing that there was a conflict of interest, since Heyliger was an attorney for several slave proprietors and had under his care a considerable number of slaves. But in spite of Bathurst's reservations, Heyliger was confirmed and served as fiscal for a number of years. P.R.O. C.O. 112/5.

111. P.R.O. C.O. 116/156.

112. In response to the governor's letter, Acting Fiscal George Bagot tried to reassure him that to punish slaves for complaining was "far from being an usual practice among managers of the Eastern District." And since the governor apparently continued to insist that the manager in question be punished, the reluctant fiscal reminded him that the matter had been before the Court of Justice, "a tribunal over whose decision your Excellency has no control. No power is vested in your Excellency to order the proceedings to be laid before you. If your Excellency has such power in this case, the principle will extend to any and every case before the court, and this, I think, will hardly be contended." Benjamin

D'Urban to George Bagot, June 7, 1824; Bagot to D'Urban, August 3, 1824. P.R.O. C.O. 116/156.

113. Ibid.

114. Proceedings of the Fiscals of Demerara and Berbice . . . , P.R.O. C.O. 116/156 and 116/138.

115. The increase in the scale of commodity production and intensification of labor exploitation robbed women of some advantages, taking from them traditional sources of authority and status. Women were particularly affected because they were the marketers. Furthermore, increased labor exploitation heightened the contradiction between production and reproduction. Morrissey, *Slave Women in the New World*, 61–62, 80. See also Beckles, *Natural Rebels*.

116. D'Urban to Bathurst, August 12, 1824, in which he makes this point and—contrary to the spirit of the "amelioration laws" of 1823—argues that it is necessary to inflict corporal punishment upon females guilty of "aggravated and repeated misconduct," P.R.O. C.O. 111/44. The fiscals' records, however, do show that more men than women were punished for crimes and misdemeanors.

117. On sexual abuses and resistance in the United States, see Darlene C. Hine, "Female Slave Resistance: The Economics of Sex," *Western Journal of Black Studies* 3 (Summer 1979):123–27; Steven Brown, "Sexuality and the Slave Community," *Phylon* 42 (Spring 1981):1–10.

118. This case ended up in the Berbice Court of Criminal Justice, and the doctors called to testify denied that the girl had been raped, since her hymen was "intact." Complaint filed by the fiscal on 31 January 1820. P.R.O. C.O. 116/139.

119. Describing work in sugar plantations in Martinique, Dale W. Tomich finds that in large plantations where the mills worked around the clock, slaves worked in seven-and-a-half-hour shifts. *Slavery in the Circuit of Sugar*, 231.

120. I have found many similar cases for Berbice. For example, some women started a fire next to the dam to drive away the sand flies so that they could nurse their infants. They were spotted by the manager, who asked if they had no work to do. The women tried to explain that they had just taken up their children who were crying. But the manager ordered them flogged. Reports of the Fiscal, January 1819–December 1823, Berbice, P.R.O. C.O. 116/138.

121. Copies of Reports from Protector of Slaves, P.R.O. C.O. 116/156. These documents referred to 1829. Had the same complaints reached the fiscal ten years earlier the cases would almost certainly have been dismissed and the slaves punished for having dared to complain.

122. After 1825, the office of fiscal was replaced by that of the protector of slaves.

123. Reports of the Fiscals, June 1819 to December 1823, P.R.O. C.O. 116/138.

124. There are many cases in the fiscals' records of this type of complaint. A very revealing case surfaced in 1822 in Berbice, when fifteen slaves went to complain that they belonged to a Mrs. Sanders and been sent to split staves, a job they were unaccustomed to. Each one was supposed to split twenty bundles per week, but they were unable to meet the quota and had been flogged. They also complained they had been forced to work on Sunday hauling up a punt. Called by the fiscal,

Mrs. Sanders's son testified that the first week the slaves had brought seventeen bundles, the second fifteen, and the third only eleven. So they had been flogged. To show his "good will," he said that when the slaves had complained of living upriver without wives, he had purchased several women. This had put a drain on his finances. He added that the "negroes" now wanted to force his mother either to sell them or to remove them from upriver, but her financial situation allowed her to do neither. He admitted that when the slaves were in the "bush," he forced them to work on Sundays, but promised to give them an extra day when they returned. P.R.O. C.O. 116/138.

125. Ibid.,

126. Ibid.,

127. In this case the fiscal did not impose any fine for lack of food because, he claimed, he had found "fine provision grounds." P.R.O. C.O. 116/156.

128. P.R.O. C.O. 116/138.

129. A few other cases resulted in severe punishments for managers. One man ordered a woman who was eight months' pregnant to be whipped. After she bore a still-born child, the manager was suspended. In the end, he was fined 2,000 guilders (equivalent to £200), sentenced to three months in jail, and dismissed from his post. Such inconsistencies reveal the judicial system's arbitrariness. Extract from the Register of the Proceedings of the Commissioners of the Court of Criminal Justice, 1819. P.R.O. C.O. 116/139.

130. "Nothing can be more keenly observant than the slaves are of that [which] affects their interests," wrote Governor D'Urban in a letter to Murray, April 1830, reproduced in Williams, ed., *Documents,* 189.

131. In the report of the fiscal for Berbice there is a complaint filed by several slaves on April 1, 1819. They said their children between 8 and 10 years old were asked to milk the goats, mind the horses, burn the coffee, clean the master's shoes, and do other work. The fiscal visited the property and found that the slaves were "dissatisfied" with a "negro" woman who was a house servant and a "favourite." P.R.O. C.O. 116/138.

132. A woman complained to the fiscal that her child had died after being given too much calomel. The case was dismissed. Ibid.

133. On January 2, 1821, Quamina from Berbice complained that he had been sold as a cooper and carpenter and was not able to pick as much cotton as the others, and for that reason had been flogged and had his back rubbed with brine. Ibid.

134. The records produced an abundance of detail and reveal the complex negotiations that went on between managers and slaves. They also show how difficult was the position of the middlemen, particularly drivers, since most were slaves and often had to punish friends and relatives. Bob, the head driver of *Belair,* for example, was forced to flog several women with the cat and lock them in the stock. Two of these women were his daughters, and one of them, Pamela, was pregnant. P.R.O. C.O. 111/43.

135. Equal attachment to this system was shown by free blacks. Typical was the case of Amelia Phippin. She had two children by her master and was taken to

England, where she became a free woman. When she returned to the colony she worked as a domestic servant. When her master died, the attorney for the estate hired her as a domestic servant. But not only did he not pay for her work, he also kept her manumission papers and refused to give her 2,000 guilders her former master had left her in his will. The protector of slaves investigated the matter and ordered the return of her papers. But Amelia claimed she never saw the money. (It was argued during the inquiry that she had already received her bequest—something she could hardly have disproved.) Returns of Complaints etc., Made to Protector, from 1 May to 31 October 1829, P.R.O. C.O. 116/156.

136. Bryan Edwards marvels at the communal sense of responsibility slaves felt toward the elderly. He says that "the whole body of negroes on a plantation must be reduced to a deplorable state of wretchedness, if, at any time, they suffer their aged companions to want the common necessaries of life, or even many of its comforts, as far as they can procure them." See Bryan Edwards, *History Civil and Commercial of the British Colonies in the West Indies* (3rd ed., 3 vols., London, 1801), 2:99.

137. In his study of peasant movements in Russia, Theodor Shanin has stressed that it is impossible to understand peasant political action without considering peasant goals. "Dreams matter," he says. "Collective dreams matter politically. That is a major reason why no direct or simple link relates political economy to political action. In between stand meanings, concepts and dreams with internal consistencies and a momentum of their own. To be sure, their structure bears testimony to the relations of power and production they are embedded in and shaped by. Patterns of thought, once established, acquire a causal power of their own to shape, often decisively, economy and politics, that is true particularly of the political impact of ideology, understood here as the dream of an ideal society in relation to which goals are set and the existing reality judged." "The Peasant Dream: Russia 1905–1907," in Raphael Samuel and Gareth Stedman Jones, eds., *Culture, Ideology, and Politics* (London, 1982), 227–43.

138. I have borrowed James Scott's concept of "public" and "hidden transcripts" (*Domination and the Arts of Resistance: Hidden Transcripts* (New Haven, 1990) and adapted it to my purposes here.

139. For a discussion of African influence in the United States, see Rawick, *The American Slave;* Wood, *Black Majority;* Lawrence W. Levine, *Black Culture and Black Consciousness: Afro-American Folk Thought from Slavery to Freedom* (New York, 1977); Sterling Stuckey, *Slave Culture: Nationalist Theory and the Foundation of Black America* (New York, 1987); Herbert G. Gutman, *The Black Family in Slavery and Freedom, 1750–1925* (New York, 1976); Tom W. Shick, "Healing and Race in the South Carolina Low Country," in Paul Lovejoy, ed., *Africans in Bondage: Studies in Slavery and the Slave Trade* (Madison, 1986), 107–24; Margaret Washington Creel, *"A Peculiar People": Slave Religion and Community Culture Among the Gullahs* (New York, 1988). The classic study of religious syncretism in Brazil is Roger Bastide, *Les Religions africaines au Brésil* (Paris, 1960). For additional references see note 140 below.

140. Attempts to recover this experience in other areas have been made by dif-

ferent historians such as Genovese, *Roll Jordan Roll;* Wood, *Black Majority;* Gutman, *The Black Family;* Monica Schuler, *"Alas, Alas Kongo"* and Afro-American Slave Culture" in Craton, ed., *Roots and Branches;* Michael Craton, *Searching for the Invisible Man: Slave and Plantation Life in Jamaica* (Cambridge, Mass., 1978); Margaret Crahan and Franklin Knight, eds., *Africa and the Caribbean: The Legacies of a Link* (Baltimore, 1979); Edward Kamau Brathwaite, *The Development of Creole Society in Jamaica, 1670–1820* (Oxford, 1971), and "Caliban, Ariel, and Unprospero in the Conflict of Creolization: A Study of the Slave Revolt in Jamaica in 1831–32," in Vera Rubin and Arthur Tuden, eds., *Comparative Perspectives on Slavery in New World Plantation Societies* (New York, 1977), 41–62; Levine, *Black Culture and Black Consciousness;* Stuckey, *Slave Culture;* John Blassingame, *The Slave Community: Plantation Life in the Antebellum South* (New York, 1972); Rawick, *From Sundown to Sunup;* Albert J. Raboteau, *Slave Religion: The "Invisible Institution" in the Antebellum South* (New York, 1978); Paul D. Escott, *Slavery Remembered: A Record of Twentieth-Century Slave Narratives* (Chapel Hill, 1979). Many anthropologists since Herskowitz have contributed to the discussion, particularly Sidney W. Mintz and Richard Price, *An Anthropological Approach to the Afro-American Past: A Caribbean Perspective* (Philadelphia: Institute for the Study of Human Issues, Occasional Papers in Social Change, 2, 1976), and Sidney W. Mintz, ed., *Slavery, Colonialism and Racism* (New York, 1974). Some have focused more on the creation of the new culture, while others have been more interested in African roots and the way the slaves' cultures were redefined under slavery. The attempt to search for African roots, however, is still very difficult because slaves came from different parts of Africa and different cultures and only a knowledge of these cultures will make it possible to identify "survivals" in the New World. As Richard Price has pointed out, this is a very risky and sometimes misleading enterprise. For a very insightful analysis, see Sally Price and Richard Price, *Afro-American Arts of the Suriname Rain Forest* (Berkeley, 1980), and Richard Price's commentary on Monica Schuler's "Afro-American Culture," in Craton, ed., *Roots and Branches,* 141–50. Seen from this perspective, slave protest and resistance would have to be redefined to identify sources of pain and conflict that have usually been neglected. In his study of the Stono Rebellion of 1739 in South Carolina, John K. Thornton shows that a knowledge of the history of the early eighteenth-century kingdom of Kongo can shed light on the slaves' motivations and actions. "African Dimensions of the Stono Rebellion," *American Historical Review* 96:4 (October 1991):1101–15. See also Oruno D. Lara, "Resistance to Slavery: From African to Black American," in Rubin and Tuden, eds., *Comparative Perspectives on Slavery in the New World Plantation Societies,* 465–81; and David Barry Gaspar, *Bondmen and Rebels: A Study of Master-Slave Relations in Antigua with Implications for Colonial British America* (Baltimore, 1985); idem, "Working the System: Antigua Slaves, Their Struggle to Live," *Slavery and Abolition* 13:3(1991):131–55.

141. Christian's case was submitted to the Court of Criminal Justice. P.R.O. C.O. 116/139.

142. The survival of this type of matrilineal kinship among the Saramaka is reported by Brother Kersten, a Moravian missionary who, referring to a potential

convert, wrote: "Grego is going with his mother's brother to Paramaribo. It is in fact, the mother's brother who is responsible for him because among the local negroes the father has no say over his children. It is always the mother's oldest brother who is responsible for the children." Quoted by Richard Price, *Alabi's World* (Baltimore, 1990), 348.

143. P.R.O. C.O. 116/139. Another story equally intriguing is that of a slave who went to complain that after his wife was whipped she gave birth prematurely to a dead child. The child had a broken arm and its body was lacerated. Because it is unlikely that such injuries would have been sustained in the mother's womb, the lacerations were almost certainly inflicted after birth. This might make sense in light of the fact that it was a common practice among some groups in West Africa to perform a ritual to exorcize the ghost of a child who died during the first week. On the other hand, it is possible that the child had been hurt during birth as the manager claimed. Here, as in analogous cases, we are left merely with tantalizing speculations. Only further research on both sides of the Atlantic can help to clarify such issues.

144. "Obeah" is discussed in the next chapter.

145. Mintz and Price, *An Anthropological Approach to the Afro-American Past.*

146. This would be analogous to the process described by the anthropologist George Foster for sixteenth-century Mexico, where traditions from certain parts of Spain came to predominate over others. Examining religious rituals, historians have found, for example, that in Jamaica and Suriname the predominant influence was from Ashanti, while in Haiti and some areas in Brazil it was Dahomean, even though these particular groups were outnumbered. See Martin L. Kilson and Robert I. Rotberg, eds., *The African Diaspora: Interpretive Essays* (Cambridge, Mass., 1976).

147. Writing of Suriname, Richard Price notices that "Although rapid religious syncretisms among slaves of diverse African provenance were an earmark of colonial Suriname's first hundred years, rituals and other performances associated with Papa, Nago, Loango, Pumbu, Komanti, and other African 'nations' (as they were often called in Afro-America) were still an important feature of late eighteenth century life among both Saramakas and plantation slaves." And he adds that by the 1760s "Papa," "Luangu," or "Komanti" rites and dances would have included people and ideas of quite varied (and mixed) African ancestry. "Nevertheless, bundles of rites or drums/dances/songs/language, which had their origin in particular African ethnicities, were kept together by eighteenth-century Saramakas (as they are still today)." *Alabi's World,* 308–9.

148. The most insightful study of this phenomenon has been done by Edward Kamau Brathwaite, who has stressed the ambiguities involved in this process and provided us with a "prismatic" rather than the usual linear view of creolization. *The Development of Creole Society in Jamaica 1770–1820,* and *Kumina* (Boston, 1972). For the impact this process of creolization had on the rebellion of 1821 in Jamaica see Brathwaite's fascinating essay, "Caliban, Ariel and Unprospero in the Conflict of Creolization." Brathwaite's metaphor, that most slaves were bound by instinct and custom, "to their mother's milk and buried navel string," is relevant here. So is his comment that Sam Sharpe, the leader of the Jamaican Rebellion,

"could never have been the Christian hero made [of] him by the missionaries because, although he was a deacon in the Baptist church, he was also, unknown and invisible to the missionaries who thought they patronized his soul, a 'ruler' in his own right in his own people's church." Brathwaite's remark also applies to the leaders of the Demerara rebellion, particularly to Quamina. "Caliban, Ariel, and Unprospero," 54.

149. James Walvin argues that "by the mid-1820s abolitionists assumed that West Indian slaves possessed those rights which, in the 1790s, the popular radicals had claimed for themselves." Abolitionists were denouncing the incompatibility between slavery and the "rights of man" and assuming that "the slaves' rights were identical to their own English rights." Walvin stresses that there had been a profound change in British society, "thirty years before, such sentiments had been denounced by Ministers and judges. Indeed, Englishmen had been transported for asking for these rights for themselves." James Walvin, "The Rise of British Popular Sentiment for Abolition, 1787–1832," in Roger Anstey, Christine Bolt, and Seymour Drescher, eds., *Anti-Slavery, Religion, and Reform* (Hamden, Conn., 1980), 155

150. Dale W. Tomich's comments on Martinique could be applied to Demerara. "For the master, the provision ground was the means of guaranteeing cheap labor. For slaves, it was the means of elaborating an autonomous style of life. From these conflicting perspectives evolved a struggle over the conditions of material and social reproduction in which the slaves were able to appropriate aspects of these activities and develop them around their own interests and needs." *Slavery in the Circuit of Sugar,* 260–61.

151. In 1824, managers were celebrating the introduction of the task-system in Demerara. They argued that it was much more productive than the usual gang system because it did not require as much supervision and the slaves were much "happier." McDonnell, *Considerations on Negro Slavery.* Eight years later, however, the protector of slaves Elliot testified before a House of Commons committee that the system had failed miserably. Select Committee on the Extinction of Slavery, P.R.O. C.O. ZHCI/1039.

152. An interesting analysis of different versions of this rebellion appears in Brackette F. Williams, "Dutchman Ghosts and the History Mystery: Ritual, Colonizer and Colonized, Interpretations of the 1763 Berbice Slave Rebellion," *Journal of Historical Sociology* 3:2 (June 1990):134–65. For a detail chronicle of the rebellion, see Roth, ed., "The Story of the Slave Rebellion on the Berbice, 1762"; Moore, "Slave Rebellions in Guyana."

153. St. Clair, who was in Demerara from 1806 to 1808, mentioned a slave plot in 1807, when the conspirators were betrayed by a slave woman who lived with a young Scotsman overseer. A piece of paper written in Arabic was found. See St. Clair, *A Soldier's Sojourn in British Guiana,* 232.

154. This percentage is confirmed by the "List of Offences Committed by Male and Female Plantation Slaves in the Colony of Demerara and Essequebo." P.R.O. C.O. 116/156.

155. *Royal Gazette,* November 13, 1819.

156. On maroons in Guyana, see Alvin O. Thompson, "Brethren of the Bush: A Study of Runaways and Bush Negroes in Guyana" (mimeo, Department of History, University of West Indies, Barbados, 1975). See also his *Colonialism and Underdevelopment in Guyana, 1580–1803* (Bridgetown, Barbados, 1987). This study contains a map of maroon communities in Guyana. See also James G. Rose, "Runaways and Maroons in Guyana History," *History Gazette* (University of Guyana, Turkeyen) 4 (January 1989):1–14; and Richard Price, *The Guyana Maroons, A Historiographical and Bibliographical Introduction* (Baltimore, 1976). An advertisement in the *Royal Gazette,* January 24, 1807, mentioned runaways harbored by wood cutters, plank sawyers, and punt makers. The notice claimed that the runaways found they had to work even harder in hiding, and so returned to their plantations.

157. P.R.O. C.O. 114/7, April 3, 1807.

158. The importance of day-to-day forms of resistance has been stressed (perhaps overmuch) by James Scott in *Weapons of the Weak: Everyday Forms of Peasant Resistance* (New Haven, 1985) and *Domination and the Arts of Resistance.* For day-to-day resistance in the Caribbean, see Tomich, *Slavery in the Circuit of Sugar;* Rebecca Scott, *Slave Emancipation in Cuba: Transition to Free Labor, 1860–1889* (Princeton, 1985); David Patrick Geggus, *Slavery, War, and Revolution: The British Occupation of Saint Domingue 1793–1798* (Oxford, 1982); Hilary M. Beckles and Karl Watson, "Social Protest and Labour Bargaining: The Changing Nature of Slaves' Responses to Plantation Life in Eighteenth-Century Barbadoes," *Slavery and Abolition* 8:3 (1987):272–93; Barbara Bush, "Towards Emancipation: Slave Women and Resistance to Coercive Labour Regimes in the British West Indian Colonies, 1790–1838," 222–43; and Richard Hart, *Slaves Who Abolished Slavery* (Mona, Jamaica, 1980).

159. Slave strikes are found in many places throughout the Caribbean and in other parts of the New World. There is evidence that in some African societies, when the people were not happy with the decisions taken by their leaders, they could bring a village to a halt. Whether these practices were merely imported or generated within is an open question. John Wray reports several cases. Particularly interesting is a case in 1812 when slaves refused to obey orders because the manager had not allowed them to attend religious services. Mary Turner has identified slave strikes in Jamaica and has shown that the Jamaican rebellion of 1831 started with a general strike. See her very insightful essay "Chattel Slaves into Wage Slaves." For a bibliography on this subject see note 107.

160. In fact, as early as the 1940s Raymond A. Bauer and Alice H. Bauer were stressing the importance of day-to-day resistance in the United States, but it is only in the past fifteen years that this has become an important issue in the historiography of slavery everywhere. Bauer and Bauer, "Day-to-Day Resistance to Slavery," *Journal of Negro History* 27 (October 1942):388–419.

161. Genovese, *Roll, Jordan, Roll,* 598.

162. Seen from this perspective the slaves' day-to-day acts of resistance acquire a new significance, particularly when we consider that their notions of rights and the strategies they developed during slavery would be crucial in the organization of

post-emancipation societies. For such continuities, see Julie Saville, "A Measure of Freedom: From Slave to Wage Laborer in South Carolina, 1860–1868" (Ph.D. dissertation, Yale University, 1986). The continuity between day-to-day forms of resistance and rebellion is stressed by Mary Turner in her comment on Hilary Beckles's "Emancipation by Law or War? Wilberforce and the 1816 Barbados Slave Rebellion," in Richardson, ed., *Abolition and Its Aftermath,* 105–10. Turner suggests (p. 109) that in order to improve our understanding of slaves' rebellions we should begin by placing them in the "context of chattel labour relations: not just the heroic moments of action but the protracted daily struggles at the point of production."

163. For an insightful analysis of the conditions necessary for people to resist oppression, see Barrington Moore, Jr., *Injustice: The Social Bases of Obedience and Revolt* (New York, 1978).

164. Compensation claims or criminal court records often used to identify slaves' complaints and resistance tend to over-represent cases of violence such as attempt to murder, theft, and arson. The use of newspapers as a main source leads by contrast to an over-representation of runaways. See for example, Gaspar, *Bondmen and Rebels,* 194–202. Missionaries' letters and diaries, on the other hand, tend to stress physical punishment. Plantation records and the records of the protector of slaves offer a more encompassing view of the many forms of protest, but like any other source they have to be used with caution. The question that such records might have over-represented physical punishment has been raised. But all sources, particularly missionaries' correspondence and governors' letters, seem to confirm that physical punishment was very frequently used in Demerara. It is possible, however, that after being defeated in 1823, slaves became even more aggressive and as a result the number of punishments increased.

165. There was some overlapping and some disparity between the totals and the number within each category. But even so, they give an admirable portrait of the state of insubordination of the slave population. Copies of Reports from Protectors of Slaves 1826–1830. The quote is from the report for the year 1828. See List of Offences Committed by Male and Female Plantation Slaves in the Colony of Demerara and Essequebo, Made Up from Return of Punishments Forwarded to Protector of Slaves, from January 1 to June 1828, P.R.O. C.O. 116/156.

166. Dale W. Tomich says that overseers were callous and unrelenting toward the slaves. He attributes such a behavior to the fact that they "were at the point of confrontation between the master's drive for surplus production and his demand for the maintenance of social control, on the one side, and the recalcitrance of the slaves, on the other. The difficulty of their situation could affect the way that they handled the slaves." *Slavery in the Circuit of Sugar,* 240. The remark applies with even more reason to drivers. For a perceptive analysis of the ambiguous position of slave drivers, see Genovese, *Roll, Jordan, Roll,* 365–88.

167. The same pattern is recorded in the returns for the six-month period ending December 31, 1827, but the proportion of women punished is higher. Of a total of 10,513 slaves punished, 6,014 were males and 4,499 females. On the other hand, from January to June 1828, some 6,092 males and 3,962 females were punished, in

a total population of 62,352 slaves, of which 34,106 were males and 28,246 were females. Females constituted about 45 percent of the total population but were accused of 39 percent of the offenses. For the period ending June 30, 1829, male offenses registered amounted to 5,666 while those of females were around 3,000. Copies of Reports from Protectors of Slaves, 8–9. See also Report from Protector A. W. Young to Sir B. D'Urban, December 1829, in Copies of the Reports from Protectors of Slaves, P.R.O. C.O. 116/156.

168. Barbara Bush, using plantation records for Grenada and Guyana from the early 1820s to the beginning of the apprenticeship period, found that women were more often "accused of insolence, 'excessive laziness,' disobedience, quarreling and 'disorderly conduct' than were male slaves." Barbara Bush, "Towards Emancipation: Slave Women and Resistance to Coercive Labour Regimes in the British West Indian Colonies, 1790–1838," 35.

169. After the rebellion, with the new regulations imposed by the British government, the limit of stripes was set at 25. In 1829, a total of 8,710 punishments was recorded, involving 5,666 males: 8 received 8 stripes; 352 from 6 to 10 stripes; 1,332 from 11 to 15; 1,108 from 16 to 20; and 2,334 from 21 to 25 stripes. (These figures are put under suspicion by other reports in which the punishment, the name of the slave and the plantation, the nature of the offense, and the nature and extent of the punishment are given. There we find slaves who were condemned to 40, 60, and even 90 stripes.) According to the official reports for 1829, there were 312 slaves punished by confinement. All women were punished by confinement or the treadmill, since the new slave regulations prohibited whipping females.

170. The treadmill was an instrument of punishment operated by one or more persons walking on the moving steps of a wheel or treading an endless sloping belt, usually driving a machine such as a pump or a small mill.

171. Quash was probably a slave living downtown instead of on a plantation; he was listed as belonging to Anthony Osborn.

172. List of Cases Appearing in the Punishment Record Returns of the Colony of Demerara and Essequebo for the Half-Year from 1st of January to 30th June 1829. Copies of Reports from Protectors of Slaves, P.R.O. C.O. 116/156.

Chapter 3.
The Fiery Furnace

1. Letter from Post, April 27, 1808, P.R.O. C.O. 114/7. For the earlier attempts to establish a mission in Demerara, see Winston McGowan, "Christianity and Slavery: Slave, Planter and Official Reaction to the Work of the London Missionary Society in Demerara, 1808–1813," a paper delivered at the 12th Conference of Caribbean Historians, University of West Indies, Trinidad, March 30-April 4, 1980.

2. This seems to contrast with Mary Turner's description of the situation in Jamaica, where evangelical missionaries had arrived much earlier. See Mary Turner, *Slaves and Missionaries: The Disintegration of Jamaican Slave Society, 1787–1834*

(London, 1982). The late arrival of missionaries in Demerara may be explained by the late incorporation of Demerara into the British empire. The British government's support for missionary work would break the colonists' resistance.

3. LMS IC, Demerara, Van Cooten's letter, September 5, 1807.

4. "Memoir of the Late Hermanus Hilbertus Post, Esq.," *Evangelical Magazine* 19 (January 1811):1–7; (February 1811):41–49.

5. *Evangelical Magazine* 19 (January 1811): 7.

6. LMS IC, Demerara, George Burder's letter, October 18, 1808.

7. *Evangelical Magazine* 19 (February 1811):42.

8. LMS IC, Demerara, Wray's letter, October 20, 1808.

9. LMS IC, Demerara, Wray's letter, May 8, 1808.

10. LMS IC, Demerara, May 2, 1808.

11. LMS Board Minutes, July 25, August 8 and 15, 1808.

12. LMS IC, Demerara, Wray's letter, June 4, 1808. See also *Royal Gazette,* May 21, 1808.

13. LMS IC, Demerara, Wray's letters, June 4 and July 10, 1808.

14. LMS Board Minutes, November 28, 1808.

15. LMS IC, Demerara, Wray's letter, December 1808, February 4 and 13, 1809.

16. LMS Board Minutes, November 28, December 12, 1808; January 30, February 13, 1809.

17. Useful here is the concept used by Althusser, Laclau, and Gõren Therborn of ideology as "interpellation." This notion suggests the active role of individuals, and re-establishes a dialectic between ideas and other human practices. See especially Therborn, *The Ideology of Power and the Power of Ideology* (2nd ed., London, 1982).

18. LMS IC, Berbice, Wray's letter, May 19, 1824.

19. LMS IC, Berbice, Wray's letter, September 1813. A document produced by the local government in 1823 showed that Davies received after August 1813 a subsidy from the government. In 1823, it amounted to 157 pounds or 2,200 local currency. P.R.O. C.O. 116/192. Elliot received 1,200 in local currency. See Guyana National Archives, Minutes of the Court of Policy, 1823, Colonial Receiver Books, 263.

20. *Royal Gazette,* April 2, 1822.

21. LMS IC, Demerara, Davies's letter, October 4, 1809.

22. LMS IC, Berbice, Wray's letter, February 28, 1814.

23. Robert Strayer finds the same interest in Africa. "Many Africans," he says, "felt that mere possession of the Bible or acquisition of the skills of literacy was effective in warding off misfortune or promoting temporal success." See Robert Strayer, "Mission History in Africa: New Perspectives on an Encounter," *African Studies Review* 19:1 (April 1976): 3.

24. LMS IC, Berbice, Wray's letter, June 16, 1814.

25. LMS Board Minutes, June 26, 1811.

26. J. A. James, *The Sunday School Teacher's Guide* (2nd American ed., from 5th English ed., New York, 1818), 43, 50, 53, 57. This belief in the correlation

between crime and ignorance was extremely widespread in the nineteenth century, although it had little basis in reality, as Harvey J. Graff shows in his essay on Canada, " 'Pauperism, Misery and Vice': Illiteracy and Criminality in the Nineteenth Century," *Journal of Social History* 2 (1977): 245–68.

27. In 1823, according to the *Evangelical Magazine* 1, New Series (July 1823): 291, there were 765,000 children attending 7,173 Sunday schools in Great Britain, under the supervision of 71,276 teachers. A year later, according to the magazine, the number of schools had increased to 7,537, teachers to 74,614, and children to 812,305. *Evangelical Magazine* 2, New Series (July 1824):286.

28. James, *The Sunday School Teacher's Guide,* 14–45, 47, 69, 80–81, 142, 174.

29. In a letter from Berbice, November 1, 1816, Wray wrote about a manager who could hardly read and write, but who, when he learned that an adult slave had been teaching children to read and pray, had forbidden him to continue doing so. According to Wray, the manager said that books had made "the negroes" in Barbados rise, and insisted that books would "ruin the country." LMS IC, Berbice, Wray's letter, November 1, 1816.

30. Patricia Hollis, *Class and Conflict in Nineteenth-Century England, 1815–1850* (London, 1973), 331–40.

31. This address had been published in the *Evangelical Magazine* 17 (February 1809):83.

32. In 1816, Wray suggested to the LMS directors the publication of a tract he had written two years before under the title "A View of the Political Benefits Which Would Accrue from the Instruction of the Negroes in the Principles of Christianity," in which he discussed several aspects of slavery in the colony, including many insurrections. LMS IC, Berbice, Wray's letter, December 13, 1816.

33. *Royal Gazette,* March 1, 1821.

34. LMS IC, Berbice, Wray's letter, July 4, 1815.

35. *A Plain Catechism, Containing the Most Important Doctrines and Duties of the Christian Religion, to Which Is Added the Duty of Children and Servants* (London, 1810), in P.R.O. C.O. 111/11. See also LMS Board Minutes, June 25, 1810.

36. LMS IC, Demerara, Appendix B 2.

37. Mary Turner notices that in Jamaica both black Baptists and Wesleyans "attracted support from freedmen and slaves conscious of the oppression they suffered and anxious to find new compensations for their disabilities and new outlets for frustration. Discontent as well as religious interest and curiosity assembled their first congregations." The intense hostility displayed to black Baptists and Wesleyans "made continued support of these churches . . . a form of opposition to the system, a tradition periodically refreshed over the years by the opposition the missionaries continued to encounter." *Slaves and Missionaries,* 197.

38. LMS IC, Berbice, Wray's letter, January 11, 1816.

39. Wray, Journal, December 19, 1808.

40. An Act for Regulating the Government and Conduct of Slaves in Dominica stipulated that "every owner, renter or director, or the attorney, agent or other representative of such owner . . . of any slave or slaves attached to plantations, shall and he is hereby required to allot to each and every such slave or slaves a sufficient

portion of land, not less than half an acre for each slave of whatever age; and also allow one day in every week over and above Sundays and the holidays . . . for the purpose of cultivating the said land in provisions." *Papers Relating to the Treatment of Slaves in the Colonies. Acts of Colonial Legislatures, 1818–1823* (House of Commons, 1824).

41. LMS IC, Berbice, Wray's letter, December 13, 1816. Cheveley also noticed that many whites were living with colored women, but did not allow them to sit at their table with them. In the Anglican Church, blacks were segregated in a gallery. Advertisements for new plays being shown in Demerara stressed that "free coloured persons only will be admitted in the back seats."

42. LMS IC, Berbice, Wray's letter, May 7, 1820.

43. LMS IC, Berbice, Wray's letter, March 28, 1822.

44. LMS IC, Demerara, Wray's letter, August 31, 1812.

45. Manumissions were not very common in Demerara. From 1808 to 1821 only 142 males and 335 females were manumitted in the colonies of Demerara and Essequibo. These figures include what were defined as ordinary and extraordinary manumissions, giving an average of 34 a year. There was a substantial increase during the years 1815 and 1816, when sugar prices were high. Of the males manumitted, only one in ten was an adult; the others were male children manumitted with their mothers. Considering that the ratio of all males to females manumitted was about 10 females for every 13 males, it is obvious that women had more opportunities than adult males. A large number of manumissions were bought by the slaves or other blacks rather than granted by whites. P.R.O. C.O. 111/37.

46. Barbara Bush contends that the relations between black women and white men have been exaggerated. "Though undoubtedly a small minority of women followed this pattern, the indications are that it was far from typical. . . . [T]he majority of women rejected rather than encouraged the sexual advances of white men." Barbara Bush, "Towards Emancipation: Slave Women and Resistance to Coercive Labour Regimes in the British West Indian Colonies, 1790–1838," in David Richardson, ed., *Abolition and Its Aftermath: The Historical Context, 1790–1916* (London, 1985), 46–47.

47. LMS IC, Berbice, Wray's letter, July 4, 1815. See also Wray, Journal, May 15, 1815.

48. LMS IC, Berbice, Wray's letter, December 16, 1823.

49. Richard Price finds similar patterns. One of the Moravians who had established himself among the Saramakas noted that each "negro" was allowed to take as many wives as he was able to support in a "respectable" way. Price concludes that the proportion of adult men who at any given time had two or more wives was close to 20 percent among the eighteenth-century Saramakas. *Alabi's World* (Baltimore, 1990), 382–83.

50. *Royal Gazette,* November 21, 1820.

51. "Creole" is used here to designate slaves born in the New World.

52. G. W. Roberts, "Movements in Slave Populations of the Caribbean During the Period of Slave Registration," in Vera Rubin and Arthur Tuden, eds., *Comparative Perspectives on Slavery in New World Plantation Societies* (New York,

1977), 145–60. *The New Times,* March 22, 1824, mentioned that in 1820 there were 24,526 male African-born slaves and 14,282 females. See also P.R.O. C.O. ZMCI/ 1039.

53. On slave conversions, see Mechal Sobel, *Trabelin' On: The Slave Journey to an Afro-Baptist Faith* (Princeton, 1979), 108–9; Donald G. Mathews, "Religion and Slavery: The Case of the American South," in Roger Anstey, Christine Bolt, Seymour Drescher, eds., *Anti-Slavery, Religion, and Reform* (Hamden, Conn., 1980), 221–22; Albert Raboteau, *Slave Religion: The "Invisible Institution" in the Antebellum South* (New York, 1980).

54. As in Africa, Christianity may have appeared as an alternative to the cleansing rituals designed to rid the community of witchcraft. For Africa, see the important essay by Robert Strayer, "Mission History in Africa: New Perspectives on an Encounter," *African Studies Review* 19:1 (April 1976):3. Similar phenomena occurred in the New World. Richard Price notices that Christianity was used by the Saramaka maroons in Suriname as protection against witchcraft. In discussing rare cases of conversion of Saramakas, Price also reveals their pragmatism. *Alabi's World,* 261, 424.

55. LMS IC, Demerara, Wray's letter, May 9, 1812.

56. Wray, Journal, September 10, 1808.

57. LMS IC, Demerara, Wray's letter, June 4, 1808. Albert Raboteau found that slaves in the southern United States also believed that when they died they would return to Africa. *Slave Religion,* 32.

58. Monica Schuler, commenting on Jamaica, rejects the missionary version of African "conversion" as inaccurate, since slaves merely adopted from Christianity "familiar elements which paralleled their own beliefs, placing the borrowed symbols, ideas, and practices within an essentially African context." Monica Schuler, *"Alas, Alas, Kongo": A Social History of Indentured African Immigration into Jamaica, 1841–1865* (Baltimore, 1980), 86. The presence of modified forms of "African" rituals in several parts of the Caribbean today lends credibility to her view.

59. On the compatibility between many African religious beliefs and Christianity, see Michael Craton, "Slave Culture, Resistance and Emancipation," in James Walvin, ed., *Slavery and British Society, 1776–1846* (Baton Rouge, 1982), 112–13.

60. Paul Bohannan and Philip Curtin, *Africa and Africans* (3rd ed., Champaign-Urbana, Ill., 1988), 191.

61. Wray, Journal, May 15, 1808. Wray reported another funeral in a letter dated October 30, 1813, from Berbice. This time it was a woman's funeral, but the ceremony was very similar. The only thing Wray added to his version of the death rites was that the slaves sent messages to their uncles, aunts, and other relatives who were dead, and emptied rum glasses on the coffin and put food as well as pipes and tobacco into the coffin. For detailed descriptions of funeral rites among the Saramakas, see Price, *Alabi's World,* 88, 104, 217, 399. Similar practices were also to be found in many Caribbean societies. See also Jerome S. Handler and Frederick W. Lange, *Plantation Slavery in Barbados: An Archaeological and Historical Investigation* (Cambridge, Mass., 1978), and Turner, *Slaves and Missionaries,* 55.

62. LMS IC, Demerara, Davies's letter, May 3, 1814. Albert J. Raboteau notes

the importance of charms among slaves in the United States, and suggests that for many the Bible itself was a charm. *Slave Religion,* 33–35.

63. LMS IC, Demerara, Davies's letter, May 3, 1814.

64. Hesketh J. Bell says that the word is probably derived from *obi,* "a word used on the East coast of Africa to denote witchcraft, sorcery and fetishism in general." See Hesketh J. Bell, *Obeah: Witchcraft in the West Indies* (London, 1893), 6.

65. LMS IC, Demerara, a copy of Wilberforce's letter, September 13, 1820.

66. There is a great similarity between the way "obeah" was handled by missionaries and the story of the benadanti reported by Carlo Ginzburg in *The Night Battles: Witchcraft and Agrarian Cults in the Sixteenth and Seventeenth Centuries* (Baltimore, 1983). A ritual that is not necessarily evil is perceived as such by the representatives of the Church, and gradually its practitioners come to play the adversarial role the Church had attributed to them.

67. Because there is some possibility that the word *Congo* was confused with *Kong,* which is the name of a group living in a different area, Wray's identification is not entirely reliable. Yet Hesketh J. Bell describes a similar Congo dance "in which the performers stood round in a ring and without moving from their places just lifted one foot from the ground, bringing it down again with a stamp in a sort of cadence, continually bowing to each other, and muttering some refrain started by one of them, clapping their hands the while." *Obeah: Witchcraft in the West Indies,* 32.

68. Richard Price describes a modified version of the use of water to heal diseases or to purify the community and protect it against evil. The herb used was called "siebie siebie" (*Scoparia dulcis*), and was also used by slaves in similar rituals in Demerara. Price, *Alabi's World,* 36.

69. The expression "trances" is mine, not Wray's.

70. Monica Schuler makes a distinction between obeah (evil practices) and myal (aiming at eradicating evil). Studying Jamaica, she describes a myal ritual to discover the sources of obeah, similar to the rite reported by Wray. *"Alas, Alas, Kongo,"* 41–42. Mary Turner also notes the similarity between myal preachers in twentieth-century Jamaica and "fanti" priests among the Akan people. *Slaves and Missionaries,* 56.

71. For the distinction between "good" and "bad" medicine, see Richard B. Sheridan, *Doctors and Slaves: A Medical and Demographic History of Slavery in the British West Indies, 1680–1834* (Cambridge, Eng., 1984), 75–77; George Way Harley, *Native African Medicine, with Special Reference to Its Practice in the Mano Tribe of Liberia* (London, 1970); M. J. Field, *Religion and Medicine of the Ga People* (London, 1961); and D. Maier, "Nineteenth-Century Asante Medical Practices," *Comparative Studies in Society and History* 21:1 (January 1979):63–81. See also Monica Schuler, "Afro-American Slave Culture," in Michael Craton, ed., *Roots and Branches: Current Directions in Slave Studies* (Toronto, 1979), 121–37; see also Edward K. Brathwaite's commentary in the same anthology, p. 155.

72. I am paraphrasing here remarks made by Paul Bohannan and Philip Curtin in *Africa and Africans,* 199.

73. Exodus 22:18.

74. "The judicial apparatus of the West India colonies," notes one scholar, "was neither uniform nor efficient. Crown colonies retained the judicial systems of their former rulers. In Saint Lucia the system was French; in Trinidad, Spanish. The legal system in British Guiana was Roman Dutch." William A. Green, *British Slave Emancipation: The Sugar Colonies and the Great Experiment, 1830–1865* (Oxford, 1976), 78.

75. The estate belonged to Winter and Co. of London, and Wray was hoping that if they could understand the benefits that would come from preaching the Gospel to the slaves, they might help the mission.

76. LMS IC, Berbice, Wray's letter, October 19, 1819.

77. It is possible to argue that the slave experience tended, in some circumstances, to alter innocent traditional rituals, conferring upon them a cruel and sinister character. For an analogous phenomenon involving South American Indians, see Michael Taussig, "Culture or Terror—Space of Death: Roger Casement's Putumayo Report and the Explanation of Torture," *Comparative Studies in Society and History* 26:3 (July 1984):467–97. See also his suggestive book, *The Devil and Commodity Fetishism in South America* (Chapel Hill, 1980). Albert J. Raboteau has found evidence of the practice of the "Water Mamma" dance in the southern United States. *Slave Religion,* 26.

78. LMS IC, Berbice, Wray's letter, January 15, 1822.

79. LMS IC, Demerara, Gravesande's letter, February 6, 1815.

80. LMS IC, Berbice, Wray's letter, June 9, 1814.

81. LMS IC, Demerara, Wray's letter, April 19, 1813.

82. LMS IC, Berbice, Wray's letter, June 9, 1814.

83. LMS IC, Berbice, Wray's letter, May 30, June 3, and July 4, 1815.

84. LMS IC, Demerara, Wray's letter, May 23, 1812.

85. LMS IC, Demerara, Wray's letter, October 12, 1812. Wray's letter, December 17, 1812.

86. That happened in 1819 at *Le Resouvenir* when the manager Van der Haas was replaced by John Hamilton.

87. As Sidney Mintz observed, "the masters' monopoly of power was constrained not only by their need to achieve certain results in terms of production and profit, but also by the slaves' clear recognition of their masters' dependence upon them." This dependence, of course, was affected by the masters' ability and readiness to inflict punishment. Sidney Mintz and Richard Price, *An Anthropological Approach to the Afro-American Past: A Caribbean Perspective* (Philadelphia, Institute for the Study of Human Issues, Occasional Papers in Social Change 2, 1976).

88. LMS IC, Demerara, Appendix B 1, Bentinck's Proclamation. Also LMS Board of Directors' Minutes, August 19, 1811, and P.R.O. C.O. 111/11.

89. That this reputation was underserved is obvious to anyone who knows the insignificant role the King played in the abolition of the slave trade. According to Robin Blackburn, Romilly, the former solicitor general, was privately outraged when the Royal Jubilee of 1809 was made the occasion to credit George III with passage of the Abolition Bill. *The Overthrow of Colonial Slavery, 1776–1848* (London, 1988), 315.

90. P.R.O. C.O. 111/11.

91. LMS Board Minutes, November 5, 1811, January 10 and 27, 1812.

92. LMS IC, Demerara, Wray's letters, January 13, April 6 and 10, and May 12, 1812.

93. Demerara and Essequibo 1812–1815, Minutes of the Court of Policy. PRO CO 114/8.

94. LMS IC, Demerara, Wray's letter, February 6, 1813.

95. LMS IC, Demerara, Wray's letter, June 18, 1821.

Chapter 4.
A True Lover of Man

1. LMS IC, Berbice, Wray's letter, August 6, 1815.

2. LMS IC, Berbice, Wray's letter, January 11, 1816.

3. LMS Board Minutes, January 8, 1816.

4. LMS Candidates' Papers, James's letter, February 5, 1816.

5. LMS Candidates' Papers, James's letter, February 23, 1816. Cf. R. W. Dale, ed., *The Life and Letters of John Angell James: An Unfinished Autobiography* (New York, 1861), 142.

6. There is a discrepancy between Smith's first biography, published in the *Evangelical Magazine* after his death, and the information given in Dale, ed., *The Life and Letters of John Angell James,* 142. There is also some confusion about whether there were two John Smiths applying at more or less the same time or just one. After consulting all the board of directors' minutes and the minutes of the committee of examination from 1812 to 1817, and comparing them with other sources, I concluded that Dale's version is probably more correct than the one provided by Smith's earlier biographer and later endorsed by others. The question is, why have John Smith's biographers (including the London Missionary Society's own) never mentioned James's prophetic letter? Is it possible that, writing immediately after Smith's trial, the first biographer thought that any allusion to James's letter might be detrimental to the LMS, or to Smith's own reputation? Smith's enemies would probably have loved to see James's letter, for it would give them even more ground for their claim that Smith had been responsible for the rebellion. For Smith's biographers, see Introduction, note 1. Stiv Jakobson's essay on Smith published in *Am I Not a Man and a Brother? British Missions and the Abolition of the Slave Trade and Slavery in West Africa and the West Indies, 1786–1838* offers an interpretation very similar to ours and relies on the same documents from the LMS, but Jakobson minimizes the role of the slaves.

7. LMS Candidates' Papers, Smith's letter, January 22, 1816.

8. Smith's biographers say he was born in 1790 but there are references in the LMS board of directors' minutes to the fact that he was twenty-four years old in 1816 when he applied to be a missionary, which would make his birth year 1792. There is also evidence showing that he was fourteen when he started his seven-year apprenticeship; it terminated in 1812–13, which also seems to indicate that he was

born in 1792. The date 1790 was given in a letter of 1816 by James Scott of Somers Town. LMS Candidates' Papers, 1816.

9. The letter from James Scott of Somers Town confirms Smith's autobiographical sketch. It says among other things that Smith's "character stands high amongst the members of the Church who had had the opportunity to be best acquainted with him," that he had been particularly useful in instructing the children in the Sunday school, and that should he leave the place his loss would be much regretted. Quoting the oldest deacon at Tonbridge, who called Smith "a very pious and sensible man," Scott went on to say that he had testified that Smith was "truly evangelical in sentiment and very zealous in the cause of God." Smith's first thoughts about being a missionary dated from 1811, when he heard a Mr. Jefferson preach a missionary sermon. LMS Candidates' Papers, 1816.

10. Letter from John Smith, St. John's Lane, Clerkenwell, 1814, in LMS Minutes of the Committee of Examination, 1814. See also LMS Board Minutes, January 17, 1814.

11. LMS Minutes of the Committee of Examination, July 26, 1815, notes a letter from Mr. James of Birmingham "respecting a young man who was desirous of becoming a Missionary to whom he had given some Instructions, but whose time was so fully occupied that Mr. James proposed to allow him four shillings a week toward the support of himself and his mother to enable him to devote two or three hours a day to improvement." The proposal was accepted. The name of the person is omitted but there is reason to believe that James was referring to Smith. According to the minutes of January 8, 1816, a letter was read from Mr. J. Smith from Middle Street, Gosport, a student supported by the Hampshire Association with "a view to his being an itinerant offering himself as a missionary." It was agreed that a letter should be written to his pastor, the Rev. Mr. James of Birmingham, "for his testimonial which if satisfactory Mr. Smith will be authorized to attend next Monday." On February 12, 1816, John Smith of St. John's Lane, Clerkenwell, came before the committee for his first examination. It was satisfactory, and it was agreed that he come again two weeks later. On February 26, Smith was examined for a second time, and the committee decided to recommend that the board of directors accept him. See also Board Minutes, Meeting of the Directors, January 8, 22, and 29, and February 12 and 26, 1816; and Dale, ed., *The Life and Letters of John Angell James,* 142.

12. LMS Board Minutes, March 6, 1816.

13. Students at Gosport received 120 lectures on theology; 30 on the Old Testament and 30 on the New; 20 on evidences of Christianity; 16 on Jewish Antiquities; 35 missionary lectures; 40 on the pastoral office; 5 on universal grammar; 5 on logic; 35 on rhetoric; 28 on ecclesiastical history; 4 lectures on dispensations before the Christian era; 11 on the different periods of the Church before Christ; and 30 on geography and astronomy. They were also drilled in composition and preaching. Some students of the "learned languages" read parts of Caesar, Sallust, Homer, the Hebrew Bible, Cicero, Ovid, Xenophon, and Homer. LMS Minutes of Committee of Examination, May 15, 1815.

14. Smith's letter to the LMS Directors, October 17, 1816, included by mistake

among LMS West Indies and British Guiana, Journals. See also Board Minutes, October 21, November 25, 1816.

15. "Letter on the Exemplary Behaviour of Ministers," *Evangelical Magazine* 6 (August 1798): 319–22. For other examples, see the magazine's "Consideration Recommended to the Missionaries," ibid. 4 (August 1796): 332–37; "A Farewell Letter from the Directors of the Missionary Society to the Missionaries Going Forth to the Heathen in the South-Sea Islands," ibid. (September 1796): 353–59; "The Dignity of the Ministerial Character" ibid. (September 1796): 362–64; "An Address from the Directors of the Missionary Society," ibid. (December 1796): 493–503; "Christian Patience," ibid. 5 (January 1797): 28–30; "An Address from the Ministers of Christ, in the Direction of the Missionary Society, to Their Brethren in the Gospel Ministry," ibid. (August 1797): 321 and 362–68; and "On the Exemplary Behaviour of Ministers, Letter II," ibid. 6 (September 1798): 360–61.

16. About the missionaries' vocation for martyrdom, see C. Duncan Rice, "The Missionary Context of the British Anti-Slavery Movement," in James Walvin, ed., *Slavery and British Society 1776–1846* (Baton Rouge, 1982), 158–59.

17. "Address from the Directors of the Missionary Society to Their Brethren in the Gospel Ministry," *Evangelical Magazine* 5 (August 1797): 365.

18. Ibid. 4 (August 1796): 322–37.

19. Ibid. 4 (August 1796): 337.

20. *Documentary Papers Produced at the Trial of Mr. John Smith, Missionary,* in *Further Papers: Copy of the Documentary Evidence Produced Before a General Court Martial* (House of Commons, 1824), 27–29. Cited hereafter as *Further Papers, Documentary Evidence.*

21. George Rudé stresses that "although the antagonism between capital and labour still remained muted, the 'meaner sort' had a strong sense of how far it was decent for the rich, or those living 'at the polite end of the town,' to flaunt their wealth in the faces of the poor. . . . This popular egalitarianism, so strongly at variance with the tenets of an aristocratic society, was no doubt fostered by the long tradition of religious freedom and democracy which the 'middling' and poor, through dissenting chapels and conventicles and tavern discussions, had inherited from the days of the great Revolution and the Good old Cause. It was linked, too, with the strong popular belief that the Englishman, as a 'free-born' subject, had a particular claim on 'liberty'." George Rudé, *Hanoverian London, 1714–1808* (London, 1971), 98–99.

22. James Walvin, "The Rise of British Popular Sentiment for Abolition, 1787–1832," in Roger Anstey, Christine Bolt, and Seymour Drescher, eds., *Anti-Slavery, Religion, and Reform,* (Hamden, Connecticut, 1980), 153. See also James Walvin, "The Public Campaign in England Against Slavery, 1787–1834," in David Eltis and James Walvin, eds., *The Abolition of the Atlantic Slave Trade: Origins and Effects in Europe, Africa, and the Americas* (Madison, Wisc., 1981).

23. P.R.O. C.O. 323/41. See also Blackburn, *The Overthrow of Colonial Slavery, 1774–1848* (London, 1988), 320–22.

24. Hilary McD. Beckles, "Emancipation by Law or War? Wilberforce and the 1816 Barbados Slave Rebellion," in David Richardson, ed., *Abolition and Its After-*

math: The Historical Context, 1790–1916 (London, 1985), 80–103; Michael Craton, "The Passion to Exist: Slave Rebellions in the British West Indies, 1650–1832," *Journal of Caribbean History* 13 (1980): 12; Michael Craton, "Proto-Peasant Revolts: The Late Slave Rebellions in the British West Indies, 1816–1832," *Past and Present* 85 (1979): 119; Michael Craton, *Testing the Chains: Resistance to Slavery in the British West Indies* (Ithaca, New York, and London, 1982).

25. Smith, Journal, January 1, 1818.

26. Bolingbroke, *Voyage to the Demerary,* 12; Richard Shomburgk's *Travels in British Guiana 1840–1844,* I:15.

27. Cheveley, Journal.

28. All these figures are for the year Bolingbroke visited Demerara. See *Voyage to the Demerary,* 71. Between 1817 and 1823 the exchange rate seems to have ranged between £80 and £100 to 1,000 guilders.

29. LMS IC, Demerara, Smith's letter, March 4, 1817.

30. LMS IC, Demerara, Smith's letter, March 4, 1817.

31. Smith, Journal, March 19 and 20, 1817.

32. LMS IC, Demerara, Smith's letter, May 7, 1817.

33. Smith, Journal, February 17, 1818.

34. Ibid., May 25, 1817.

35. Isaiah 9:17 reads: "Therefore the Lord shall have no joy in their young men, neither shall have mercy on their fatherless and widows; for every one is an hypocrite and an evildoer, and every mouth speaketh folly. For all this his anger is not turned away, but his hand is stretched out still." It is impossible to know precisely what Smith said, or whether he confined himself to 9:17. But it is easy to understand the impact his words may have had, particularly if he also used 9:19, which has a promise of war and violence: "Through the wrath of the Lord of hosts is the land darkened, and the people shall be as the fuel of the fire: no man shall spare his brother."

36. Smith, Journal, July 9, 1818. Slaves transformed missionary messages everywhere. For Jamaica, see Mary Turner, *Slaves and Missionaries: The Disintegration of Jamaica Slave Society, 1787–1834* (Champaign-Urbana, Ill., 1982), 95.

37. Smith, Journal, February 9, 1821.

38. Ibid., June 14, 1818.

39. LMS IC, Smith's letter, February 24, 1817.

40. LMS IC, Demerara, Smith's letter, October 13, 1817; Smith, Journal, June 22, December 10, 1817; November 14, 1821.

41. Smith, Journal, July 4, 1818.

42. The Moravians who went to live among the Saramaka maroons in the eighteenth century had similar experiences. Richard Price notices that "Differing missionary and Saramaka notions regarding property and its role in social relations served as a privileged symbolic idiom in the negotiation of power throughout the history of the mission." *Alabi's World* (Baltimore, 1990), 99. Apparently, in African stateless societies a man could collect his debts by going to the compound of his debtor and removing the property (a goat, for example, or something equivalent) that was owed to him. According to this tradition Romeo would have had the right

to "appropriate" Smith's chickens. About this African tradition, see Paul Bohannan and Philip Curtin, *Africa and the Africans* (3rd ed., Champaign-Urbana, Ill., 1988), 166. The reputation of being incorrigible thieves that the slaves had among whites may in part be attributed to this tradition. For an interpretation that sees theft as a form of resistance, an expression of the slaves' moral economy, see Alex Lichtenstein, "That Disposition to Theft, with Which They Have Been Branded: Moral Economy, Slave Management and the Law," in Paul Finkelman, ed., *Rebellions, Resistance, and Runaways Within the Slave South* (New York, 1989), 255–82.

43. Smith, Journal, October 7, 1818.

44. In keeping with congregationalist practice, religious services were open to all, but church meetings were only for communicants.

45. Smith, Journal, May 10, 1817.

46. Ibid., September 20, 1821.

47. Ibid., June 15, 22, 1817.

48. Ibid., October 1, 1821.

49. Ibid., August 1, 1817.

50. Ibid., August 6, 1821.

51. Ibid., April 1, 1821. The way that Smith records Dora's story, saying that her son had been "stolen" from her is revealing of his antislavery point of view.

52. Ibid., June 30, 1818.

53. Ibid., January 13, 1822.

54. Ibid., March 23, 1818.

55. Bristol's deposition in Smith's trial, in *Copy of the Proceedings,* House of Commons, 1824, 13.

56. Donald Mathews remarks that "rooted in the reformed traditions of both the Continent and Great Britain, evangelicalism brought an intense personal piety to hundreds of thousands of people in the English-speaking world before the end of the eighteenth century." Donald G. Mathews, "Religion and Slavery: The Case of the American South," in Anstey, Bolt, and Drescher, eds., *Anti-Slavery, Religion and Reform,* 208. See also Donald G. Mathews, *Religion in the Old South* (Chicago, 1977).

57. Smith, Journal, March 13, 1817.

58. Ibid., August 10, 1817.

59. Ibid., September 28, 1817.

60. Ibid., March 15, 1818.

61. Ibid., July 19, 1818. Exodus 3:7–8 could indeed be applied to the slaves as to any oppressed people in the world, particularly because it contained a promise of deliverance: "And the Lord said, I have surely seen the affliction of my people which are in Egypt, and have heard their cry by reason of their task-masters; for I know their sorrows; And I am come down to deliver them out of the hand of the Egyptians, and to bring them up out of that land unto a good land and a large, unto a land flowing with milk and honey."

62. Ibid., October 16, 1817.

63. Ibid., November 2, 1817.

64. When an old slave named Hannah died on January 3, 1818, Smith wrote in his diary: "She lived unnoticed, and unnoticed died. The princess Charlotte had

many attendants, but Hannah none except those kind angels who are sent to minister to the heirs of salvation. If as some have said, to love God be an earnest of heaven, I believe Hannah had that earnest." Smith, Journal, January 3, 1818.

65. In one of his letters, Smith mentioned that the slaves had gathered £190. LMS IC, Demerara, Smith's letter, October 14, 1818. In another, dated November 29, 1822, he noted £130 *s* 13.

66. Smith, Journal, May 11, 1818.

67. Ibid., March 30, 1817.

68. For a description of how Saramaka maroons handled a case of adultery, see Price, *Alabi's World,* 118.

69. Smith, Journal, April 6, 1817.

70. Ibid., April 11, 1817.

71. LMS IC, Demerara, Smith's letter, May 7, 1817.

72. Smith, Journal, July 9, 1818.

73. Ibid., May 13, August 11, 1817.

74. The Smiths' maid Charlotte said that "Mr. Smith flogged Cooper once very much with a horse-whip for impudence to Mrs. Smith." "Statement of the Negro Woman Charlotte," in *Further Papers,* 52.

75. Smith, Journal, October 19, 1819.

76. Ibid., April 15, 1821.

77. Ibid., April 29, 1821.

78. Ibid., June 3, 1821.

79. Ibid., October 6, 1822. "Cohabit" and "ravish" were Smith's renderings of what the slaves told him.

80. Ibid., March 5, 1820. The same problem was faced by missionaries in the United States. Albert J. Raboteau, *Slave Religion: The "Invisible Institution" in the Antebellum South* (New York, 1980), 185.

81. Smith, Journal, September 7, 1821.

82. Ibid., October 9, 1821.

83. Robert Dirks questions estimates based on planters' reports. He calculates a daily caloric intake of 1,500 to 2,000. "Resource Fluctuations and Competitive Transformations in West Indian Slave Society," in Charles Laughlin, Jr., and Ivan A. Brady, eds., *Extinction and Survival in Human Populations* (New York, 1978), 137–39. For a more optimistic calculation, see Kenneth F. Kiple and Virginia A. Kiple, "Deficiency Diseases in the Caribbean," *Journal of Interdisciplinary History* 11:2 (Autumn 1980): 197–215.

84. On diet, clothing, and housing in other areas of the Caribbean, see Richard B. Sheridan, *Doctors and Slaves: A Medical and Demographic History of Slavery in the British West Indies, 1680–1834* (Cambridge, Eng., 1985), 135–41, 163–78, and chap. 3, *passim.*

85. Smith's letter, reproduced in the Rev. Edwin Angel Wallbridge, *The Demerara Martyr: Memoirs of Reverend John Smith, Missionary to Demerara, with Prefatory Notes Containing Hitherto Unpublished Historical Matter by J. Graham Cruickshank,* Vincent Roth, ed. (Georgetown, 1943), 56–72.

86. Smith, Journal, September 3, 1817.

87. Ibid., April 6, 1817.
88. Ibid., May 4, 1817.
89. Ibid., September 7, 1817.
90. Ibid., September 13, 1817.
91. Ibid., September 13 and 14, 1817.
92. Ibid., September 28, 1817.
93. Ibid., September 18, October 7, 1817.
94. Ibid., April 29 and March 29, 1818.
95. Ibid., March 29, 1818.
96. Ibid., April 19, 1817, March 21, 1819.
97. Ibid., September 3, 1817. Four months later, on Christmas Day, he wrote, "As usual the whip was cracking upon the negroes' backs from 12 to 1 o'clock." Ibid., December 25, 1817.
98. Ibid., March 2, 1818.
99. The *Rule on the Treatment of Servants and Slaves,* reissued in 1784 by the Dutch, stipulated: "In case a slave has misbehaved in such manner that necessity obliges him to be punished it shall be done moderately and not in cruelty or passion. Such punishment from the master to his slave may not be extended to more than 25 lashes in order that all misfortunes may be avoided and such punishment may not be inflicted before the slave shall have been laid down flat on his belly and tied between or to four posts." It also said that if a more severe punishment was required the slave should be sent to the fortress, where he would be punished after a formal sentence had been obtained. Apparently the number of strokes authorized by law had since increased.
100. Smith, Journal, July 12, 1818.
101. Ibid., August 10, 1818.
102. Ibid., October 13, 1821.
103. LMS IC, Demerara, Smith's letter, March 1818.
104. Smith, Journal, June 28, 1821.
105. LMS IC, Demerara, Smith's letter, February 18, 1823.
106. Throughout the Caribbean missionaries had the same complaints. Sheridan, *Doctors and Slaves,* 167.
107. Smith, Journal, May 10 and August 21, 1817.
108. Ibid., February 12, 1819.
109. Ibid., November 5, 1820.
110. Ibid., August 12, 1822.
111. LMS IC, Demerara, Smith's letter, March 29, 1820.
112. Smith, Journal, March 5, 1818.
113. LMS IC, Demerara, Smith's letter, June 4, 1818.
114. Smith, Journal, April 18, 1818.
115. Ibid., May 14, 1818.
116. Smith's annual stipend was guaranteed by Post's will, and he could be reasonably confident of it. But there were also many other little perquisites such as milk, vegetables, fresh fish, fuel, which the manager could easily vary or withhold as he pleased.

117. LMS IC, Demerara, Smith's letter, October 13, 1817.

118. LMS IC, Demerara, Smith's letter, March 19, 1820.

119. LMS IC, Demerara, Davies's letter, March 2, 1819.

120. Smith, Journal, December 10, 1817: "My journey was rendered extremely unpleasant by my companion violently slandering the character of Mr. . . . " Also December 5 and December 26: "I never liked the company of Mr. . . . and Mrs. . . . [T]hey are always scandalizing the Missionary Society."

121. Ibid., September 23, 1821.

122. Ibid., November 12, 1821.

123. Ibid., December 5, 1817.

124. Ibid., November 25, 1822.

125. Ibid., December 31, 1821.

126. Ibid., March 26, 1821.

127. In February of 1823, he boasted that the average congregation was of 800 persons. LMS IC, Demerara, Smith's letter.

128. Smith, Journal, September 1, 1822.

129. Donald G. Mathews, "Religion and Slavery: The Case of the American South," in Anstey, Bolt, and Drescher, eds., *Anti-Slavery, Religion, and Reform,* 222. See also Donald G. Mathews, *Slavery and Methodism: A Chapter in American Morality, 1780–1845* (Princeton, 1965) and *Religion in the Old South* (Chicago, 1977); Mechal Sobel, *Trabelin' On: The Slave Journey to an Afro-Baptist Faith* (Westport, Conn., 1979); Eugene D. Genovese, *Roll, Jordan, Roll* (New York, 1974), 161-284; Mary Turner, *Slaves and Missionaries: The Disintegration of Jamaican Slave Society, 1784–1834* (Champaign-Urbana, Ill., 1982).

130. Smith, Journal, February 18, 1820.

131. Ibid., March 18, 1821.

132. Ibid., September 1822.

133. On smallpox, see Sheridan, *Doctors and Slaves,* 249–67.

134. A few months earlier, Governor Murray had issued a proclamation declaring that smallpox had been brought to the colony "by a negro from St. Vincent." Murray had required all persons to inoculate their families and slaves with cow pox as quickly as possible. *Royal Gazette,* June 12, 1819.

135. Smith, Journal, October 31, November 9, 1819.

136. Smith's description contrasts sharply with the idealized picture offered by managers. Fanny Kemble, who lived in the United States on a Georgia plantation in the 1830s, gave an equally devastating account of the "hospital" she found on her husband's plantation. Fanny Kemble, *Journal of a Resident on a Georgian Plantation in 1838–39* (rpt. Chicago, 1969), cited by Sheridan, *Doctors and Slaves,* 289.

137. Smith, Journal, November 21, 29, 30, December 4, 1819.

138. Ibid., October 8, 1822.

139. Ibid., January 27, 1823.

140. *Royal Gazette,* January 3, 1815.

141. Ibid., May 8, 1821.

142. Ibid., December 4, 1821.

143. The membership of the Scottish Kirk included Lachlan Cuming, James

Johnstone, Charles Edmonstone, Alexander Simpson, George Buchanan, Colin Macrae, Alexander Grant, W. M. Munro, Peter Grant, M. Hyndman, Gavin Fullarton, John Fullarton, Donald Campbell, Charles Grant, Michael McTurk, and many others. J. Graham Cruickshank, *Pages from the History of the Scottish Kirk in British Guiana* (Georgetown, 1930).

144. *Royal Gazette,* January 28, February 1, 1822.

145. LMS IC, Demerara, Mercer's letter, February 22, 1820.

146. LMS IC, Demerara, several letters, in particular Smith's of February 28, 1823. Also Smith, Journal, where from September 3, 1822, onward there are several references to his fruitless attempts.

147. Smith, Journal, September 11, 13, 22, 24, 1822.

148. Ibid., October 21, 1822.

Chapter 5.
Voices in the Air

1. Song, quoted in Lawrence Levine, *Black Culture and Black Consciousness: Afro-American Folk Thought from Slavery to Freedom* (New York, 1977), xiii.

2. *The Liberator,* September 3, 1831, quoted in John B. Duff and Peter Mitchell, eds., *The Nat Turner Rebellion: The Historical Event and the Modern Controversy* (New York, 1971), 41.

3. This chapter is based primarily on four sets of records of the many trials that grew out of the events of 1823. Three sets were the records hastily produced by the Colonial Office in response to requests from the House of Commons in the spring of 1824, which the House immediately ordered published: (1) *Demerara . . . A Copy of the Minutes of the Evidence on the Trial of* John Smith, *a Missionary in the Colony of Demerara, with the Warrant, Charges, and Sentence:*—VIZ *Copy of the Proceedings of a General Court Martial* . . . (House of Commons, 1824), which will be cited as *Proceedings;* (2) *Demerara: Further Papers . . . Respecting Insurrection of Slaves . . . with Minutes of Trials* (House of Commons, 1824), which will be cited as *Further Papers Respecting Insurrection;* and (3) *Demerara: Further Papers . . . Copy of Documentary Evidence Produced Before a General Court Martial* . . . (House of Commons, 1824), which will be cited as *Further Papers: Documentary Evidence.* The fourth set consists of manuscripts and other documents in the Public Record Office at Kew, P.R.O. C.O. 111/39–46, particularly untitled correspondence and reports on the rebellion and the trials, in P.R.O. C.O. 111/43. I have also used a a second collection labeled "Negro Trials," in P.R.O. C.O. 111/44 and 111/53. Also used for cross reference was the London Missionary Society's collection of documents, *Report of the Proceedings Against the Late Rev. John Smith* (London, 1824). Particular passages from all these records are cited here only when specific and salient language or information seemed relevant to a reconstruction of what was happening. Any sort of clear understanding of the trial records, however—and certainly any grasp of the experience of the individual slaves who were tried

or who testified—must depend on a careful cross-referencing to many other sources. I have relied particularly on the journals and letters of the missionaries, on other Colonial Office documents in the Public Record Office, and on the two Demerara newspapers, the *Royal Gazette* and the *Guiana Chronicle,* and on Joshua Bryant, *Account of an Insurrection of the Negro Slaves in the Colony of Demerara Which Broke Out on the 18th of August, 1823* (Demerara, 1824), which will be cited hereafter as Bryant, *Account of an Insurrection.*

4. Smith, Journal, April 13, 1817.

5. *Further Papers Respecting Insurrection,* Examination of Bristol, 26–28; Deposition of Paris, 30–31.

6. Smith, Journal, December 15, 1820.

7. *Further Papers Respecting Insurrection,* 26.

8. *Royal Gazette,* April 2, 1822. Also P.R.O. C.O. 114/9.

9. Smith, Journal, August 12, 1822.

10. The importance of rumor in slave rebellions is stressed by Michael Craton, "Slave Culture, Resistance and the Achievement of Emancipation in the British West Indies, 1783–1838," in James Walvin, ed., *Slavery and British Society, 1776–1846* (Baton Rouge, 1982), 105–6.

11. *Further Papers: Documentary Evidence,* 6, 7.

12. The Rev. Wiltshire Staunton Austin, minister of the Established Church, who resided in Georgetown. For a short autobiography see P.R.O. C.O. 111/144.

13. Smith, Journal, June 3, 1823.

14. Ibid., June 3, 1823.

15. Ibid., June 25, 1823.

16. In "A Conversation of Rumors: The Language of Popular Mentalités in Late-Nineteenth-Century Colonial India," Anand A. Yang stresses the role rumors had in rebellions in Bengal. He argues that—as utterances of, by, and for the people—rumors flowed through channels of everyday life, channels of communication created by regular networks of face-to-face relations among neighbors, friends, kinfolk, or even mere "nodding acquaintances who were to be encountered in the local marketplace." Yang also calls attention to the importance of studying rumors as "an entrance to the history of mentalités." *Journal of Social History* 20 (1987):485–505. See also his "Sacred Symbol and Sacred Space in Rural India: Community Mobilization in the Anti-Cow Killing Riot of 1893," *Comparative Studies in Society and History* 22 (1980):576–96. The importance of rumors had been stressed much earlier by Georges Lefebvre, in his now-classic *The Great Fear of 1789: Rural Panic in Revolutionary France* (1st American ed., trans. Joan White, New York, 1973). A similar point has been made by Michael Adas in his study on peasants in Southeast Asia: "From Avoidance to Confrontation: Peasant Protest in Precolonial and Colonial Southeast Asia," *Comparative Studies in Society and History* 23 (1981):271–97.

17. Hansard's Parliamentary Debates, New Series, IX, 256–360.

18. William Green, *British Slave Emancipation: The Sugar Colonies and the Great Experiment, 1830–1865* (Oxford, 1976), 102. See also Lowell Joseph Ragatz,

The Fall of the Planter Class in the British Caribbean, 1763–1833: A Study of Social and Economic History (London, 1928), 411.

19. Edwin Angel Wallbridge, *The Demerara Martyr: Memoirs of the Reverend John Smith, Missionary to Demerara, with Prefatory Notes . . . by J. Graham Cruickshank*, Vincent Roth, ed. (Georgetown, 1943), 93–94. For details, see Ragatz, *The Fall of the Planter Class in the British Caribbean*, 408–15.

20. Demerara and Trinidad were in a particularly vulnerable position because of their status as crown colonies. The old British colonies like Barbados and Jamaica enjoyed more autonomy.

21. Parliamentary Papers, XXIV (1824), 427. These reforms were to be imposed on the crown colonies and only recommended to the others. See also Michael Craton, James Walvin, and David Wright, eds., *Slavery, Abolition and Emancipation: A Thematic Documentary* (London, 1976), 300–303.

22. Guyana National Archives, Minutes of the Court of Policy, August 1823, 1129. See also P.R.O. C.O. 112/5.

23. *Further Papers Respecting Insurrection,* 25, 28, 39, 46–49.

24. Ibid., 46–48. For the consulting of oracles before wars in Africa, see Cyril Daryll Forde and P.M. Kaberry, eds., *West African Kingdoms in the Nineteenth Century* (London, 1967).

25. LMS IC, Berbice, John Wray's letter, December 13, 1823. Wray thought Jack Gladstone was twenty-five, but during the trial Jack himself said that he was thirty. The advertisement offering a reward of 1,000 guilders for the apprehension of Quamina and Jack appeared in the *Berbice Royal Gazette* of September 3, 1823.

26. This letter was later appended to the trial documents. *Further Papers: Documentary Evidence,* 8.

27. Wray said Jack had two children. LMS IC, Berbice, Wray's letter, September 4, 1823. An article published in *The Colonist,* May 3, 1824, said he had four. LMS Odds Colonial Newspapers. The article in *The Colonist* is very revealing. *The Colonist,* refuting stories that had appeared in the *New Times* in London and in the *Morning Chronicle,* reported: "Eight years ago, Jack, in the instability of his temporary connections then called a girl wife, who afterwards left him for the 'use' of the neighbouring proprietor, and to these days he never exhibited any symptoms of ill will toward that gentleman. . . . He was afterward married to a different woman by the late missionary Smith and by her has had four children." Jack's first wife was Susanna.

28. Smith, Journal, December 16, 1821.

29. Ibid., June 1, 1817.

30. *Further Papers Respecting Insurrection,* 43.

31. Jack's testimony, ibid., 39, 43–44, and "Prisoner Jack's Statement in Defence," ibid., 76–79; see also Susanna's and Seaton's testimony, ibid., 28, 43, and 47.

32. Ibid., 47–48.

33. Some slaves later testified that Hamilton had promised to give them arms.

34. Later in the trial Paris denied that he had ever said any of this. *Further Papers Respecting Insurrection,* 41. See also P.R.O. C.O. 111/45.

35. Smith, Journal, July 5, 1823, and *Further Papers Respecting Insurrection,* 27.

36. Romeo's deposition, *Proceedings,* 10.

37. Ibid., 48.

38. Ibid., 49.

39. In the records the word Coromantee also appears as Curamantee, Cromantee, Coramantyn, Coromantine, Koromantee, and sometimes even Komanti.

40. It is possible that those slaves were Popo, a group who may have shared the same language. See the linguistic map of Western Africa in Michael Kwamena-Poh, John Tosh, Richard Waller, and Michael Tidy, *African History in Maps* (New York and London, 1982).

41. In several earlier uprisings there had been an alliance between plantation slaves and the "bush negroes." This was true in 1795, 1804, and 1810. In 1795 there were at least eight large bush settlements in the hinterland of Demerara. In 1804, Cudjoe, one of the leaders of an uprising, who was recognized as a "King" by the Akan slaves, was said to be in contact with the Bush people, and in 1810 it was said that the Bush Negroes of the Abary-Mahaicony area intended to instigate a revolt among the slaves on both the East and the West Coast of Demerara. See Alvin O. Thompson, *Brethren of the Bush: A Study of Runaways and Bush Negroes in Guyana, c. 1750–1814* (Barbados, Department of History, University of West Indies, 1975), 26–35; *Colonialism and Underdevelopment in Guyana, 1580–1803* (Barbados, 1987), 142; and "Some Problems of Slave Desertion in Guyana, 1750–1814" (University of West Indies Occasional Papers, Barbados, 1976).

42. *Royal Gazette,* April 21, 1806, fretted that "Bush negroes" and runaways were coming to the market on Sundays.

43. The best description of the Saramakas and their interaction with plantation slaves and white colonists and missionaries is given in Richard Price, *Alabi's World* (Baltimore, 1990); see also his *The Guiana Maroons (Bush Negroes): A Historical and Bibliographical Introduction* (New Haven, 1972). Thompson, *Colonialism and Underdevelopment in Guyana,* contains a map of maroon communities revealing their relative proximity to plantations. See also James G. Rose, "Runaways and Maroons in Guyana History," *History Gazette* (University of Turkeyen) 4 (January 1989): 1–14.

44. The *Royal Gazette,* February 1, 1815, has an advertisement for a runaway slave called "Goodluck," who called himself Sammy de Groot. He was said to be about five feet, eight inches tall, "a good looking man with whiskers, speaking Dutch and English pretty well." Since Goodluck does not seem to be a common name, it is possible that the runaway is the same person that appears in the records as Jack's friend.

45. Smith, Journal, January 8, 1820. Also Wallbridge, *The Demerara Martyr,* 83.

46. It is possible that instead of being a Congo, the slave was from Kong, one of the many groups that apparently shared a language with other Akan groups.

47. *Further Papers Respecting Insurrection,* 59.

48. For the revolt in Grenada, see Michael Craton, *Testing the Chains: Resistance to Slavery in the British West Indies* (Ithaca and London, 1982), 180–94.

49. This also appears in Bristol's testimony as reported by Alexander McDonnell in *Consideration on Negro Slavery with Authentic Reports, Illustrative of the Actual Condition of the Negroes in Demerara* (London, 1825), 242. Apparently, before the news about the amelioration laws had arrived in the colony, the *Dochfour*'s coopers, who "used to make two puncheons out of dressed staves every day," had "only turned out one, to see if their master would be satisfied."

50. During the Parliamentary debate about Smith's trial, it was said that Susanna was Jack's wife. Other evidence seems to confirm this, though not explicitly. See T. C. Hansard, *Parliamentary Debates, from the Thirtieth Day of March to the Twenty-fifth of June, 1824* (vol. XI, London, 1825).

51. *Further Papers Respecting Insurrection,* 67, 77.

52. The leading role played by skilled workers has been pointed out by many historians of slavery. Dale W. Tomich argues that the privileged position these slaves enjoyed "may have created an ambivalent, but not necessarily conservative response to the slave system. . . . Indeed the contradiction between individual dignity and self-worth, on the one hand, and slavery status, on the other may have been experienced more palpably by these slaves." Dale W. Tomich, *Slavery in the Circuit of Sugar: Martinique and the World Economy, 1830–1848* (Baltimore, 1990), 227. A similar point has been made by Eugene Genovese for the United States in *Roll, Jordan, Roll: The World the Slaves Made* (New York, 1974), and by Orlando Patterson for Jamaica in *An Analysis of the Origins, Development, and Structure of a Negro Slave Society in Jamaica* (London, 1967).

53. Barbara Kopytoff, in a stimulating essay "Religious Change Among the Jamaican Maroons: The Ascendance of the Christian God Within a Traditional Cosmology," *Journal of Social History* 20 (Fall 1986–87): 463–85, analyzes the interplay between African traditional cults and Christianity among maroons in Jamaica. She stresses that maroon religion, like maroon society, was a creole composite from its beginning in the late seventeenth century, and also argues that there was an incompatibility between the Protestant European Christianity preached to the maroons and traditional maroon religious ideas. This became apparent when the Church Missionary Society catechists lived among the maroons.

54. See, for example, the work of Barbara Bush: "Towards Emancipation: Slave Women and Resistance to Coercive Labour Regimes in the British West Indian Colonies, 1790–1838," in Richardson, ed., *Abolition and Its Aftermath,* 27–54; "Defiance and Submission: The Role of Slave Women in Slave Resistance in the British Caribbean," *Immigrants and Minorities* 1 (1982): 16–38, particularly 23–31; "The Family Tree Is Not Cut," Women and Cultural Resistance in Slave Family Life in the British Caribbean," in G. Y. Okihiro, ed., *Resistance: Studies in African Caribbean, Latin American, and Afro-American History* (Boston, 1984); and *Slave Women in Caribbean Society, 1650–1838* (Bloomington, 1986); also Lucille Mathurin, *The Rebel Woman in the British West Indies During Slavery* (Kingston, 1975); Marietta Morrissey, *Slave Women in the New World: Gender Stratification in the Caribbean* (Lawrence, Kan., 1989); Rhoda E. Redcock, "Women and Slavery in the Caribbean: A Feminist Perspective," *Latin American Perspectives* 12 (Winter 1985): 63–80. For a comparison with the United States, see Deborah Gray White, *Ar'n't*

I a Woman? Female Slaves in the Plantation South (New York, 1985); Elizabeth Fox-Genovese, *Within the Plantation Household: Black and White Women of the Old South* (Chapel Hill, 1988) and "Strategies and Forms of Resistance," in Okihiro, ed., *Resistance: Studies in African Caribbean,* 143–65; Angela Davis, "Reflections on the Black Women's Role in the Community of Slaves," *The Black Scholar* 3 (December 1971):3–15.

55. There is at least one piece of evidence that this was the case. At *Foulis,* Daniel called a meeting on Sunday night before the rebellion and ordered the women to leave. All men were then asked to kiss the Bible as a form of oath. *Further Papers Respecting Insurrection,* 12. See also Mathurin, *The Rebel Woman,* 21. Much of the planning for the famous Baptist War of 1831 in Jamaica took place at religious meetings after the women had left. Similarly, in the slave conspiracy in Antigua in 1736 slaves made their pledge after the women left. See David Barry Gaspar, *Bondmen and Rebels: A Study of Master-Slave Relations in Antigua, with Implications for Colonial British America* (Baltimore, 1985), 243. Elizabeth Fox-Genovese mentions that West African traditions did not encourage female participation. *Within the Plantation Household,* 53, note 8.

56. *Further Papers Respecting Insurrection,* 58.

57. Bryant, *Account of an Insurrection,* 68; *Further Papers Respecting Insurrection,* 97–98.

58. Susanna's role is similar to that of Nancy Grigg in Barbados. Nancy told the slaves that they were to be freed on Easter Monday, but for that to happen they had to follow the example of Haiti and fight for it. *The Report from a Select Committee of the House of Assembly Appointed to Inquire into the Origin, Causes and Progress of the Late Insurrection* (Barbados, 1818), 29–31, quoted by Michael Craton, "The Passion to Exist: Slave Rebellions in the British West Indies, 1650–1832," *Journal of Caribbean History* 13 (1980): 16. For details see Craton, *Testing the Chains,* 254–66. See also H. McD. Beckles, "Emancipation by Law or War? Wilberforce and the 1816 Barbados Slave Rebellion," in David Richardson, *Abolition and Its Aftermath: The Historical Context, 1790–1916* (London, 1985), 80–103.

59. There is no doubt that the African experience was crucial in many slave rebellions. Gwendolyn Midlo Hall, in her study of Louisiana in the eighteenth century, has emphasized the importance of slaves from the Senegambia area in the articulation of rebellions. She attributes this capacity for leadership to their language and culture. Ms. Hall was kind enough to allow me to see her book manuscript, "The Afro-Creole Slave Culture of Louisiana: Formation During the Eighteenth Century." Michael Craton has also given evidence of the significant presence of Coromantee in the leadership of rebellions in the Caribbean since the seventeenth century. "The Passion to Exist," 1–20. See also Craton, *Testing the Chains.* Monica Schuler has stressed the importance of the Akan in promoting rebellions. See, in particular, "Akan Rebellions in the British Caribbean," *Savacou* 1:1 (June 1970): 8–32, and "Ethnic Slave Rebellions in the Caribbean and the Guianas," *Journal of Social History* 3 (Summer 1970): 374–85. Slaves identified as Coromante, Koromante, "Cromantee," "Coromantyn," or "Coromantine" were also leaders of important maroon groups, such as those in Jamaica and in Antigua. For the leadership

role played by Coromantee in the conspiracy of 1736 in Antigua, see Gaspar, *Bondmen and Rebels,* 234–38.

60. Richard Hart observes that Coromantee have been identified with the Akan, a name used to describe "linguistically and ethnographically" several groups in West Africa, including Ashante and Fante. But as it has always been noticed, these descriptions are precarious because any slave from different origins, such as Ga, Adagme, or Ewe, leaving from ports in West Africa might receive this identification. Hart notes that the term Coromantee was probably taken originally from the Fante coastal settlement where the English built their first trading post on the Gold Coast in 1631, now called Kromantine. *Slaves Who Abolished Slavery* (2 vols., Jamaica, 1985).

61. *Further Papers Respecting Insurrection,* 67.

62. For a comparison between day names as they were used by Akan, Fante, and Jamaicans, see Hart, *Slaves Who Abolished Slavery,* 2:9. Also Schuler, "Akan Slave Rebellions," 28–29. Peter Wood (relying on an article published by David De Camp, "African Day Names in Jamaica," *Language* 43 (1967): 139–49) gives a list of similar names. *Black Majority: Negroes in Colonial South Carolina, from 1670 through the Stono Rebellion,* 185.

63. Gold Coast trade had been important in the first half of the century when Akan wars had devastated the region. About 321,000 slaves were exported from the Gold Coast. After mid-century, the volume of exports from this area had declined and the trade was concentrated on the Fante coast. The main posts of operation were the English castle at Anambo and the Dutch port of Kromantine. There was a notable expansion of the slave trade in the 1780s and 1790s when the Asante attempted to occupy the Fante coast, and also as a consequence of the struggles among the Akan for mastery of the interior. It is thus possible that many slaves in Demerara came from this area during this period. See Paul Lovejoy, *Transformations in Slavery: A History of Slavery in Africa* (Cambridge, Eng., 1983), 56, 81, 97. See also Schuler, "Akan Slave Rebellions in the British Caribbean," 8–33.

64. Cuffy's plot in Barbados in 1675 was led by Coromantee, who also played an important role in the 1753 rebellion in Antigua. The rebellion of 1760 in Jamaica has also been attributed to Akan-speaking slaves. Igbos and Coromantee slaves were leaders in the Jamaica uprising of 1776, and in many others. Craton, *Testing the Chains,* 115–79, *passim.* See also Schuler, "Akan Slave Rebellions in the British Caribbean," 8–31, and "Ethnic Slave Rebellions in the Caribbean and the Guianas," 374–85; Gaspar, *Bondmen and Rebels,* 234–38. For a description of the Coromantee, see Bryan Edwards, *History, Civil and Commercial, of the British Colonies in the West Indies* (3 vols., London, 1801), 2:74–86.

65. Paul Bohannan and Philip Curtin stress that descent groups may contain several million people and use the sanction of kinship obligations to bind their members to the "right" course of action. They suggest that unilinear descent groups were and still are the basis for most extra-familial social organizations in Africa. They form political groups, religious congregations, and even production and land-owning units. "It is loyalty to the descent group, as well as to the family, that is

under discussion when Africans talk about their 'obligations' to 'their' people." *Africa and the Africans* (3rd ed., Champaign-Urbana, Ill., 1988), 123.

66. Among the many reports published by the African Institution is one by Henry Meredith containing answers to "Queries relative to Africa, as they respect that District of the Gold Coast, called the Agooma Country." The region, located near the Equator, had the sea on the south and the countries of Akoon, Adgumakoon, Assin, Akim, and Akra as its boundaries. It covered an area described as twenty miles in length and fifteen miles in breadth. In this area the "Fantee language" was spoken and the day names were identical to those found in Demerara, where there were many slaves from the Gold Coast.

67. Joseph Greenberg, in the latest edition of *Languages of Africa,* says that the number of languages spoken in Africa may be above 800. The Greenberg classification contains five major language groups. The largest of them, which also covers the most extensive geographical area, Greenberg calls the Niger Kardofan group. Most people belonging to a linguistic family and sub-families would find no difficulty in communicating with each other. *The Languages of Africa* (3rd ed., Bloomington, 1970). See also his *Studies in African Linguistic Classification* (New Haven, 1955).

68. William C. Suttles, Jr., "African Religious Survivals as Factors in American Slave Revolts," *Journal of Negro History* 56 (April 1971) 97–104, cited in Gaspar, *Bondmen and Rebels,* 322.

69. Among the Angolans, for example, in order to bypass lineage and to allow for cooperation with other groups there were several other institutions, including secret societies, uniting people across their social boundaries. Skill or profession provided a network that served to transmit skills and knowledge. Curing cults and witchcraft provided another vehicle through which people could abandon temporarily their primary loyalty. Mbundu neighborhoods held regular circumcision camps in which the young men of a locality joined together, regardless of their position in descent group genealogies. Secret societies called *kilombos* united warriors by co-opting unlimited numbers of alien males. A male, no matter what his ethnic origins, became a fully qualified warrior by demonstrating his personal ability to fight and by completing the kilombo initiation rites. They excluded all women from their ceremonies. Joseph C. Miller, *Kings and Kinsmen: Early Mbundu States in Angola* (Oxford, 1976), 45–52, 151–237.

70. Edward Kamau Brathwaite remarks that Paul Boble (one of the leaders of the Jamaica slave rebellion of 1831 and a pastor in the Baptist black church), Toussaint in Haiti, and Sam Sharpe all "carried the myal title of Daddy (Dada)" and that an "Afro-myal movement" underlay the more liberal-reformist creole concern with justice and land. He finds similar trends in the 1763 slave rebellion in Berbice, in Barbados in 1816, and in the Nat Turner rebellion in the United States in 1831. "The African Presence in Caribbean Literature," in Sidney W. Mintz, ed., *Slavery, Colonialism, and Racism* (New York, 1974), 77–78.

71. Among the Igbos there were hundreds of patrilineal clans. But government consisted of two basic institutions, the Ama-ala Council of Elders and the village assembly of citizens. Any adult male could sit on the council. In routine matters

the elders ruled by decree and proclamation, but when their decisions were likely to produce disputes the Ama-ala could assemble in the ward square. At the assembly the elders laid the issues before the people and every man had a right to speak. The village assembly was considered the Igbo man's birthright. Ibibios were famous for their secret societies. Age-set loyalty could also be as strong as the bonds of lineage. See J. B. Webster, A. A. Boahen, M. Tidy, *The Revolutionary Years: West Africa Since 1800* (London, 1980), 87–107. See also Forde and Kaberry, eds., *West African Kingdoms in the Nineteenth Century,* 9–26.

72. The respect due to elders was common in many African societies. Among the Yoruba, for example, the junior persons had to show respect for the senior by obeisance. Forde and Kaberry, eds., *West African Kingdoms in the Nineteenth Century,* 51.

73. Barry Gaspar finds that in "Akan society, the young men are regarded as a distinct social group called *mmrantie* (plural of *abrantie*). They may chose their own leader, but they owe allegiance to the state ruler and can be called upon by the elders for various services of importance to the state; the elders, however, maintain their distance from the young men, who have frequently posed a political threat to traditional Akan rulers." Gaspar, *Bondmen and Rebels,* 238.

74. On this subject, see the works of Michael Craton, Edward Kamau Brathwaite, Richard Price, Sidney Mintz, Monica Schuler, and Franklin Knight, as cited in Chapter 2, note 140.

75. For an admirable example of oral history, see Richard Price, *First Time: The Historical Vision of an Afro-American People* (Baltimore, 1983).

76. See Monica Schuler, *"Alas, Alas, Kongo": Social History of Indentured African Immigration into Jamaica, 1841–1865* (Baltimore, 1980). See also Franklin W. Knight and Margaret E. Crahan, *Africa and the Caribbean: The Legacies of a Link* (Baltimore, 1979); Philip Curtin, ed., *Africa Remembered: Narratives by West Africans from the Era of the Slave Trade* (Madison, Wisc., 1967); Michael Craton, *Searching for the Invisible Man: Slaves and Plantation Life in Jamaica* (Cambridge, Mass., 1978).

77. Experts on African history have called attention to the techniques, such as secret societies and age-grading, used to counteract ethnic rivalries. See, for example, Ade Ajayi and Michael Crowder, eds., *History of West Africa* (rev. ed., 2 vols., New York, 1976), 2:194–97, cited by Monica Schuler, "Afro-American Slave Culture," in Michael Craton, ed., *Roots and Branches: Current Directions in Slave Studies* (Toronto, 1979), 122.

78. *Further Papers Respecting Insurrection,* 33.

79. Ibid., 42.

80. For a similar assessment of slavery in the United States, see Paul D. Escott, *Slavery Remembered: A Record of Twentieth-Century Slave Narratives* (Chapel Hill, 1979).

81. Several scholars interested in riots have recently stressed the importance of networks. John Bohsted observes that many sociologists have turned away from explanations of social movements which focused exclusively on mental components of collective action and have turned to the analysis of the organizational strengths or

weaknesses which either enable people to act on their grievances or prevent them from acting. "Gender, Household, and Community Politics: Women in English Riots 1790 1810," *Past and Present* 120 (August 1988), 88–122. See also John Bohsted and Dale E. Williams, "The Diffusion of Riots: The Patterns of 1766–1795 and 1801 in Devonshire," *Journal of Interdisciplinary History* 19, 1 (Summer 1988): 1–24, and John Bohsted, *Riots and Community Politics in England and Wales, 1790–1810* (Cambridge, Eng., 1983).

82. Curiously enough, no reference by blacks to the Haitian rebellion was recorded; the only references were made by whites, who were constantly haunted by the specter of Haiti.

83. It is worth noticing that the practice of holding a public meeting before a war was common in some parts of Africa.

84. Later that night Sandy met Paris, Telemachus, and Joseph at *Bachelor's Adventure,* and Paris warned them again if they followed Sandy's plan they would be killed. In his deposition, Barson, a slave from *Paradise,* told a story that is indicative of the degree to which the slaves had appropriated their masters' culture. He said that he had been at the chapel on the seventeenth, and had seen Quamina, Jack, and others talking. Jack had a paper in his hand, which he appeared to be reading to them. Gilbert told him that Jack said the slaves would be free very soon, and that even the white people knew about it. Barson had his doubts, but Gilbert reassured him by saying that it must be true "because it was on the paper."

85. *Further Papers Respecting Insurrection,* 77.

86. In Dominica, an act of the House of Assembly, June 1, 1821, required every owner, renter, or attorney to allot each slave a sufficient portion of land—not less than half an acre for every slave of whatever age—and also allow one day in every week, "over and above Sundays" and holidays, so that slaves could cultivate their crops. *Papers Relating to the Treatment of Slaves in the Colonies,* VIZ., *Acts of Colonial Legislatures 1818–1823* (House of Commons, 1824), 23.

87. Typical was the story reported in his trial by a slave from *Foulis.* He said that on his return from *Success,* Daniel called a slave meeting and told them the news. He ordered the women to leave, and all slaves present took an oath by kissing the Bible. The practice of taking oaths before a war has been identified in other rebellions in the Caribbean. Michael Craton noted that in a rebellion in Antigua in 1735 oaths administered "in at least seven different places, were sealed with a draft of rum mixed with grave dirt and cock's blood, and included pledges to kill all whites, to follow the leaders without questioning, to stand by each other, and to observe secrecy on pain of death. In some cases the oath was made with the hand on a cockerel, and in one case the leader Secundi 'called to his Assistance a Negro Obiahman, or Wizard, who acted his Part before a great number of slaves.' " Craton finds such oaths similar to those practiced in Ghana ("The Passion to Exist," 7). In Demerara the slaves used the Bible instead. As Monica Schuler and others have noticed, slaves often used the Bible as a talisman and saw missionaries as sorcerers.

88. This information was given by Archibald Brown[e], who reported a con-

versation he had had with Jack after his arrest. P.R.O. C.O. 111/45. It should be noticed that Brown also appears as Archibald Browne in other documents.

89. Estimates of the number of slaves involved vary from as few as eight to 12,000. See P.R.O. C.O. 111/44; *Further Papers Respecting Insurrection,* 15; Bryant, *Account of an Insurrection,* 110–16.

90. This was a common characteristic of slave rebellions everywhere. Barry Gaspar notes that "Often assigned leadership roles among the slaves by whites," these privileged slaves played an important role in conspiracies. "Particularly well situated to play such roles were slave drivers, who, by the very nature of their work, stood astride the worlds of slaves and slaveowners." *Bondmen and Rebels,* 232. The importance of artisans and drivers in the plotting of slave rebellions has been stressed by Michael Craton in *Testing the Chains,* and by many others who have studied slave rebellions in the Caribbean and elsewhere.

91. In his trial, Duke denied all the charges against him and stressed that "he had not touched property belonging to white men." *Further Papers Respecting Insurrection,* 83.

92. According to Robert Moore, Berbice slaves in rebellion in 1763 joined the Djukas. Robert Moore, *Slave Rebellions in Guyana* (mimeo., University of Guyana, 1971).

93. *Guiana Chronicle,* June 9, 1823.

94. Smith, Journal, March 19, 1819.

95. *Royal Gazette,* May 24, 1821.

96. In 1827, four years after the rebellion, the governor noticed that one out of three slaves was punished every year in Demerara—and those were only the recorded cases.

97. *Royal Gazette,* January 25, 1821, recorded Alexander Simpson's mortgage on *Montrose* and another on *Le Reduit* and *Wittentung.*

98. See the *Royal Gazette,* October 16, 1821. The *Royal Gazette,* December 13, 1821, gives the amount of sugar imported into Great Britain from 1807 to the beginning of 1821 and its average price each year. Prices rose between 1808 and 1811 from 33*s* 11*d* to 46*s* 10*d*, then down in 1812 to 36*s* 5*d* and up again to reach a peak in 1815 at 73*s* 4½*d.* From that point, prices went down to 30*s* 2½*d* in 1821, the lowest point. (See Fig. 1.) Figures on sugar production for Demerara show that the amount almost doubled between 1815 and 1820, from 18,607,091 pounds to 35,128,197; coffee and cotton shrank almost 50 percent. See *Royal Gazette,* March 21, 1816, and February 22, 1821. A House of Commons investigation concluded that sugar produced in "Demerary" in 1814 amounted to 2,300 hogsheads, and in 1820 the amount had increased to 5,200. During this period the population of slaves seems to have remained pretty stable, varying between 72,000 and 75,000 for Demerara and Essequibo. The *Royal Gazette,* January 12, 1819, gives data for 1818 for Demerara and Essequibo: 43,683 and 18,725 slaves "attached to estates," for a total of 62,418. Only 12,793 slaves were reported as "belonging to individuals" in both areas taken together. The same year, Demerara produced 23,000,821 pounds of sugar, Essequibo 29,069,228. Demerara and Essequibo also produced similar amounts of rum and molasses. Given the much small number of slaves in Esse-

quibo, planters there had clearly moved much closer to a sugar monoculture. But Demerara planters were still heavily committed to coffee and cotton. In 1818, Demerara produced more than nine million pounds of coffee, and Essequibo well under one million. And Demerara's cotton production was fifteen times that of Essequibo. See also *Royal Gazette,* July 21, 1821. Another document, published March 21, 1822, in the *Royal Gazette*—a petition from Jamaica to the crown about the state of the British market—calculated a going price of £57 per ton of sugar of moderate quality. Of this, £27 went for customs, £15 to pay freight and other charges, and only £15, "not amounting to three elevenths of the gross produce, remains to the colonist to reward his establishments, and the profit upon the large fixed capital he necessarily employs." The *Royal Gazette,* June 22, 1822, reproduced a letter from Joseph Marryat describing the critical situation of the West Indian planters. It gave as an example the depreciation of property in Demerara, where an estate purchased seven years earlier for £40,000 sold for £13,400 in the spring of 1822.

Chapter 6.
A Man Is Never Safe

1. George Lamming, *The Pleasures of Exile* (London, 1960), 121.

2. Alexander Simpson's deposition, in *Demerara: Further Papers . . . Respecting Insurrection of Slaves . . . with Minutes of Trials* (House of Commons, 1824), 38. Cited hereafter as *Further Papers Respecting Insurrection.* See also a slightly modified version in Joshua Bryant, *Account of an Insurrection of the Negro Slaves in the Colony of Demerara Which Broke Out on the 18th of August, 1823* (Demerara, 1824), 2. Cited hereafter as Bryant, *Account of an Insurrection.*

3. Bryant, *Account of an Insurrection,* 2–27.

4. Ibid., 4–5.

5. John Stewart's deposition, *Further Papers Respecting Insurrection,* 79–80.

6. *Further Papers: Documentary Evidence,* 7, and *Further Papers Respecting Insurrection,* 68. Also *Demerara . . . A Copy of the Minutes of the Evidence on the Trial of* John Smith, *a Missionary in the Colony of Demerara, with the Warrant, Charges, and Sentence:*—VIZ. *Copy of the Proceedings of a General Court Martial . . .* (House of Commons, 1824), 27. Cited hereafter as *Proceedings.*

7. *Demerara: Further Papers . . . Copy of Documentary Evidence Produced Before a General Court Martial . . .* (House of Commons, 1824), 7. Cited hereafter as *Further Papers: Documentary Evidence.*

8. Jane Smith's affidavit in the London Missionary Society, *Report of the Proceedings Against the Late Rev. J. Smith, of Demerara, Minister of the Gospel, Who Was Tried Under Martial Law, and Condemned to Death, . . . Including the Documentary Evidence Omitted in the Parliamentary Copy* (London, 1824), 191. Cited hereafter as LMS, *Report of the Proceedings.*

9. John Smith to Mercer, August 20, in ibid., 182. Also cited in Edwin Angel Wallbridge, *The Demerara Martyr: Memoirs of Reverend John Smith, Missionary to*

Demerara, with Prefatory Notes . . . by J. Graham Cruickshank, Vincent Roth, ed. (Georgetown, 1943), 104.

10. Smith to the Secretary of the LMS, August 21, 1823, LMS, *Report of the Proceedings,* 183–84. Also quoted in Wallbridge, *The Demerara Martyr,* 109. See also Smith's letter to the Rev. Mr. Mercer, August 20, 1823, ibid., 104.

11. Depositions of John Bailey and John Aves, *Proceedings,* 24–25; *Further Papers Respecting Insurrection,* 28–30.

12. *Further Papers Respecting Insurrection,* 30.

13. *Proceedings,* 43.

14. Jane Smith's affidavit, November 13, 1823, in LMS, *Report of the Proceedings,* 191–194.

15. Elizabeth's deposition, in *Proceedings,* 29.

16. John Smith to the Secretary of the London Missionary Society, August 21, 1823, in LMS, *Report of the Proceedings,* 183–84. Also quoted in Wallbridge, *The Demerara Martyr,* 108–11.

17. Thomas Nurse's deposition, in *Proceedings,* 29–31. See also John Smith to the First Fiscal, August 22, 1823, in LMS, *Report of the Proceedings,* 181–82. Also reproduced in Wallbridge, *The Demerara Martyr,* 111–13.

18. Council for World Mission, London Missionary Society Archives, Journals, West Indies and British Guiana, 1809–1825, Smith, Journal, July 1, 1819, and October 8, 1822. Cited hereafter as Smith, Journal.

19. Smith, Journal, December 4, 1819.

20. Cited by John Wray in a letter of July 31, 1824, to the London Missionary Society. LMS IC, Berbice.

21. *Proceedings,* 30.

22. Ibid., 66.

23. John Smith to V. A. Heyliger, First Fiscal, August 22, 1823, in LMS, *Report of the Proceedings,* 181–82; also in Wallbridge, *Demerara Martyr,* 111–12.

24. Extract from a letter from Demerara forwarded to the London Missionary Society by a Rev. W. Brown, a minister from Belfast, December 13, 1823. LMS IC, Berbice.

25. This is the version given by Bryant, but the letter Governor Murray sent to Earl Bathurst on August 24 said the slaves had demanded unconditional emancipation. Bryant, *Account of an Insurrection,* 5. Murray to Bathurst, August 24, 1823, in *Further Papers Respecting Insurrection,* 6. See also P.R.O. C.O. 111/53.

26. Murray to Bathurst, August 24, 1823, in *Further Papers Respecting Insurrection,* 5–6. See also P.R.O. C.O. 111/53.

27. Bryant, *Account of an Insurrection,* 5–6.

28. *New Times* (London), March 22, 1824; LMS, Odds. Different sources give slightly different figures. The Blue Book for 1823 shows whites: 2,009 males and 250 females; free blacks: 1,336 males and 1,773 females; slaves: 41,025 males and 33,126 females. P.R.O. C.O. 116/192.

29. Barry W. Higman, *Slave Populations of the British Caribbean, 1807–1834* (Baltimore, 1984), 85–95.

30. According to estimates at the time, about 13,000 slaves were involved.

"Schedule A, Exhibiting the Number of Estates Whose Negroes Were Engaged in the Rebellion," in *Further Papers Respecting Insurrection,* 15.

31. Bryant, *Account of an Insurrection,* 14.

32. Joshua Bryant, the source here, and the first to write a history of the rebellion, was a member of this battalion.

33. Murray to Bathurst, August 24, 1823, in *Further Papers Respecting Insurrection,* 6.

34. Cheveley, Journal.

35. Cheveley, Journal; Bryant, *Account of an Insurrection,* 33–36. Bryant uses "Chevely" but in the LMS Archives the name appears as John C. Cheveley.

36. *Further Papers Respecting Insurrection,* 6–7.

37. Ibid., 7.

38. Michel Foucault, *Discipline and Punish: The Birth of the Prison* (trans. Alan Sheridan, New York, 1979).

39. Cheveley, Journal.

40. In the *Royal Gazette,* January 13, 1821, there was a petition of John Hopkinson, Esq., for the "Negro man" Dublin.

41. Bryant, *Account of an Insurrection,* 46–47.

42. Ibid., 56–59.

43. Cheveley, Journal.

44. *Further Papers Respecting Insurrection,* 32, 83, 85. Bryant, *Account of an Insurrection,* 41.

45. Bryant, *Account of an Insurrection,* 50.

46. *Further Papers Respecting Insurrection,* 18.

47. *Royal Gazette,* September 3, 1823.

48. Bryant, *Account of an Insurrection,* 78.

49. Osnaburg is a type of canvas common throughout the Caribbean. It was originally from Osnabruck, Hanover. Cyril Hamshere, *The British in the Caribbean* (Palo Alto, Calif., 1972), 127.

50. *Further Papers Respecting Insurrection,* 98.

51. Accra, one of the leaders of the 1763 rebellion in Berbice, committed suicide after his defeat. According to Moore this was an Akan tradition. Robert Moore, *Slave Rebellions in Guyana* (Mimeo., University of Guyana, 1971).

52. Richard, the head driver who had had such an important role in the seizing of the manager of *Success,* was found many months later, many miles away.

53. Cheveley, Journal. Cheveley may have misread the signs. The crowd may in fact have been greeting the prisoners and making fun of the soldiers.

54. Cheveley, Journal.

55. Extract from a letter from Demerara forwarded to the London Missionary Society by a Rev. W. Brown, a minister from Belfast, December 13, 1823. LMS IC, Demerara.

56. Cf. Foucault, *Discipline and Punish,* 50.

57. Referring to England, Douglas Hay says that "The constitutional struggles of the seventeenth [century] had helped to establish the principles of the rule of law: that offences should be fixed, not indeterminate; that rules of evidence should

be carefully observed; that the law should be administered by a bench that was both learned and honest" (p. 32). Law had become a power with its own claims, and "equality before the law" a matter of principle. But he also shows how the use of pardon "allowed the rulers of England to make the courts a selective instrument of class justice, yet simultaneously to proclaim the law's incorruptible impartiality, and absolute determinacy." "Property, Authority and the Criminal Law," in Douglas Hay, Peter Linebaugh, John G. Rule, E. P. Thompson, Cal Winslow, eds., *Albion's Fatal Tree: Crime and Society in Eighteenth-Century England* (New York, 1975), 17–63.

58. Philip Corrigan and Derek Sayer, *The Great Arch: English State Formation as Cultural Revolution* (Oxford, 1985), 155–56.

59. *Second Report of the Committee of the Society for the Mitigation and Gradual Abolition of Slavery* (London, 1825), 4.

60. LMS IC, Berbice, Wray's letter, July 31, 1824, citing an article in the *Royal Gazette*, 1808.

61. *Guiana Chronicle*, January 7, 1824.

62. An order of the King-in-Council of June 30, 1821, provided that no punishment should be inflicted in the colonies other than those which were allowed in England. In "free-labor" societies, of course, the realities of class imposed their own limits on the ideal of equal justice for all. The fairness of judicial procedures is still an issue in contemporary societies. Poor and black people are still struggling to make the liberal promise of equality before the law a reality.

63. *Royal Gazette*, May 8, 1821.

64. For more detail, see Chapter 1.

65. For a comparison with other areas, see Elsa V. Goveia, "The West Indian Slave Laws of the Eighteenth Century," *Revista de Ciencias Sociales* 4 (March 1960): 75–105; J. Thorsten Sellin, *Slavery and the Penal System* (New York, 1976); Paul Finkelman, *The Law of Freedom and Bondage: A Casebook* (New York, 1965); Leon A. Higginbotham, *In the Matter of Color: Race and the American Legal Process* (New York, 1978); Helen T. Catterel, *Judicial Cases Concerning American Slavery and the Negro* (5 vols., Washington, D.C., 1926–37). For an examination of the contradictory nature of the slave law, see David B. Davis, *The Problem of Slavery in Western Culture* (New York, 1966), 244–61; Eugene Genovese, *Roll, Jordan, Roll: The World the Slaves Made* (New York, 1974), 25–49.

66. Ralph from plantation *Success*, Duke from *Clonbrook*, Kinsale from *Bachelor's Adventure*, Cudjoe, from *Porter's Hope*, Gilbert from *Paradise*, Smith from *Friendship*, Quamine from *Haslington*, Cuffy from *Annandale*, Zoutman from *Beter Wervagting*, Primo and Quaco from *Chateau Margo*, and many others. For the slave trials, see: Minutes of Slaves' Depositions, P.R.O. C.O. 111/53; "Negro Trials," P.R.O. C.O. 111/44 and 45; *Further Papers Respecting Insurrection*; and *Further Papers: Documentary Evidence*.

67. *Further Papers Respecting Insurrection*, 66–99. There is a list of "Cases of Conviction of Whites and Other Persons on Slave Evidence Between the Years 1774 and 1824, Extracted from the Records of Criminal Sentences of the Court of Justice of Demerara and Essequebo," in P.R.O. C.O. 111/44.

68. *Royal Gazette,* April 6, 1819.

69. *Papers Relating to the Treatment of Slaves in the Colonies; Acts of Colonial Legislatures; 1818–1823* (House of Commons, 1824), 8. Cited hereafter as *Papers Relating to the Treatment of Slaves.*

70. *Papers Relating to the Treatment of Slaves,* 32–33.

71. *Royal Gazette,* March 13, 15, 1821.

72. Peter Wood notices that West African naming practices gave "considerable latitude for giving the same person different names." He also called attention to possible confusions between names that appear to have originated in the European tradition but could also have derived from Africa. A name sounding like Cato, for example, was found among the Bambara, Yoruba, and Hausa. Another that sounded something like Hercules, often pronounced and spelled Heckles, was a Mende name meaning "wild animal." Wood, *Black Majority: Negroes in Colonial South Carolina, from 1670 Through the Stono Rebellion* (New York, 1975), 183–85.

73. Others had their names written in two different ways, probably due to different pronunciations: Ankey, a mulatto woman residing at *Le Resouvenir,* also appears as Antje; and Tully, a slave who lived in town, appears also as Taddy. The use of African names side-by-side with English ones may be indicative of the slaves' resistance to giving up their traditions. Similar naming practices were followed in the United States. See Henning Cohen, "Slave Names in Colonial South Carolina," in Paul Finkelman, ed., *The Culture and Community of Slavery* (New York, 1989), 50–56, and, in the same collection, Cheryl Ann Cody, "There Was No 'Absalom' on the Ball Plantations: Slave Naming Practices in the South Carolina Low Country, 1720–1865," 15–49; and John C. Inscoe, "Carolina Slave Names: An Index to Acculturation," 163–90.

74. Brathwaite remarks that in traditional societies names were so important that a change could transform a person's life. The use of multiple names was a defensive strategy to avoid evil spells. One could add that in a slave society the use of different names could also help to confuse masters and public authorities. Edward Kamau Brathwaite, "The African Presence in Caribbean Literature," in Sidney Mintz, ed., *Slavery, Colonialism, and Racism* (New York, 1974), 91. Similarly, Hesketh J. Bell remarks that slaves often had several names. *Obeah: Witchcraft in the West Indies* (2nd rev. ed., London, 1893), 11.

75. Those who were officers in the militia were identified as such, rather than by profession. That was the case of Lieut. Thomas Nurse, who was a merchant in Georgetown, and of Michael McTurk, who, instead of being identified as a planter, appears as "a captain of the First Company, Second Battalion of the militia, a burgher-captain, living on plantation *Felicity.*"

76. For further confirmation, he referred to Gilles's previous owner and manager, H. B. Fraser, and to John McLean of *Vrees en Hoop.*

77. In a letter written to John Wray, January 2, 1824, Jane Smith reported that her servant Charlotte had heard the "negroes" of *Le Resouvenir* say that as the prisoners were on their way to town, they were told that if they said all they could against the parson they would be saved. LMS IC, Berbice, Wray's letter, January 2, 1824.

78. *Further Papers Respecting Insurrection,* 76–79.

79. Ibid., 78–79.

80. Ibid., 74.

81. Ibid., 30, 31, 54.

82. Ibid., 29.

83. LMS IC, Berbice, Wray's letter, March 29, 1824. See also a declaration of Richard Padmore, jailer in Demerara, stating that Paris had told Austin that Captain Edmonstone had directed him to tell those lies and promised that if he threw all the blame on Parson Smith and Mr. Hamilton, he would probably "get clear." The jailer says that later Paris admitted to him that he had lied to Austin about Edmonstone. The jailer was obviously in connivance with the authorities. The day he was to die, Paris persisted in telling Austin that Captain Edmonstone had instructed him to tell lies. Since the same issue was raised apropos of Jack Gladstone's trial, this last version is probably correct. P.R.O. C.O. 111/44 and 45.

84. In *Discipline and Punish,* Foucault observed that in the nineteenth century public executions disappeared in different European countries. But, in spite of the renewed efforts of people like Romilly, Mackintosh, and Fowell Buxton, England was "one of the countries most loathe to see the disappearance of public execution, perhaps because of the role of model that the institution of jury, public hearings, and respect of habeas corpus had given to her criminal law; above all, no doubt because she did not wish to diminish the rigour of her penal laws during the great social disturbances of the years 1780–1820." *Discipline and Punish,* 14. Thus it is not surprising that public executions continued in places like Demerara.

85. Bryant, *Account of an Insurrection,* 61–62.

86. Ellick of *Coldingen;* Attila of *Plaisance;* France of *Porter's Hope;* Billy of *Ann's Grove;* Harry of *Triumph;* and Quintus of *Beter Verwagting.*

87. Telemachus, Scipio, and Jemmy of *Bachelor's Adventure;* Lindor and Picle of *La Bonne Intention;* Beffaney of *Success;* Tom of *Chateau Margo;* Paul of *Friendship;* and Quamina of *Noot en Zuyl.*

88. Telemachus and Jemmy at *Bachelor's Adventure;* Lindor, at *La Bonne Intention;* and Paul, at *Friendship.*

89. Bryant, *Account of an Insurrection,* 66.

90. Foucault, *Discipline and Punish,* 49.

91. *Royal Gazette,* June 20, 1820.

92. Bryant, *Account of an Insurrection,* 69–73, describes the scenes vividly. More details can be found in P.R.O. C.O. 111/53.

93. The list of slaves executed can be found in P.R.O. C.O. 111/45. See also Bryant, *Account of an Insurrection,* 109–11.

94. P.R.O. C.O. 111/53, and "List of Slaves Punished from August 1823 to January 1824," P.R.O. C.O. 111/45.

95. Ralph and Quaco from *Success;* Primo and Quaco from *Chateau Margo;* Duke from *Clonbrook;* Kinsale from *Bachelor's Adventure;* Cudjoe from *Porter's Hope;* Inglis of *Foulis;* Adonis from *Plaisance;* Gilbert from *Paradise;* Smith from *Friendship;* Quamine from *Haslington;* Nelson from *New Orange Nassau;* and Zoutman from *Beter Verwagting.*

96. *Further Papers Respecting Insurrection,* 65–66.

97. In a letter to Governor Murray, written almost a year after the revolt, the attorney for *Success* praised Jack Gladstone for his behavior and called him an extraordinary man. He stressed that prior to the revolt Jack Gladstone was liked "by us all for his good behavior, intelligence and usefulness," and even during the insurrection he had distinguished himself for his forbearance and humanity to whites. The attorney also mentioned that "Mr. Gladstone" had interceded with the government on Jack's behalf. He concluded his letter by expressing his hope that at a future time Jack Gladstone might be permitted to return to the plantation and to his family. P.R.O. C.O. 111/44. See also P.R.O. C.O. 111/53.

98. Most managers seemed to be happy to see their slaves back. When the governor informed them that the King had remitted the slaves' sentences and asked them whether they would like to have their slaves back, and under "what mark of disgrace," most managers and attorneys responded positively. "I should be most glad were the man Duke restored to the estate in consideration for his family who are quiet and well disposed," responded Hugh Rogers. Van Watenshoodt from *Plaisance* said he had lost seven of his finest people and this had considerably weakened his small gang; thus "the restoration of the negro Adonis cannot but prove of benefit to the estate." This was the tone of most of the letters. A few managers or slaveowners requested that their slaves be sent back but put in light chains for a while. And even fewer declared they did not want the slaves back for fear that they would instigate others to rebel. These slaves were sent to the colony workhouse. The most striking letter was written by the attorney for *Success,* who requested that Beffany and Ralph wear chains on their legs as a mark of disgrace for being violent characters and acting with "marked *atrocity*" toward Stewart and the overseer by putting them in the stocks. P.R.O. C.O. 111/44.

99. Harry from *Triumph* had led the attack at *Mon Repos* and had set the house on fire. Lindor had given orders to set fire to *La Bonne Intention.* Thomas had attacked George Mason, the manager from *Chateau Margo,* with his cutlass. Paul of *Friendship* had been the first to enter his master's house, collared him and tried to wrestle a gun from his hand. Quamina from *Noot en Zuyl* was seen at *Elizabeth Hall* advising the slaves to do away with the whites. Beffaney and others had struck the overseer of *Success* with their cutlasses and released Jack Gladstone.

100. *Chateau Margo, La Bonne Intention, Beter Verwagting, Triumph, Mon Repos, Good Hope, Lusignan.*

101. *Coldingen, Non Pareil, Enterprise, Paradise, Foulis, Porter's Hope, Enmore, Golden Grove, Haslington.*

102. LMS IC, Berbice, Smith to Wray, June 25, 1823.

103. LMS IC, Berbice, Wray's letter, December 16, 1823.

104. LMS IC, Berbice, Wray's letter, August 6, 1815.

105. LMS IC, Berbice, Wray's letter, August 6, 1823.

106. LMS IC, Berbice, Wray's letters, July 17 and August 27, 1823.

107. LMS IC, Berbice, Wray's letter, August 27, 1823.

108. LMS IC, Berbice, M. S. Bennet, Berbice's fiscal, to Wray, September 1, 1823.

109. LMS IC, Berbice, Wray's letters, September 1, 2, and 3, 1823.

110. LMS IC, Berbice, Wray's letter, September 4, 1823.

111. LMS IC, Berbice, Wray's letter, September 4, 1823.

112. LMS IC, Berbice, Wray's letter, undated, probably written after September 29 and before October 6.

113. LMS IC, Berbice, Wray's letter, October 6, 1823.

114. LMS IC, Berbice, Wray's letter, October 9, 1823.

115. LMS IC, Berbice, Wray's letter, October 6, 1823.

116. The exact words were, "How much more safe was Mr. Campbell among the savages of Africa than we are among our own countrymen." LMS IC, Berbice, Wray's letter, November 15, 1823.

Chapter 7.
A Crown of Glory That Fadeth Not Away

1. The proceedings were published in two forms, one by the House of Commons and the other by the London Missionary Society. There are some differences between the two. The Missionary Society and its supporters felt the differences significant enough to justify a new and more complete edition. What follows relies on both versions: *Demerara . . . A Copy of the Minutes of the Evidence on the Trial of* John Smith, *a Missionary in the Colony of Demerara, with the Warrant, Charges, and Sentence:*—VIZ *Copy of the Proceedings of a General Court Martial. . . .* (House of Commons, 1824), which will be cited as *Proceedings;* and the London Missionary Society, *Report of the Proceedings Against the Late Rev. J. Smith, of Demerara, Minister of the Gospel, Who Was Tried Under Martial Law, and Condemned to Death, . . . Including the Documentary Evidence Omitted in the Parliamentary Copy* (London, 1824), which will be cited as LMS, *Report of the Proceedings.* See also "Court Martial of Missionary John Smith," P.R.O. C.O. 111/42 and "Missionary John Smith's Case," P.R.O. C.O. 111/53. See also 111/39 and 111/43.

2. *Proceedings,* 3–4.

3. Blue Book: Public Functionaries' Salaries. P.R.O. C.O. 116/192.

4. *Proceedings,* 8. The significance that the Old Testament had for slaves everywhere has been noted by many historians. See, for example, Albert J. Raboteau, *Slave Religion: The "Invisible Institution" in the Antebellum South* (New York, 1980), 311.

5. Romeo told the court that Jack Gladstone was not a deacon, although he did help sometimes to teach the catechism. He was a "wild fellow" and did not come regularly to the chapel.

6. The passage reads: "41: And when he was come near, he beheld the city, and wept over it. 42: Saying, If thou hadst known, even thou, at least in this thy day, the things which belong unto thy peace! but now they are hid from thine eyes."

7. This statement could invalidate Jane Smith's statement that she, not her husband, had sent for Quamina. Her statement, however, was confirmed by other

witnesses. It is possible, as Smith suggested, that Romeo was lying, but it is also possible that both Smiths, independently of each other, had sent for Quamina.

8. It is extraordinary how well some slaves could remember the texts they read in the Bible the day before the rebellion. Joe, who was called after Azor to testify, gave the following interpretation to Luke 41–42: "I do not recollect what the text was but there are some words in the chapter I know. The parson said, the Lord Jesus Christ sent a disciple into a certain village, and you will see a colt tied there, bring it unto Me, and if the master of the colt should ask you what you are going to do with it, you must say the Lord hath need of it; and they brought it to the Lord, and laid some raiment on it; and He rode it to Jerusalem, and He rode it to the top of a mountain where He could see Jerusalem all over; and He wept over Jerusalem, and said, if they had known their peace, that is to say, if the people knew what belonged to them, they would believe in Him; now their trouble would come upon them. So far I can make out; I cannot remember anything more." *Proceedings,* 10.

9. *Proceedings,* 36.

10. Privately, however, Davies informed the LMS that Smith's manners and stubbornness had provoked the ire of the whites. LMS IC, Demerara, Davies's letter, May 1824.

11. "Minute and Process Verbal of an Inquest Held This 6th Day of February 1824 on the Body of John Smith." P.R.O. C.O. 111/44. The inquest found that Smith "was worn away to a skeleton," and that two days before his death had lost the power of speech and of deglutition (swallowing).

12. For details, see Cecil Northcott, *Slavery's Martyr: John Smith of Demerara and the Emancipation Movement, 1817–1824* (London, 1976).

13. P.R.O. C.O. 111/39. See also *The Colonist,* December 1, 1823.

14. Governor D'Urban to Lord Bathurst, June 7, 1824. P.R.O. C.O. 111/44. See also Austin's testimony, in Report from the Select Committee on the Extinction of Slavery Throughout the British Dominion. . . . 1832. P.R.O. C.O. ZHCI/1039.

15. In a letter sent to G. Burden from Downing Street, July 20, 1824, Adam Gordon said that the resentment excited against Austin by the publication of an extract from one of his letters, in which he freely reflected on the community as well as vindicated the conduct of Mr. Smith, had led Lord Bathurst "to consider that the return of Mr. Austin to Demerary would not be attended with advantage to the Community or with comfort for himself." Gordon said that Lord Bathurst intended to propose that Austin be employed in some other colony. P.R.O. C.O. 112/12. See also P.R.O. C.O. 111/44.

16. *Guiana Chronicle,* October 27, 1824.

17. Joshua Bryant, *Account of an Insurrection of the Negro Slaves in the Colony of Demerara Which Broke Out on the 18th of August, 1823* (Demerara, 1824), 98–106.

18. *Guiana Chronicle,* February 27, 1824.

19. *The Wesleyan Methodist Magazine* (3d Series, 3 (January 1824):50–53; ibid., (February 1824):108; and *The Times,* December 12, 1823. See also Lowell Joseph

Ragatz, *The Fall of the Planter Class in the British Caribbean, 1763–1833* (New York, 1928), 431–33.

20. *The Christian Observer, Conducted by Members of the Established Church for the Year 1824* 24 (April 1824):221–226. For the impact on missions in Jamaica, see Mary Turner, *Slaves and Missionaries: The Disintegration of Jamaican Slave Society, 1757–1834* (Champaign-Urbana, Ill., 1982), 112–26.

21. P.R.O. C.O. 111/43.

22. *Evangelical Magazine and Missionary Chronicle* New Series, 1 (November 1823):473.

23. Ibid. (November 1823):473 and (December 1823):513.

24. Elliot's letter to the LMS, October 18, 1823, excerpted in the *Evangelical Magazine,* New Series, 2 (January 1824):29–31.

25. *Evangelical Magazine,* New Series, 2 (February 1824):73.

26. See a letter from the LMS to Earl Bathurst, December 16, 1823, requesting that because of Smith's health he be sent home, whatever the outcome of his trial. P.R.O. C.O. 111/43.

27. *Evangelical Magazine,* New Series, 2 (April 1824):163.

28. *Wesleyan Methodist Magazine* 3rd Series, 3 (January 1824):42, ibid. (April 1824):189.

29. Ibid. (June 1824):403–404 and 414.

30. *The Baptist Magazine* printed the minutes of a meeting of the Protestant Society for the Protection of Religious Liberty, where several cases of persecution of itinerant preachers were reported. The Society complained that the power of the Anglican clergy was increasing. No bells could be tolled, no monuments erected, no vestries held, without the consent of the clergyman. "On the magisterial bench, the number of clergymen was considerable; and . . . when they predominated at the quarter-sessions, the evil was great to Protestant Dissenters, and the good not great to anybody else." The Society resolved that it "would continue to regard the right of every man to worship God according to his conscience, as an invaluable sacred, and unalienable right; and all violations of that right, by monarchs or by multitudes, by penal laws or lawless violence, by premiums for conformity or exclusion for non-conformity, as unjust, and oppressive, inexpedient and profane." *Baptist Magazine for 1823* 15, 6 (June 1823):250–51. See also ibid. 16, 3 (March 1824):81; ibid. 16, 4 (April 1824):168; ibid. 16, 7 (July 1824):302–3.

31. *Christian Observer* 24, 1 (January 1824):18, 47, 64–68.

32. Ibid. 2 (Febuary 1824):87–93.

33. Ibid. 3 (March 1824):153–63.

34. Ibid. 3 (March 1824):160–61.

35. For an analysis of the role abolitionist rhetoric had in legitimizing British social order, see David Brion Davis's "Reflection on Abolitionism and Ideological Hegemony," *American Historical Review* 92:4 (October 1987): 797–812. For a critique of Davis's interpretation, see John Ashworth, "The Relationship Between Capitalism and Humanitarianism," ibid., 813–28; and Thomas L. Haskell, "Convention and Hegemonic Interest in the Debate over Antislavery: A Reply to Davis and Ashworth," ibid., 829–50. Davis, examining British self-congratulatory rhetoric

of the post-abolitionist period, says that Buxton's culture, by annihilating its supposed "antithesis," acquired the attributes of omnipotence and universality. *Slavery and Human Progress* (New York, 1984), 125. For a critique of the ideological and hegemonic character of abolitionism, see also Seymour Drescher, *Capitalism and Antislavery: British Mobilization in Comparative Perspective* (New York, 1987), 144–45.

36. See David Eltis, "Abolitionist Perceptions of Society After Slavery," in James Walvin, ed., *Slavery and British Society, 1776–1846* (Baton Rouge, 1982), 198. Patricia Hollis reveals tensions between the working class and the antislavery movements in "Anti-Slavery and British Working-Class Radicalism in the Years of Reform," in Roger Anstey, Christine Bolt, and Seymour Drescher, eds., *Anti-Slavery, Religion, and Reform* (Hamden, Conn., 1980), 294–318. For a different view, see Betty Fladeland, " 'Our Cause Being One and the Same': Abolitionists and Chartism," in Walvin, ed., *Slavery and British Society,* 69–99. In *Capitalism and Antislavery* (pp. 143–45), Drescher argues that abolitionism was "a harbinger, or a catalyst of other popular mobilization, rather than a deflector of social conflict. . . . Abolition was not charity at a distance but a premature strike against the aristocracy at home." And he says that before the end of the Napoleonic wars, William Cobbett stood virtually alone in his denunciation of the "negrophile hypocrisy of British abolitionism," but toward the 1830s a growing number of workers considered abolitionism a diversion from "the real miseries of white slaves at home."

37. *Christian Observer* 24, 6 (June 1824): 359.

38. Ibid. 8 (August 1824): 398–99, 479–87.

39. George Rudé speaks of "the popular belief that the Englishman, as a 'free born' subject, had a particular claim on 'liberty.' " *Hanoverian London, 1714–1808* (London, 1971), 99.

40. Wray mentioned that people sang it in the street and it was not uncommon for slaves to sing it. LMS IC, Berbice, Wray's letter, December 15, 1823. Derek Jarret says of "Rule Britannia" that "it contained both the sturdy defiance of the freedom-loving island race and also the arrogant imperialism of the most rapidly developing commercial industrial nation in the world." *England in the Age of Hogarth* (London, 1986), 40.

41. Betty Fladeland notes that "The more the abolitionists publicized the horrid conditions of slavery, the more likely it was that British workers would see themselves as caught in similar circumstances. By the 1830s leaders of working-class movements realised that it would be to their advantage to capitalise on the achievement of the abolitionists in awakening sympathy for the downtrodden, and to do this by copying abolitionist strategy. Workers' cries of enslavement, in turn, forced abolitionist attention to miseries at home." " 'Our Cause Being One and the Same,' " 69. Seymour Drescher also argues that accounts of slavery served to intensify rather than to screen the image of working-class privation of rights. Slavery served as a metaphor. See Seymour Drescher, "Cart Whip and Billy Roller, Or, Anti-Slavery and Reform: Symbolism in Industrializing Britain," *Journal of Social History* 15:1 (Fall 1981):3–24. Drescher expands on this argument in "The Historical Context of British Abolition," in David Richardson, ed., *Abolition and*

Its Aftermath: The Historical Context (London, 1985), 3. For women's participation in the campaign for emancipation, see James Walvin, "The Propaganda of Anti-Slavery," in Walvin, ed., *Slavery and British Society,* 61–63.

42. LMS Odds.

43. Ibid.

44. There were two hundred such abolitionist associations in Britain. See *New Times,* June 26, 1824. LMS Odds.

45. LMS Odds.

46. *John Bull,* June 7, 1824, in LMS Odds. See also *John Bull,* October 26 and December 14, 1823; January 25 and March 29, 1824. LMS Odds. For the slaves' and Smith's trials, see *John Bull,* February 1 and April 26, 1824. LMS Odds.

47. T. C. Hansard, *The Parliamentary Debates, from the Thirtieth Day of March to the Twenty-fifth Day of June, 1824* (vol. XI, London, 1825), 961–1076, 1167–68.

48. Ibid., 1206–1315.

49. C. Duncan Rice says that "Smith came to play much the same role in British anti-slavery propaganda as the murdered newspaper editor Elijah P. Lovejoy in America." *The Rise and Fall of Black Slavery* (New York, 1975), 253.

50. "Report from the Select Committee on the Extinction of Slavery Throughout the British Dominions with the Minutes of Evidence. Appendix and Index" (House of Commons, August 1832). P.R.O. C.O. ZHCI/1039.

51. On November 22, 1831, the King by an order in council promulgated the Consolidation Slave Ordinance, extending to slaves practically all civil rights but freedom. Two years later, in June 1833, the House of Commons passed five resolutions, including immediate measures for the abolition of slavery, but establishing seven and a half years of apprenticeship, with compensation to be paid to masters. With this proviso, an Act of August 21, 1833, granted freedom to all slaves starting August 1, 1834. Slaveholders received £51 17s for each slave. The failure of the apprenticeship scheme became immediately obvious. Blacks' resistance to compulsory labor intensified. Ironically, five years later, Michael McTurk moved in the Court of Policy to bring the whole system to an end and grant unconditional freedom to all slaves, while Peter Rose voted against McTurk's motion. But the motion prevailed and on August 1, 1838, the slaves were made free. A. R. F. Webber, *Centenary History and Handbook of British Guiana* (Georgetown, 1931), 163–93. See also Henry G. Dalton, *The History of British Guiana* (2 vols., London, 1855), 1: 412, 429; and *Almanack and Local Guide of British Guiana, Containing the Laws, Ordinances and Regulations of the Colony, the Civil and Military Lists, with a List of Estates from Corentyne to Pomeroon Rivers* (Demerara, 1832), 223–91.

INDEX